Structure and Properties of Cell Membranes

Volume I

A Survey of Molecular Aspects of Membrane Structure and Function

Editor

Gheorghe Benga, M.D., Ph.D.

Head, Department of Cell Biology
Faculty of Medicine
Medical and Pharmaceutical Institute
Cluj-Napoca, Romania

CRC Press
Taylor & Francis Group
Boca Raton London New York

CRC Press is an imprint of the
Taylor & Francis Group, an **informa** business

First published 1985 by CRC Press
Taylor & Francis Group
6000 Broken Sound Parkway NW, Suite 300
Boca Raton, FL 33487-2742

Reissued 2018 by CRC Press

Library of Congress Cataloging in Publication Data
Main entry under title:

A Survey on molecular aspects of membrane structure and
function.

(Structure and properties of cell membranes; v. 1)
Includes bibliographies and index.
1. Cell membranes. 2. Molecular biology. I. Benga,
Gheorghe, II. Series. [DNLM: a. Cell Membrane—
physiology. 2. Cell Membrane—ultrastructure.
3. Biological Transport. QH 601 S9285]
QH601.S785 1985 574.87'5 84-19943
ISBN 0-8493-5764-0

A Library of Congress record exists under LC control number: 84019943

Publisher's Note
The publisher has gone to great lengths to ensure the quality of this reprint but points out that some imperfections in the original cop may be apparent.

Disclaimer
The publisher has made every effort to trace copyright holders and welcomes correspondence from those they have been unable contact.

ISBN 13: 978-1-315-89786-8 (hbk)
ISBN 13: 978-1-351-07696-8 (ebk)

Visit the Taylor & Francis Web site at http://www.taylorandfrancis.com and the
CRC Press Web site at http://www.crcpress.com

PREFACE

In recent years it has become apparent that many essential functions of living cells are performed by membrane-associated events. Membranes are highly selective permeability barriers that impart their individuality on cells and organelles (Golgi apparatus, mitochondria, lysosomes, etc.) by forming boundaries around them and compartmentalizing specialized environments. By receptor movement and responses to external stimuli, membranes play a central role in biological communication. Owing to various enzymes attached to or embedded into membranes they are involved in many metabolic processes. The two important energy conversion processes, photosynthesis in chloroplasts and oxidative phosphorylation in mitochondria, are carried out in membranes. To understand all these processes, which are essential for living organisms, it is necessary to understand the molecular nature of membrane structure and function.

The main purpose of this book is to provide in-depth presentations of well-defined topics in membrane biology, focusing on the idea of *structure-function relationships at the molecular level*. The book consists of three volumes.

Volume 1 covers general aspects of structure-function relationships in biological membranes. Attention has been paid both to protein and lipid components of cell membranes regarding the interactions between these components, mobility of proteins and lipids, as well as to the physiological significance of membrane fluidity and lipid-dependence of membrane enzymes. Since some molecular components of the plasma membrane appear to function in concert with some component macromolecules of basement membranes a review of this expanding topic has been included.

Volume 2 is devoted to models and techniques which allow molecular insights into cell membranes. After the first chapter describes quantum chemical studies of proton translocation, several chapters present the most extensively used model systems (monomolecular films, planar lipid bilayers, and liposomes) in relation to biomembranes as well as the reconstitution of membrane transport systems. The description of some biophysical techniques (X-ray, spin labeling ESR, NMR) is focused on their use in studying stucture-function relationships in cell membranes. The remaining chapters in this volume are devoted to the physiological significance of surface potential of membranes and of the dietary manipulation of lipid composition.

Volume 3 covers transport at the molecular level in selected systems. The first chapters present basic kinetics and pH effects on membrane transport, while subsequent chapters focus on the effect of membrane lipids on permeability in prokaryotes or on Ca^{2+} permeability. Three chapters describe structure-function relationships in mitochondrial H^+-ATPase, cytochrome oxidase, and adenine nucleotide carrier. The last chapter is devoted to exocytosis, endocytosis, and recycling of membranes, which are distinct, albeit overlapping, cellular processes.

From this survey it is obvious that by application of biochemical and biophysical techniques it is possible to explain membrane phenomena at the molecular level in a meaningful way. Moreover, it is now clear that the study of cell membranes at the molecular level is important for understanding the alterations leading to abnormal cells or the understanding of drug and pesticide action. The multidisciplinary approach of research in this area and the permanent need for information regarding the recent advances require new books on cell membranes. The present collection of reviews is by no means a comprehensive treatise on all aspects of "membranology", rather a sampling of the status of selected topics. The volumes, providing contributions for reference purposes at the professional level, are broadly aimed at biochemists, biologists, biophysicists, physicians, etc., active investigators working on cell membranes and hopefully will also be of great help to teachers and students at both the undergraduate or postgraduate levels.

THE EDITOR

Dr. Gheorghe Benga, M.D., Ph.D., is the Head of the Department of Cell Biology at the Medical and Pharmaceutical Institute, Cluj-Napoca, Romania. He is also heading the Laboratory of Human Genetics of the Cluj County Hospital.

In 1967, Dr. Benga received an M.D. with academic honors from the Medical and Pharmaceutical Institute. After 3 years of internship (1966 to 1969) in basic medical sciences (biochemistry, microbiology), he studied for a Ph.D. in medical biochemistry from 1969 to 1972 under Prof. Ion Manta, Department of Biochemistry, at the same Institute. In 1972, Dr. Benga received a B.Sc. in Chemistry and in 1973, an M.Sc. in Physical Chemistry of Surfaces from the University of Cluj.

From 1972 to 1978, Dr. Benga was Lecturer and Senior Lecturer in the Department of Biochemistry at the Medical and Pharmaceutical Institute. In 1974 he was awarded a Wellcome Trust European Travelling Fellowship and spent 1 year in England as a postdoctoral research worker under Prof. Dennis Chapman, Department of Chemistry and Biochemistry, University of Sheffield and Chelsea College University of London. In 1978, Dr. Benga was appointed to head the newly formed Department of Cell Biology at the Medical and Pharmaceutical Institute. He is currently teaching cell biology to medical students.

In addition to his other duties, Dr. Benga has spent several 1- to 3-month periods as a Visiting Scientist at many British and American universities and in 1983 was a Visiting Professor at the University of Illinois at Urbana-Champaign.

Dr. Benga has attended several international courses on biomembranes and has presented numerous papers at international and national meetings, as well as guest lectures at various universities and institutes in Romania, England, the U.S., the Netherlands, and Switzerland. He has taken an active part in the organization of three international workshops on biological membranes (1980, Cluj-Napoca — Romanian-British; 1981, Cluj-Napoca — Romanian-American; 1982, New York City — American-Romanian) and has published over 80 papers to date.

Dr. Benga is the author of several text books of cell biology for medical students and of the book, *Biologia moleculară a membranelor cu aplicaţii medicale*, published by Editura Dacia, Cluj-Napoca, 1979. He is the co-author of *Metode biochimice în laboratorul clinic,* Editura Dacia, 1976; co-editor of *Biomembranes and Cell Function,* New York Academy of Sciences, 1983; and co-editor of *Membrane Processes: Molecular Biology and Medical Applications,* Springer-Verlag, New York, 1984.

His major interests in the field of biological membranes include the characterization of molecular composition and functional properties of human liver subcellular membranes, the molecular interactions (lipid-protein, lipid-sterol, and drug effects) in model and natural biomembranes, and the investigation of water diffusion through red blood cell membranes.

Dr. Benga is President of the Cluj-Napoca Section of the Romanian National Society of Cell Biology and Vice-President of this Society. He is on the board of the Subcommission of Biochemistry of the Romanian Academy and is on the editorial board of *Clujul Medical.*

CONTRIBUTORS

L. I. Barsukov
Shemyakin Institute of Biorganic Chemistry
U.S.S.R. Academy of Sciences
Moscow, U.S.S.R.

Gheorghe Benga
Head, Department of Cell Biology
Faculty of Medicine
Medical and Pharmaceutical Institute
Cluj-Napoca, Romania

L. D. Bergelson
Professor
Head, Lipid Department
Shemyakin Institute of Biorganic Chemistry
U.S.S.R. Academy of Sciences
Moscow, U.S.S.R.

Giovanna Parenti Castelli
Associate Professor of Biochemistry
Istituto di Chimica Biologica
University of Bologna
Bologna, Italy

Lewis D. Johnson, M.D.
Professor
Department of Pathology
University of South Carolina
Columbia, South Carolina

Giorgio Lenaz
Professor of Biochemistry
Istituto Botanico
University of Bologna
Bologna, Italy

Reiner Peters
Max-Planck-Institut für Biophysik
Frankfurt, West Germany

Gregory B. Ralston
Senior Lecturer in Biochemistry
Department of Biochemistry
University of Sydney
Sydney, New South Wales
Australia

Fritiof S. Sjöstrand
Professor Emeritus of Molecular Biology
Department of Biology
University of California
Los Angeles, California

Charles G. Wade
Manager, Magnetics Technical Support
IBM Instruments, Inc.
San Jose, California

John M. Wrigglesworth
Senior Lecturer
Department of Biochemistry
Kings College
University of London
London, U.K.

STRUCTURE AND PROPERTIES OF CELL MEMBRANES

Gheorghe Benga

Volume I

The Evolution of Membrane Models
Protein-Protein Interactions in Cell Membranes
Lateral Mobility of Proteins in Membranes
Lateral Diffusion of Lipids
Topological Asymmetry and Flip-Flop of Phospholipids in Biological Membranes
Membrane Fluidity: Molecular Basis and Physiological Significance
Lipid Dependence of Membrane Enzymes
Protein-Lipid Interactions in Biological Membranes
Basement Membrane Structure, Function, and Alteration in Disease

Volume II

Basic Kinetics of Membrane Transport
pH Effects on Membrane Transport
The Effect of Membrane Lipids on Permeability and Transport in Prokaryotes
The Influence of Membrane Lipids on the Permeability of Membranes to Ca^{2+}
Molecular Aspects of Structure-Function Relationship in Mitochondrial H^+-ATPase
Molecular Aspects of the Structure-Function Relationship in Cytochrome *c* Oxidase
Molecular Aspects of the Structure-Function Relationships in Mitochondrial Adenine Nucleotide Carrier
Exocytosis, Endocytosis, and Recycling of Membranes
The Surface Potential of Membranes: Its Effect on Membrane-Bound Enzymes and Transport Processes

Volume III

Quantum Chemical Approach to Study the Mechanisms of Proton Translocation Across Membranes Through Protein Molecules
Monomolecular Films as Biomembrane Models
Planar Lipid Bilayers in Relation to Biomembranes
Relation of Liposomes to Cell Membranes
Reconstitution of Membrane Transport Systems
Structure-Function Relationships in Cell Membranes as Revealed by X-Ray Techniques
Structure-Function Relationships in Cell Membranes as Revealed by Spin Labeling EPR
Structure and Dynamics of Cell Membranes as revealed by NMR Techniques
The Effect of Dietary Lipids on the Composition and Properties of Biological Membranes

TABLE OF CONTENTS

Volume I

Chapter 1

THE EVOLUTION OF MEMBRANE MODELS

Fritiof S. Sjöstrand

TABLE OF CONTENTS

I. THE BEGINNING OF AN EVOLUTION

Before electron microscopy made it possible to observe membranes, it was theoretically assumed that membranes are present at the surfaces of the cells and the cell nucleus to account for the difference in the composition of the media on the two sides of these boundaries. The membranes were conceived of as forming barriers that controlled the exchange between these media, passively due to the physical properties of the membranes.

The observation by Overton[1] in 1899 that substances that are soluble in nonpolar solvents permeated the surface of plant protoplasts faster than water-soluble substances led him to propose that the surface of cells is impregnated by a substance that might be similar to a fatty oil.

This observation drew the attention to the lipids as possible basic components of the cell surface, and it led to the first membrane model proposed by Gorter and Grendel[2] in 1925, according to which a single lipid bilayer covers the cell surface (Figure 1).

The observation by Danielli and Harvey[3] in 1935 that the surface tension at the cell surface is too low to correspond to that of a lipid surface led to the membrane model of Danielli and Davson[4] in which a layer of protein molecules was added to each side of the lipid bilayer. The proteins would act as surfactants reducing the surface tension. In this second membrane model, the lipid bilayer is thus sandwiched between two protein layers that were originally assumed to consist of globular proteins, while later the polypeptide chains were considered to be unfolded (Figure 2).

The role of the proteins was to stabilize the lipid bilayer, and they were not considered to play any specific role by contributing to the permeability properties of the membrane. Addition of possible charge interactions between the proteins and the lipids made the model appear somewhat more sophisticated.

The further evolution of membrane models reminds one in certain aspects of the evolution of year models of automobiles. Like the automobile manufacturers who have realized that it is commercially expedient to change the shapes of automobiles only gradually, because too drastic changes have led to catastrophic declines in sales, we find that only gradual changes in the membrane models have been generally accepted, while extensive modifications have been ignored.

Electron microscopic analysis of the structure of cells led to the discovery of membranes as existing as discrete structural entities. It was found that membranes are not confined to the boundaries of cells and of the cell nucleus, but that they constitute the dominant structural components of practically all cell organelles in the cytoplasm. The cytoplasmic membranes were thus discovered starting with the membranes of the outer segment disks in photoreceptors by Sjöstrand,[5,6] followed by the discovery of membranes in the chloroplasts by Steinmann,[7] in the mitochondria by Palade[8] and by Sjöstrand,[9] in the Golgi apparatus by Dalton and Felix[10] and by Sjöstrand and Hanzon,[11] and in the basophilic cytoplasm, the ergastoplasm, that was found to correspond to the endoplasmic reticulum observed in whole cells in tissue cultures by Porter and Kallman.[12]

The cytoplasmic membranes were proposed by Sjöstrand[9] in 1953 to play a more competent role functionally than to form a barrier. This point of view was expressed by proposing that "the intracellular cytoplasmic membranes may represent a fundamental principle of organization of the cytoplasm."

However, the generally accepted concept was that membranes function as barriers, and the function of the cytoplasmic membranes including the mitochondrial membranes is to delimit compartments. The interest was, consequently, focused on the lipid bilayer structure of membranes.

For many years, the triple-layered pattern observed in electron micrographs of cross-sections of membranes was accepted as confirming the Danielli-Davson membrane model,

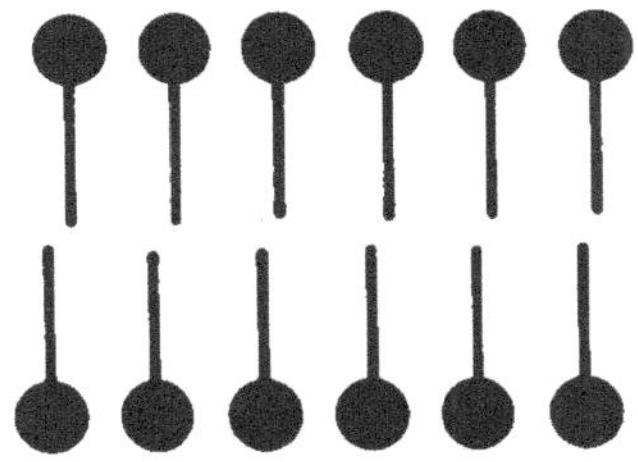

FIGURE 1. The lipid bilayer model for the plasma membrane proposed by Gorter and Grendel.[2]

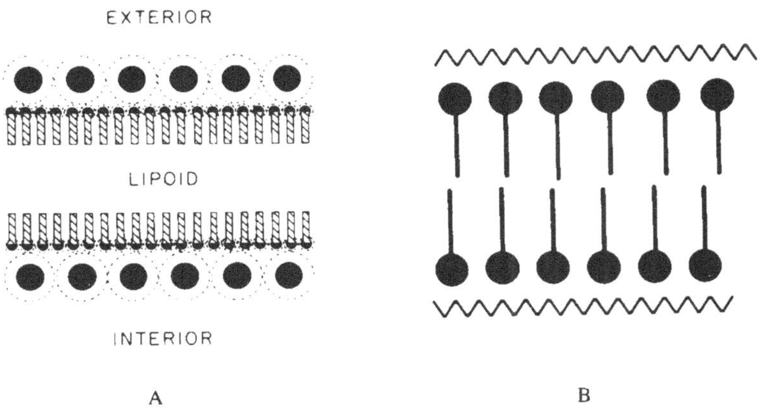

FIGURE 2.. (A) The original model for the plasma membrane proposed by Danielli and Davson;[3] (B) the modified model published in 1952 by Davson and Danielli. (From Davson, H. A. and Danielli, J. F., *The Permeability of Natural Membranes*, Cambridge University Press, London, 1952. With permission.)

and the universal presence of this pattern led to the concept that all membranes are structurally identical. For as many years, the proteins that constitute the major part of the mass of most membranes were ignored except those that contributed to the sandwich structure. They could easily be ignored because they were lost during the preparation of the specimens, although this was not acknowledged.

With the advent of high-resolution electron microscopic analysis of the structure of cells, membrane models could be based on direct observations. However, not all membrane models that were proposed from now on were based on such observations. Some were, instead, deduced on theoretical ground. Of those that were based on observations, it is possible to distinguish those in which the observations have been interpreted objectively without any preconceived concept regarding the structure of membranes and those in which the interpretation has been based on a dogmatically assumed basic structure of membranes that applies universally to all membranes.

We can, therefore, distinguish between different approaches that we will represent by different branches of our evolutionary tree. This tree also has a sucker.

II. THE FIRST BRANCH

The first branch is contributed by membrane models that have been developed on the basis of an objective evaluation of observations. On this branch, we find a model that represents a first attempt to incorporate proteins in the structure that were not associated

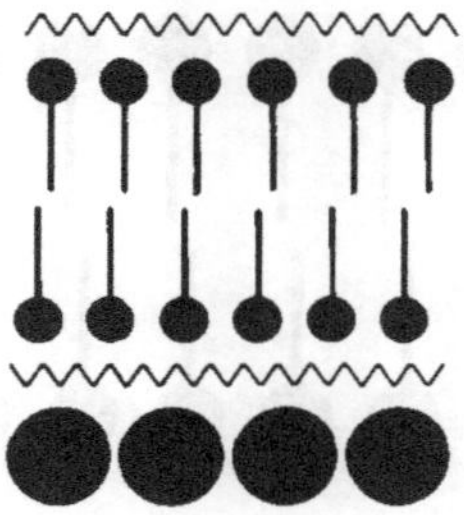

FIGURE 3. A layer of globular protein molecules was added by Sjöstrand[13] in 1960 to the cytoplasmic surface of the Danielli-Davson model to account for electron microscopic observations.

with stabilizing the sandwich structure. The plasma membrane was found to be asymmetric with a thicker stained layer located at the cytoplasmic surface. The thickness of this layer could be accounted for by an accumulation of proteins at this surface.[13] As shown in Figure 3, this layer was assumed to consist of globular protein molecules.

During the 1960s and 1970s, the proposed membrane models reflect more concern about the role the proteins may play in the structure of membranes. It eventually became obvious to some researchers that the technique to prepare tissues for electron microscopic analysis that had been so useful in making it possible to discover the cytoplasmic membranes was not satisfactory for an analysis of the membrane structure at a molecular level, because the preparatory technique led to an extensive denaturation of the proteins.

It had thus become obvious that the triple-layered structure of membranes as revealed in electron micrographs could be caused by a denaturation of the membrane proteins, because this pattern appeared irrespective of the lipid content of membranes and, in the case of mitochondrial membranes, even after extraction of practically all lipids.[14] This pattern could, therefore, not be used as a criterion for the existence of a continuous lipid bilayer if it was not known that such a bilayer existed.

Attempts were now made to improve the preparatory techniques with the aim of limiting sufficiently the extent to which the proteins were denatured to make the protein molecules retain their globular conformation. Several preparatory methods were developed and refined during the 1960s and 1970s that made it possible to collect new information regarding the structure of membranes.

An intermediate step in this development involved a slight modification of the conventional technique that, however, led to pictures of membranes showing a particulate structure instead of a triple-layered pattern. Cytoplasmic membranes seemed to consist of a layer of globules that were first interpreted to consist of lipids but later were considered to be globular proteins.[15,16] A membrane model was proposed according to which globular protein molecules were embedded in a lipid bilayer in such a way that they extended across the bilayer and were exposed at both surfaces of the membrane (Figure 4).

This observation, which later could be shown to involve a fixation artifact, was encouraging because it seemed to reveal how a globular protein molecule would appear in an electron microscopic picture, and it showed that a sufficiently high resolution could be achieved on thin sections to make it possible to observe individual protein molecules when in their globular conformation. The observation, therefore, became the starting point for a more serious attack on the problem of preserving cells without extensive denaturation of proteins in connection with the embedding for thin sectioning.

This work was motivated by the conviction that "membranes may represent a fundamental

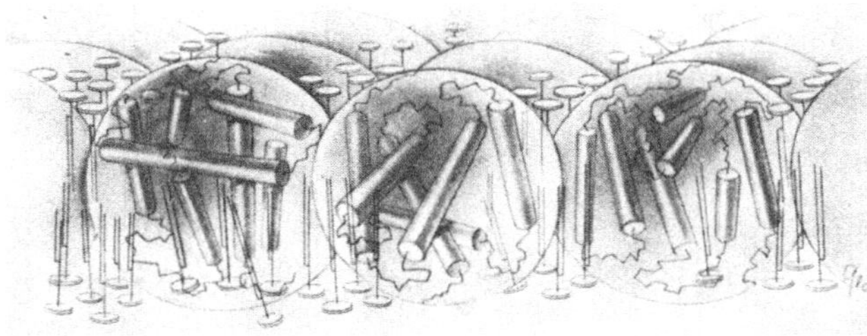

FIGURE 4. Globular protein molecules penetrating a lipid bilayer according to a model proposed by Sjöstrand in 1964. (From Sjöstrand, F. S., *Regulatory Functions of Biological Membranes*, Järnefelt, J., Ed., Elsevier, Amsterdam, 1968. With permission.)

principle of organization of the cytoplasm'' and that this ''fundamental principle'' involved a structural organization at a molecular level that offered special conditions for the functioning of complex enzyme systems, like the respiratory chain in mitochondria. The work aimed at finding out what these conditions were, which was conceived of as being of a most basic biological significance.

Membranes were considered to function both as barriers and as sites for metabolic processes, the barrier function dominating in some types of membranes while the metabolic functions dominated in others, such as the crista membrane in mitochondria. This membrane would then be suitable for the analysis since structural features related to a barrier function would be swamped by those associated with the metabolic functions. To study the structurally simplest membrane would have been meaningless because it would have been a pure barrier type of membrane.

The two most useful methods that were developed during the 1960s, the low denaturation embedding techniques developed by Sjöstrand and Barajas[17] and the freeze-fracturing technique developed by Moore and Mühlethaler,[18,19] contributed structural information that extensively deviated from that obtained earlier.

It was shown in 1968 that, in the embedded tissue, the crista membrane measured almost 150 Å in thickness instead of 40 to 50 Å in conventionally prepared tissues. Since the average diameter of polypeptides can be assumed to be 40 to 50 Å, they must form a three-dimensional aggregate within the membrane.[17] In 1970, a three-dimensional model illustrating this concept was published and is shown in Figure 5.[20]

This membrane structure represented a too radical change to become accepted. It was considered impossible that a membrane could be that thick.

It was of importance that the observations made on material embedded according to the new techniques could be confirmed by analyzing freeze-fractured and freeze-sectioned tissues.[21,22] When properly executed, the freeze-fracturing technique exposes the cell structure to a minimum of conditions that can cause a structural perturbation. The correctness of the determined thickness of the crista membrane could even be confirmed by analyzing negatively stained whole isolated unfixed mitochondria.

There were no possibilities earlier to incorporate the biochemical information regarding the respiratory chain in the membrane model, because the complexes distinguished on the basis of biochemical analysis were not structurally defined. This situation has now changed by the purification of, for instance, cytochrome oxidase and cytochrome reductase sufficiently to allow a determination of the size and the shape of these complexes. It is then of interest that these dimensions require that the crista membrane has the observed thickness of 120 to 130 Å. Cytochrome oxidase, furthermore, is associated with the lipid bilayer in a very asymmetric way, which requires that the bilayer is located at the surface of the membrane. The deduced membrane model can thus accommodate these components (Figure 6).

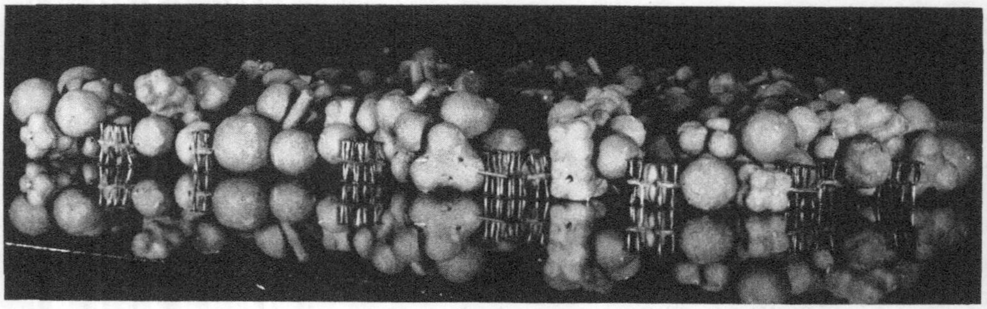

FIGURE 5. Three-dimensional model of the crista membrane in mitochondria which in 1968 was deduced on the basis of observations made on low denaturation embedded material by Sjöstrand and Barajas.[17] (Reproduced from Sjöstrand, F. S. and Barajas, L., *J. Ultrastruct. Res.*, 32, 293, 1970. With permission.)

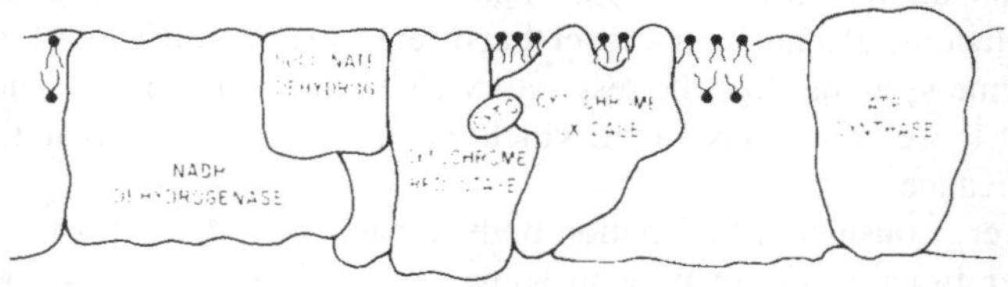

FIGURE 6. Schematic drawing of part of the crista membrane structure to illustrate that the dimensions of the membrane, as determined on low denaturation embedded tissues, can accommodate the complexes of the respiratory chain, the dimensions and shapes of which have been determined by electron microscopic analysis of single-layer crystals.

An attempt was now made to deduce what kind of environments could exist within the crista membrane. These deductions led to the conclusion that the interior of this membrane is mixed polar and nonpolar, that there is a nonpolar barrier formed by both lipids and proteins at the matrix surface of the membrane, and that the membrane represents a structure characterized by a low water activity. This would, however, require that the membrane be equipped with a water-translocating mechanism, a prediction that led to designing experiments that showed conclusively that the crista membrane has a high capacity for translocating water.[23,24]

It was concluded that the crista membrane offered environments for enzyme-catalyzed reactions that differ extensively from an aqueous environment. The mitochondrion can be considered to consist of two phases, one aqueous phase in the matrix and one semisolid phase in the cristae.[25] The latter phase can be conceived of as a biological variant of a solid state.

It was also recognized that the crista membrane is an extreme case of a metabolically active membrane and that membranes designed for a barrier function have a different structure. This was demonstrated by the analysis of four different types of membranes in addition to the crista membrane: the inner and outer surface membranes of mitochondria, the membrane of the outer segment disks in photoreceptors, and the plasma membrane of the outer segment of photoreceptors.[26-28] The deduced membrane models are shown in Figures 7 to 10.

The barrier type of membrane as illustrated by the disk membrane and the analyzed type of plasma membrane are characterized by the presence of a lipid bilayer that is penetrated only to a limited extent by proteins and a second protein layer associated with one surface

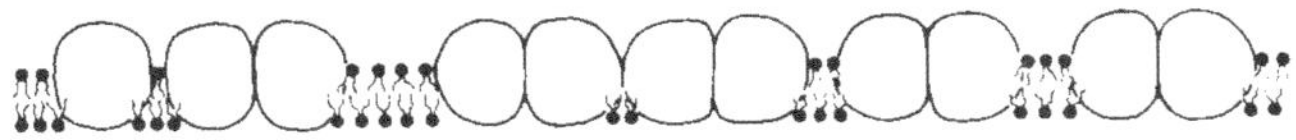

FIGURE 7. The inner surface membrane in mitochondria. The thickness of this membrane is half that of the crista membrane. Lower surface of membrane faces the matrix. (Modified after Sjöstrand, F. S., *Membrane Fluidity in Biology,* Vol. 1, Aloia, R. C., Ed., Academic Press, New York, 1983, 83.

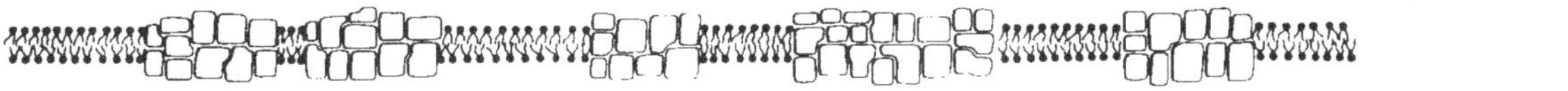

FIGURE 8. The outer surface membrane in mitochondria. The mitochondria offer favorable conditions for comparing the structure of different membranes in the same preparation and in the same pictures. The models of the three mitochondrial membranes are based on observations made on low-denaturation embedded tissues, freeze-sectioned tissues, freeze-fractured tissues, and on negatively stained whole unfixed mitochondria. The observations revealed extensive differences in the structure of these membranes, refuting the concept of one basic structure that is common to all membranes. (From Sjöstrand, F. S., *Biol. Cell.*, 39, 217, 1980. With permission.)

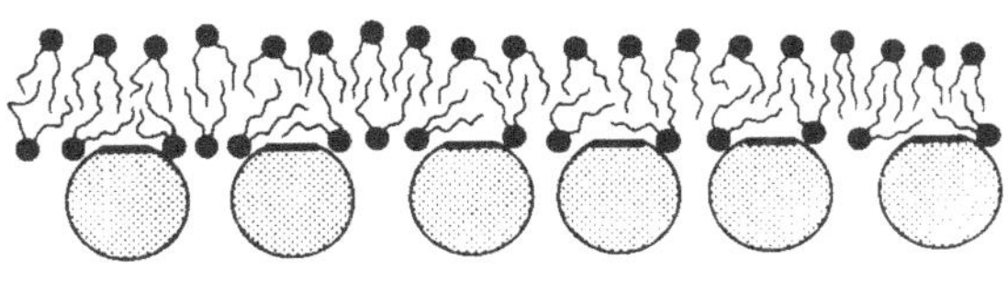

FIGURE 9. The membrane of the double membrane disks in the outer segment of photoreceptors. The rhodopsin molecules are represented by circles with a heavy straight line indicating the retinal. They are assumed to be associated hydrophobically with the lipid bilayer through the hydrophobic region at the site of the retinal. The liquid state of the lipid bilayer makes the lipid molecules expose partially their hydrocarbon chains and, also, makes the rhodopsin molecules diffuse in the plane of the membrane. The layer of rhodopsin molecules is facing the center of the disk. (Redrawn from Sjöstrand, F. S. and Kreman, M., *J. Ultrastruct. Res.*, 65, 195, 1978.)

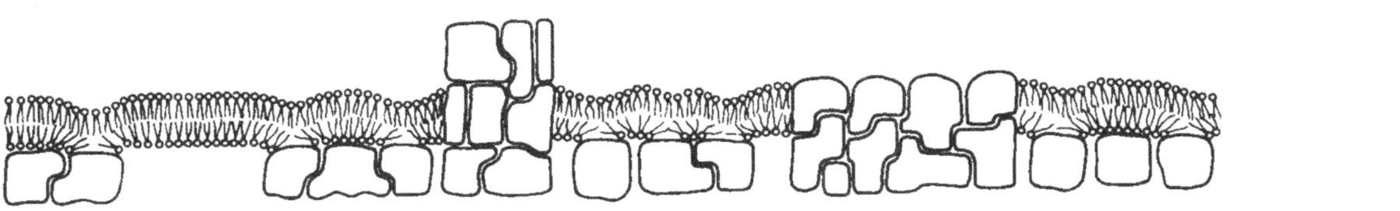

FIGURE 10. The plasma membrane of the outer segment of photoreceptors. The molecular organization is similar to that of the disk membrane. However, large multimolecular complexes permeate the plasma membrane. The drawing aims at illustrating various possible modes of interactions between the molecules in the membrane.(From Sjöstrand and Kreman, 1979.) This model is not intended to represent the plasma membrane structure universally. The diversity of the structure of membranes and the observed differences in the structure of the plasma membrane covering different parts of the same cell make it necessary to analyze each type of plasma membrane to determine its structure. Some common structural features might then be revealed but cannot be anticipated. (From Sjöstrand, F. S. and Kreman, M., *J. Ultrastruct. Res.*, 66, 254, 1979. With permission.)

of the lipid bilayer. This means that the capability of the lipid bilayer to form a barrier for charged particles is only to a limited extent reduced by the presence of permeating protein molecules that are known to impair the barrier properties of lipid bilayers.

These four membrane models were deduced mainly on the basis of analyzing freeze-fractured tissues. Information regarding the thickness of the membranes had been obtained

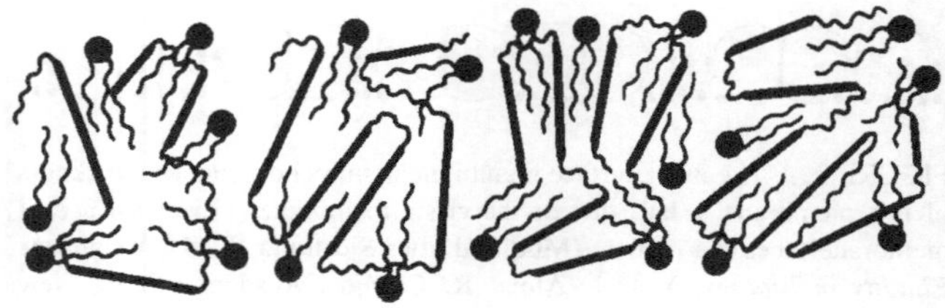

FIGURE 11. Benson's membrane model consisting of a single layer of lipoprotein complexes. It was proposed in 1968.

from embedded material and was of crucial importance in the case of the disk membrane because it allowed a unique interpretation of the pictures of freeze-fractured outer segments.

For the interpretation of pictures of freeze-fractured material, it is necessary to know the location of the fracture planes which cannot be predicted. The location was established on the basis of observations and was simplified in the case of the crista membrane that, thanks to its thickness, was frequently fractured obliquely, exposing the interior of the membrane. Such a location of a fracture plane is easy to determine, while a fracture passing parallel to a membrane offers considerable difficulties in establishing its location relative to the membrane. In the case of the outer segment disks, the periodic structure of the outer segments offered particularly favorable conditions for locating the fracture plane.

The analysis of freeze-fractured material was also supplemented by an analysis of isolated intact disks, outer segments, and mitochondria after freeze-drying and surface replication.[29] This allowed observing the surface of these membranes that was exposed. The deductions made when analyzing the freeze-fractured membranes could then be confirmed.

III. THE SECOND BRANCH

We now turn to the second branch on the evolutionary tree where we find models that have been developed on the basis of a hypothetical deduction based on indirect evidence or on the results of pure hypothetical reasoning.

The model proposed by Benson in 1967 was based on the observation that lipids in plant cell membranes seemed to associate with the proteins in a specific way to form lipo-protein complexes with a precisely defined composition. The lipid bilayer structure was completely disposed of and the membrane model consisted of a single layer of protein molecules to which lipids were bound (Figure 11).[30,31]

The second model on this branch is that of Singer and Nicolson[32] that, presented in 1972, was based on a qualitative thermodynamic reasoning. The partially nonpolar nature of protein molecules could make it thermodynamically favorable for protein molecules to associate with the lipid bilayer when they exposed nonpolar amino acid side chains extensively at their surface. These molecules would become firmly bound to the bilayer by a thermodynamic force (Figure 12).

The introduction of protein molecules in the lipid bilayer was performed very cautiously according to this model. Single protein molecules were proposed to penetrate the bilayer either halfway or completely. These firmly bound proteins were referred to as the "integral proteins", and only these proteins were considered to be of structural significance. All other proteins, constituting the "peripheral proteins", were assumed to be loosely bound to the membrane surfaces, and they did not contribute to the structural integrity of the membrane. They were, therefore, conveniently left to the reader to do what he or she wanted to do with them. An imaginary reader with some artistic, creative talent could very well design a model with a structure corresponding to that of a real membrane by playing with the "peripheral

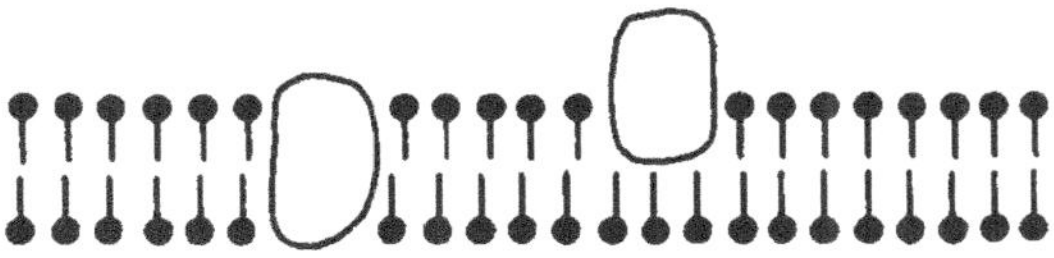

FIGURE 12. The Singer-Nicolson membrane model proposed in 1972.

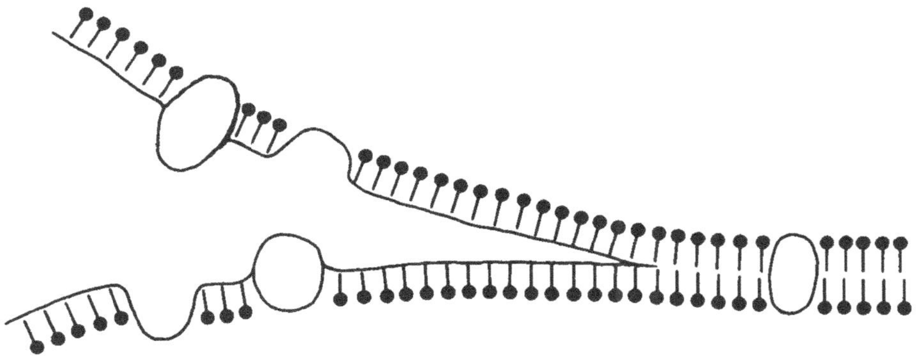

FIGURE 13. Drawing to illustrate the dogma on which pictures of freeze-fractured material are generally interpreted. The dogma simplifies the analysis and the presentation of the observations, because the researcher only has to put the established labels on the fracture faces, count particles, and measure their diameters. All particles are by definition intramembrane particles. The dogma excludes that any other particles can be observed.

proteins''. Unfortunately, however, he or she would not know when this would be the case. The instructions were missing for how to use the building kit.

The Singer-Nicolson membrane model was proposed with limited aim to present the structure of membranes. The model primarily aims at illustrating a thermodynamically possible association of proteins with a lipid bilayer to form a mechanically stable structure due to thermodynamic forces. However, several models, including that of Wallach and Gordon,[33] in which protein molecules had been assumed to penetrate a lipid bilayer, had already been proposed, and the distinction between integral and peripheral proteins was identical to that of firmly bound and loosely bound proteins proposed for the crista membrane by Sjöstrand and Barajas[17] in 1968 on the basis of actual observations.

However, the Singer-Nicolson membrane model introduced the changes sufficiently gradually for the model to sell well, and it has been generally accepted, which might be highly beneficiary for the further evolution of concepts regarding the structure of membranes, or it may have a detrimental effect on the sales of models that deviate.

IV. THE THIRD BRANCH

Along the third branch of the evolutionary tree, we have models that have been developed on the basis of observations made on freeze-fractured material, but where the interpretation of the information has been based on a dogma reflecting a preconceived opinion about membrane structure. The membrane is thus assumed to consist of a continuous lipid bilayer, and this bilayer is cleaved through its middle. The fracture plane is, therefore, always located in the middle of membranes (Figure 13).

What makes this approach highly unscientific is the fact that no observations can be made that disagree with the preconceived membrane structure since the interpretation is based on it. Instead, all observations appear to confirm the assumed structure.

The consequence of this situation is that no progress is possible. Evolution along this branch has come to a standstill fixed by a dogma that excludes the freedom of questioning that is the basis for progress in science. The same interpretation of the pictures is repeated in a monotonous way with a lipid bilayer being split through its middle revealing intramembrane particles. The only differences that distinguish membranes are the number of particles per unit area, the size of the particles, and possible regular arrays of particles. No other aspects of the membrane structure can ever be revealed because no other views of the membrane are possible.

This fixation to a dogma will doom the species on this branch to extinction.

V. THE SUCKER

A sucker on the evolutionary tree can also be recognized. The attempts to describe the lipid bilayer as structurally highly organized to allow a close packing of the molecules to guarantee strong "cohesion" between them can be considered to belong to the era when the attention was still focused on the lipid bilayer.

The demonstration that biological lipid bilayers are only 40 to 50 Å thick or even thinner and that they are in a liquid, highly disordered state means that this sucker has been cut off.

VI. A VISION

We have seen how the models describing certain features of the molecular structure of membranes have undergone an evolution that has eventually lent support to the concept that membranes "may represent a fundamental principle of organization of the cytoplasm", with the membrane structure itself establishing special conditions for metabolic functions. The early predicted diversity of the structure of membranes has also been shown to be correct.[34] A comparison of the structure of the outer surface membranes and the crista membrane in mitochondria reveals extreme structural differences.

Future research on the nature of the intramembrane environments in metabolically active membranes and the work on model systems that aim at mimicking these environments are likely to contribute to the development of an understanding of the role the membrane structure plays in cell metabolism. It is likely to lead to the development of structural biochemistry or the biochemistry of the semisolid state. This way the study will become directed towards the structurally intact cellular components in which the interactions between enzymes and between enzyme systems are retained.

In membranes with a mixed barrier and metabolic function and in which the barrier function dominates, it is likely that the metabolic functions are confined to regions within which the conditions for molecular interactions are similar to those in the crista membrane. The structural characteristics of this membrane may, therefore, be of general significance.

The building of membrane models has reached a fascinating stage in which the proposed structure can eventually be translated into its possible functional significance at a level above that of mechanical stability. We look forward to one day when it will be impossible to describe the basic metabolic functions of cells without including the structure as a factor which is at least as important as the chemical composition. Presently, the structure too closely follows the role of decorative ornaments or is used for pictorial symbolism in biochemical presentations, a symbolism that has little to do with reality.

REFERENCES

1. **Overton, E.,** Ueber die allgemeinen osmotischen Eigenschaften der Zelle, ihre vermutlichen Ursachen und ihre Bedeutung für die Physiologie, *Naturforschende Gesellschaft zu Zürich, Vierteljahrsschrift,* Zürich, 1899, 88.
2. **Gorter, E. G. and Grendel, F.,** Bimolecular layers of lipids on the chromocytes of blood, *Proc. K. Acad. Wetensch. (Amsterdam),* 29, 314, 1926.
3. **Danielli, J. F. and Harvey, E. N.,** The tension at the surface of mackerel egg oil, with remarks on the nature of the cell surface, *J. Cell. Comp. Physiol.,* 5, 483, 1935.
4. **Danielli, J. F. and Davson, H. A.,** A contribution to the theory of permeability of thin films, *J. Cell. Comp. Physiol.,* 5, 495, 1935.
5. **Sjöstrand, F. S.,** The ultrastructure of the retinal rods of the guinea pig eye, *J. Appl. Phys.,* 19, 1188, 1948.
6. **Sjöstrand, F. S.,** An electron microscope study of the retinal rods of the guinea pig eye, *J. Cell. Comp. Physiol.,* 33, 383, 1949.
7. **Steinmann, E.,** An electron microscope study of the lamellar structure of chloroplasts, *Exp. Cell Res.,* 3, 367, 1952.
8. **Palade, G. E.,** The fine structure of mitochondria, *Anat. Rec.,* 114, 427, 1952.
9. **Sjöstrand, F. S.,** Electron microscopy of mitochondria and cytoplasmic double membranes, *Nature (London),* 171, 30, 1953.
10. **Dalton, A. J. and Felix, M. D.,** Cytological and cytochemical characteristics of the Golgi substance of epithelial cells of the epididymis — in situ, in homogenates and after isolation, *Am. J. Anat.,* 94, 171, 1954.
11. **Sjöstrand, F. S. and Hanzon, V.,** Membrane structures of cytoplasm and mitochondria in exocrine cells of mouse pancreas as revealed by high resolution electron microscopy, *Exp. Cell Res.,* 7, 393, 1954.
12. **Porter, K. R. and Kallman, F. L.,** Significance of cell particulates as seen by electron microscopy, *Ann. N.Y. Acad. Sci.,* 54, 882, 1952.
13. **Sjöstrand, F. S.,** Morphology of ordered biological structures, *Radiat. Res. Suppl.,* 2, 349, 1960.
14. **Fleischer, S., Fleischer, B., and Stoeckenius, W.,** Fine structure of whole and fragmented mitochondria after lipid depletion, *Fed. Proc. Fed. Am. Soc. Exp. Biol.,* 24, 296, 1965.
15. **Sjöstrand, F. S.,** A new ultrastructural element of the membranes in mitochondria and some cytoplasmic membranes, *J. Ultrastruct. Res.,* 9, 340, 1963.
16. **Sjöstrand, F. S.,** The structures of cellular membranes and cell membrane contacts in the nervous system, in *Biochemistry and Pharmacology of Basal Ganglia; Proc. 2nd Symp. of the Parkinson's Disease Information and Research Center,* Costa, E., Cote, L. J., and Yahr, M. D., Eds., Raven Press, New York, 1966, 17.
17. **Sjöstrand, F. S. and Barajas, L.,** Effect of modifications in conformation of protein molecules on structure of mitochondrial membranes, *J. Ultrastruct. Res.,* 25, 121, 1968.
18. **Moore, H., Mühlethaler, K., Waldner, H., and Frey-Wyssling, A.,** A new freezing ultramicrotome, *J. Biophys. Biochem. Cytol.,* 10, 1, 1961.
19. **Moore, H. and Mühlethaler, K.,** Fine structure of frozen-etched yeast cells, *J. Cell Biol.,* 17, 609, 1963.
20. **Sjöstrand, F. S. and Barajas, L.,** A new model for mitochondrial membranes based on structural and on biochemical information, *J. Ultrastruct. Res.,* 32, 293, 1970.
21. **Sjöstrand, F. S. and Cassell, R. Z.,** Structure of inner membranes in rat heart muscle mitochondria as revealed by means of freeze-fracturing, *J. Ultrastruct. Res.,* 63, 111, 1978.
22. **Sjöstrand, F. S. and Bernhard, W.,** The structure of mitochondrial membranes in frozen sections, *J. Ultrastruct. Res.,* 56, 233, 1976.
23. **Candipan, R. and Sjöstrand, F. S.,** *J. Ultrastruct. Res.,* in press.
24. **Candipan, R. and Sjöstrand, F. S.,** *J. Ultrastruct. Res.,* in press.
25. **Sjöstrand, F. S.,** The structure of mitochondrial membranes: a new concept, *J. Ultrastruct. Res.,* 64, 217, 1978.
26. **Sjöstrand, F. S. and Cassell, R. Z.,** The structure of the surface membranes in rat heart muscle mitochondria as revealed by freeze-fracturing, *J. Ultrastruct. Res.,* 63, 138, 1978.
27. **Sjöstrand, F. S. and Kreman, M.,** Molecular structure of outer segment disks in photoreceptor cells, *J. Ultrastruct. Res.,* 65, 195, 1978.
28. **Sjöstrand, F. S. and Kreman, M.,** Freeze-fracture analysis of structure of plasma membrane of photoreceptor cell outer segments, *J. Ultrastruct. Res.,* 66, 254, 1979.
29. **Sjöstrand, F. S.,** Low temperature techniques applied for CTEM and STEM analysis of cellular components at a molecular level, *J. Microsc.,* 128, 279, 1982.
30. **Benson, A. A.,** Plant membrane lipids, *Annu. Rev. Plant Physiol.,* 15, 1, 1964.
31. **Benson, A. A.,** On the orientation of lipids in chloroplast and cell membranes, *Am. Oil Chem. Soc. J.,* 43, 265, 1966.

32. **Singer, S. J. and Nicolson, G. L.,** The fluid mosaic model of the structure of cell membranes, *Science,* 175, 720, 1972.
33. **Wallach, D. F. H. and Gordon, A. S.,** Lipid protein interactions in cellular membranes, *Fed. Proc. Fed. Am. Soc. Exp. Biol.,* 27, 1263, 1968.
34. **Sjöstrand, F. S.,** Electron microscopy of cells and tissues, in *Physical Techniques in Biological Research,* Vol. 3, Pollister, A., Ed., Academic Press, New York, 1966, 169.

Chapter 2

PROTEIN-PROTEIN INTERACTIONS IN CELL MEMBRANES

Gregory B. Ralston

TABLE OF CONTENTS

I. INTRODUCTION

Although the lipids of biological membranes provide a permeability barrier and a fluid boundary to the cell, the membrane proteins are responsible for specific recognition phenomena, enzymic activities, specific transport, transmembrane signaling, and structural elements that contribute to the morphology of the cell.[1,2] Through the introduction of high-resolution electrophoresis in the presence of SDS, coupled with sensitive protein detection methods, the individual polypeptides of a variety of membranes have been characterized and catalogued. However, many of the proteins of biomembranes do not float independently in an indifferent lipid bilayer. Rather, there exist complex and specific interactions between proteins. Many of these interactions may be essential for the function of the membrane. Some of these interactions may be strong and relatively fixed, either thermodynamically or kinetically.[3] Others may be weaker, transient, and modulated by environmental conditions or interaction with ligands.[4]

For many years, some of the components of the mitochondrial respiratory chain and the chloroplast photosynthetic systems have been envisaged as well-defined protein complexes.[5] Recent evidence indicates that the inner membrane matrix compartment of mitochondria and chloroplast thylakoids may have a packed protein structure, in which integral proteins embedded in the membrane interact not only with other integral proteins, but also with a highly organized protein structure in the matrix space.[6,7] There is abundant evidence for the existence of many other membrane proteins in oligomeric states. Some of these exist as homopolymeric oligomers comprised of a single polypeptide species (Table 1); others as heteropolymeric complexes comprised of several different polypeptide chains (Table 2). These tables are by no means exhaustive, but simply reflect those proteins whose structures are either best known, or the subject of considerable current investigation.

II. HOMOPOLYMERS

Biological membranes are fundamentally asymmetric.[8,9] The integral proteins have a preferred transmembrane orientation, presumably reflecting the initial biosynthesis and membrane insertion mechanism. The enormous kinetic barrier to transmembrane flip-flop ensures that most protein molecules retain their initial orientation throughout the life of the cell. In addition, all subunits of a homo-oligomer will be immersed to the same depth in the bilayer.[10,11] This fixed orientation and the planar, amphipathic nature of the lipid bilayer place considerable limitations on the ways that membrane proteins might self-associate.

A. Symmetry

The preferred type of association for homopolymeric membrane proteins is that of cyclic symmetry, with the axis of rotational symmetry passing normally through the membranes (Figure 1).[12,13] This type of association preserves the transmembrane orientation of the subunits and provides maximum stabilization through protein-protein interactions, as well as through interactions between the proteins, lipids, and aqueous environment.[13] Furthermore, the formation of oligomers with areas of protein-protein contact within the lipid bilayer reduces the contact between the hydrophobic surface of the protein and the hydrocarbon chains of the lipid, lowering some of the restrictions to protein folding that would be necessary to generate a larger hydrophobic protein surface.

Dimers, trimers, and in fact all odd-numbered, closed homopolymeric oligomers possess only cyclic symmetry.[4,14] Thus, it is perhaps not surprising to find dimers as a common structural unit for membrane proteins (see Table 1). More unusual is what appears to be a significant number of trimeric structures — including the porins,[15-17] bacteriorhodopsin,[18] carboxylesterase,[19] and viral hemagglutinin[20] (Table 1). Not long ago, on the basis of existing knowledge of soluble proteins, trimers were considered uncommon oligomeric states for

Table 1
HOMO-OLIGOMERIC TRANSMEMBRANE PROTEINS

Protein	Source	Oligomer	Ref.
ADP/ATP carrier	Mitochondria	Dimer	30
Band 3 protein	Erythrocytes	Dimer (+ tetramer?)	23—27
Porin	*E. coli*	Trimer	15,22
Porin	*S. cerevisiae* mitochondria	Trimer	16
Porin	Rat liver mitochondria	Trimer	17
Bacteriorhodopsin	*H. halobium*	Trimer	18
Carboxyl esterase	Microsomes	Trimer	19
Hemagglutinin	Influenza virus	Trimer	20
Neuraminidase	Influenza virus	Tetramer	21
Ca^{2+}-ATPase	Sarcoplasmic reticulum	Tetramer	67
Gap junction protein	Rat hepatocytes	Hexamer	97,98

Table 2
HETERO-OLIGOMERIC TRANSMEMBRANE PROTEINS

Protein	Source	Oligomer	Ref.
$Na^+ + K^+$-ATPase	Plasma membrane	$(\alpha\beta)_2$	31—33
Acetylcholine receptor	*Torpedo, Electrophorus*	$\alpha_2\ \beta\gamma\delta$	34, 35
Cytochrome *c* oxidase	Beef heart mitochondria	$(I, II, III, IV, V, VI, VII)_2$	38, 39
ATPase-F_1	Bacteria	$\alpha_3\beta_3\gamma\delta\epsilon$	75, 76

proteins.[14] In the case of hemagglutinin from influenza virus, the trimeric nature of the oligomer has been verified unambiguously, for the extramembranous domain of the protein, through the use of X-ray crystallography.[20]

One common oligomeric state for soluble proteins, the tetramer with dihedral symmetry, is an unlikely structure for integral membrane proteins because each subunit would not be able to maintain the same transmembrane orientation. Recently, Colman et al.[21] have demonstrated, for the first time, cyclic tetramer formation in the extramembrane domains of influenza virus neuraminidase (Figure 2). Klingenberg[13] has suggested that the common occurrence of oligomers among membrane transport proteins may reflect a common functional property. The interactions between subunits of a cyclic oligomer are minimal along the symmetry axis because identical residues face each other across this axis. This may facilitate the formation of a transmembrane channel between the subunits, a channel that may be important for transport. In the context of oligomers in lipid membranes, such channels are also likely to be polar, in order to favor protein-protein interactions over protein-lipid interactions.

That such channels represent the transport path in specific translocation mechanisms has yet to be verified. The evidence is still incomplete and controversial.[11] In the particular case of porin, each subunit of the trimer appears to carry its own separate pore or channel, clearly resolved in reconstructed images from electron micrographs of two-dimensional sheets.[22]

B. Indefinite Self-Association?

While the formation of oligomers with cyclic symmetry is undoubtedly an important process in biological membranes, there are no *a priori* grounds to discount subsequent association of such oligomers *via* an indefinite or isodesmic mechanism (Figure 3). Indeed, such a mechanism may underly the current uncertainty concerning the oligomeric state of the red cell band 3 protein.[23-27]

The band 3 protein is believed to occur in the erythrocyte membrane as a dimer.[23,24]

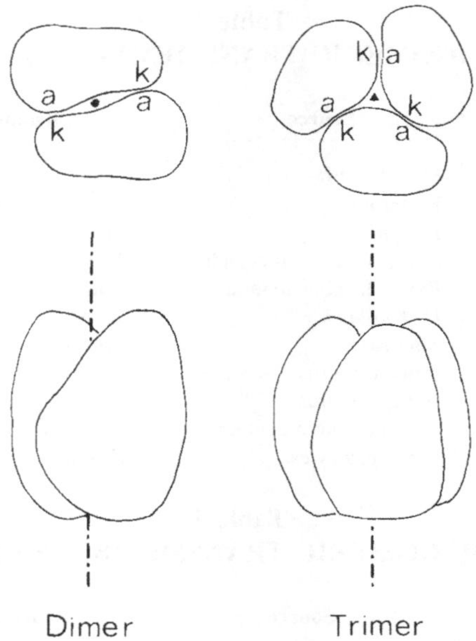

FIGURE 1. Examples of a dimer and trimer showing rotational symmetry. The top diagram shows a view along the symmetry axis, while the lower diagram shows a view along the plane of the membrane. The oligomers are stabilized by interactions between complementary sites *a* and *k* on each monomer.

However, there is mounting evidence that at least some of the band 3 population also exists as a tetramer,[25-27] and hexamers and higher oligomers may also exist.[27] How many of these are artifactual aggregates has yet to be established. However, evidence for the reversible dimer-tetramer interconversion in Triton® X-100 solution has recently been obtained,[26] and it appears that the association behavior of band 3 must include at least the tetramer as well as dimer and monomer.

Kyte[11] has pointed out that the one protein cannot form both cyclic dimers and cyclic tetramers without considerable change in shapes. In addition, symmetry dictates that any further association of a cyclic dimer cannot result in a closed structure. Instead, further association is likely to proceed indefinitely (Figure 3). In the case of the band 3 protein, this type of behavior is the most likely explanation of the conflicting reports of oligomer size. This interpretation is supported by the fact that both membrane-inserted and cytoplasmic domains of band 3 protein are separately capable of dimerization.[28,29] Thus, there are two inherently different self-association sites in band 3, and unless the angle between these sites in the plane of the membrane is zero, indefinite self-association beyond the dimer is expected.[4]

III. HETEROPOLYMERS

While many proteins appear to exist in the membrane as oligomers of a single polypeptide, there are many examples of oligomers involving different proteins (Table 2). The (Na^+, K^+)-ATPase appears to comprise two chains, associated to form an $\alpha\beta$ protomer[10] which further associates to the $(\alpha\beta)_2$ dimer.[31] This structure has been verified recently from kinetic evidence[32] and from an electron microscope study of quasi-crystalline membrane arrays.[33]

One of the best-characterized heteropolymer integral proteins, the acetylcholine receptor (from the electric ray, *Torpedo*,[34] or the electric eel, *Electrophorus*[35]) appears in electron micrographs as a ring-like structure, but has no true rotational symmetry.[36] This oligomer

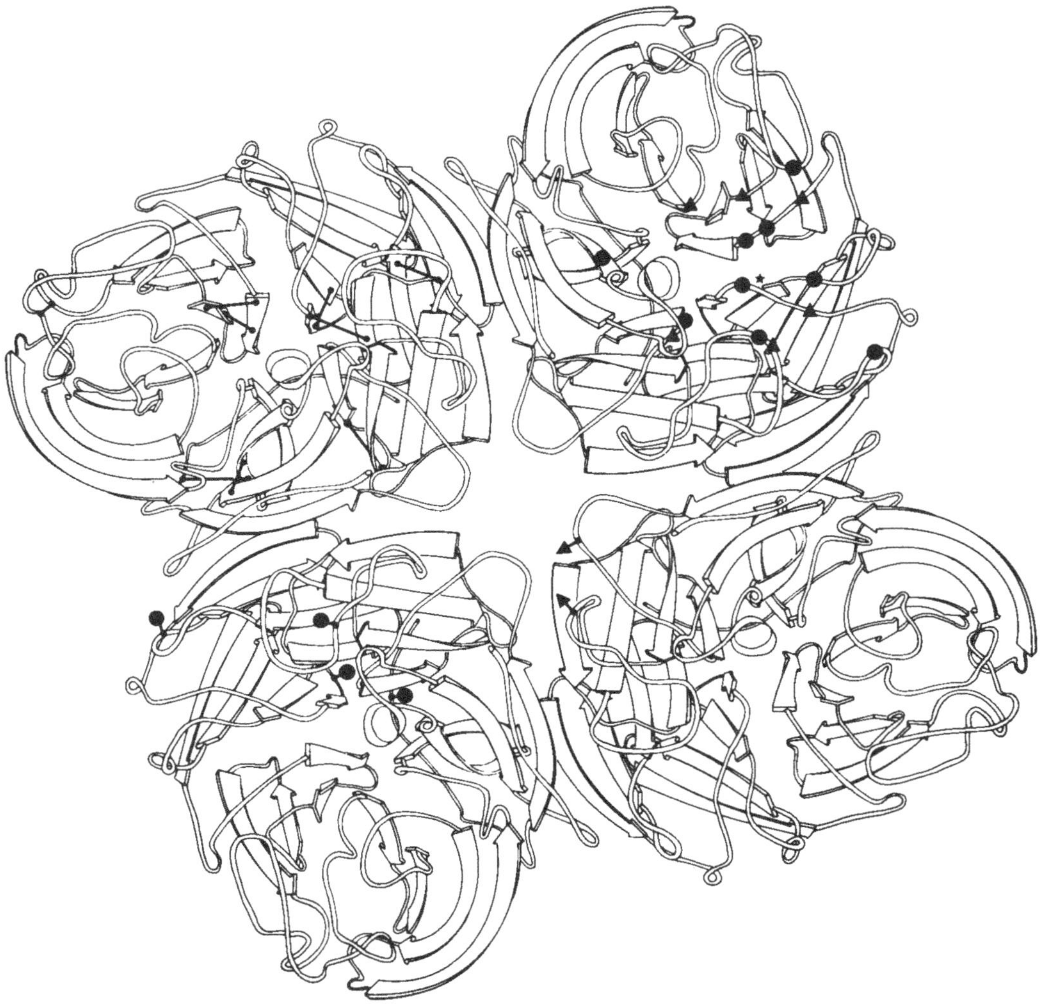

FIGURE 2. Diagram of the tetramer of influenza virus neuraminidase, viewed from above, down the axis of fourfold rotational symmetry. (Reprinted from Varghese, J. N., Laver, W. G., and Colman, P. M., *Nature (London)*, 303, 38, 1983. With permission.)

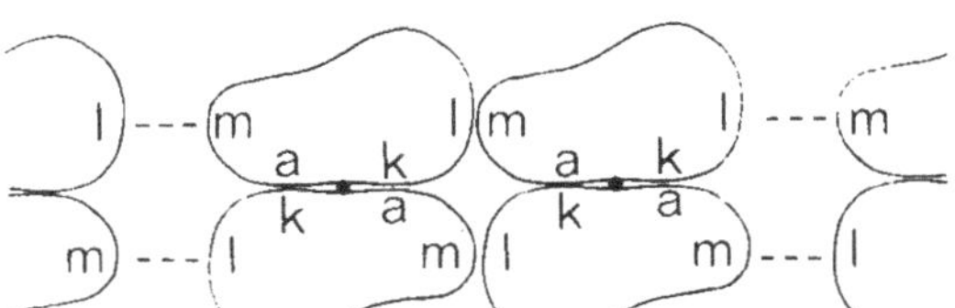

FIGURE 3. Diagram showing subsequent indefinite (isodesmic) self-association of a cyclic dimer. Association of *l* and *m* sites between two dimers does not result in a closed structure, but allows further additions to continue at each end of the oligomer.

is comprised of five polypeptides of four different types. Pairs of the ring-like receptors can be linked via disulfide bridges on the δ subunit to form dimers of rings or occasionally oligomeric chains.[37] These disulfide-linked oligomers can be dissociated by dithiothreitol, and their functional significance is unknown.

The electron transfer chains of mitochondria and chloroplasts comprise relatively stable protein complexes which can be isolated as functional oligomers.[38,39] The cytochrome oxidase (aa_3 complex) is thought to comprise at least six polypeptide chains, and the protomer undergoes a further association to a dimer.[40-43] However, with large complexes of this nature, comprising many different polypeptide chains, there is enormous difficulty in determining which polypeptides are part of the functioning complex, and which simply represent "artefactual togetherness".[38,44] In the case of cytochrome oxidase, in particular, there is also apparently wide variation between species in the number of polypeptides in the functioning molecule.[43,45,46] Georgevich et al.[47] have attempted to define the subunit composition on the basis of the minimal structure able to transfer electrons from cytochrome *c* to oxygen, coupled to the generation of a proton gradient. According to this definition, some components recognized by others as subunits of cytochrome *c* oxidase appear to be dissociable without loss of function. On the other hand, subunit III, although relatively easily dissociated,[38,47] seems to be an essential component. Enzyme depleted in component III, although capable of performing electron transfer, cannot couple this transfer to a proton gradient.[47]

IV. TRANSIENT ASSOCIATION

While the oligomers so far considered have been sufficiently stable to be isolated, relatively weak or transient association may underly some important membrane functions.[48-50] Binding of an external ligand may alter the association state in a manner that is well understood for soluble proteins, and which depends on the number of ligand molecules bound, the protein and ligand concentration, and whether a particular association state preferentially binds ligand.[51-53]

Transient, ligand-induced association underlies the mobile receptor theory for the action of some hormones,[54] according to which binding of hormone to a receptor induces an interaction between the hormone-receptor complex and adenylate cyclase, as well as with a GTP-binding regulatory protein.[55-59]

The interactions between proteins in lipid bilayers is also influenced by the lipid component, depending upon the protein/lipid ratio (or protein concentration in the bilayer) as well as the nature of the lipid. At high dilution, bovine rhodopsin is functional and appears to be monomeric, as judged from low-angle X-ray scattering.

The rotational diffusion of bovine rhodopsin also reflects these changes. Under these conditions, the rotational correlation time of approximately 10 μsec is also consistent with the existence of the monomer.[50] At higher concentrations, the correlation time increases to 20 μsec, consistent with dimer formation, both in artificial vesicles[50] and in rod outer-segment membranes.[62] These values are unchanged on bleaching,[63] suggesting that the association does not have major functional significance.[63] The residence time for rhodopsin in these transient dimers has been estimated at approximately 10^{-5} sec.[50] Below the phase transition, the correlation time rises to approximately 10^{-4} sec, independent of lipid/protein ratios,[49,50] and indicative of cluster formation as protein and lipid segregate.[64] Rhodopsin also appears to undergo transient interaction with the GTP-binding protein, transducin, during the light-induced cyclic nucleotide cascade in retinal rod outer segments.[61]

V. APPROACHES TO STRUCTURE

The study of protein interactions in membranes, particularly those involving integral proteins, poses a dilemma. Those techniques with thermodynamic authority, such as light scattering, osmometry, and sedimentation equilibrium, cannot be applied to proteins in intact membranes, but require prior solubilization and isolation of the components under study.

On the other hand, attempts to study protein interactions within the intact membrane must involve relatively indirect methods, ambiguity, and difficulty of interpretation. Only by a combination of both approaches has a clear and unambiguous picture of some protein interactions emerged. In some favorable cases information obtained from detailed genetic studies has also been helpful in the elucidation of oligomer composition.

A. Nonionic Detergent Solubilization

Nonionic detergents, exemplified by Triton® X-100,[65] are capable of solubilizing integral membrane proteins, often without loss of function,[65] and may lead to the isolation and purification of well-defined protein oligomers with integral stoichiometry. Benign solubilization is not always achieved, however, and frequently an extensive search may be needed to find a detergent that leaves function unimpaired. Occasionally, the association state of a membrane protein may depend on the particular detergent used.[65-68]

1. Subunit Composition and Stoichiometry

Once a membrane protein has been extracted and purified in solution of nonionic detergents, it may be analyzed by dissolution and electrophoresis in the presence of sodium dodecyl sulfate (SDS), in order to determine the composition. In favorable cases such analysis can also lead to reliable estimates of molecular weight and stoichiometry. However, determination of both molecular weight and stoichiometry are not trivial tasks. The well-characterized water-soluble proteins may not be valid calibration standards for the examination of integral membrane proteins, due to different detergent-binding behavior of the two classes of protein. Furthermore, the almost ubiquitous presence of carbohydrate residues in integral membrane proteins leads to anomalous electrophoretic mobility and variable staining intensity with commonly used protein stains. These confuse both molecular weight and stoichiometry determination.[69] Subunit III of cytochrome *c* oxidase shows considerably less staining intensity than the other subunits,[38,47] and in the case of a complex protein like this, ambiguity in either molecular weight or stoichiometry can confuse the issue badly.

Perhaps the only reliable method for determination of the molecular weight of subunits is by means of sedimentation equilibrium in SDS.[44,69] Such a determination requires an accurate measure of detergent binding to the protein, and this degree of reliability may be a luxury that comes at a price few can afford in terms of the quantity of purified material required. Thermodynamically reliable measurements can also be obtained from membrane osmometry.[70] Transport experiments, such as sedimentation velocity or gel filtration,[71] even in the presence of SDS, are unreliable, because of the influence of particle shape on the hydrodynamic behavior. A careful combination of both gel filtration and sedimentation velocity, however, may lead to valid results. Tanford and colleagues[44,69] have cautioned against the use of density gradient sedimentation velocity experiments for determination of sedimentation coefficients because of the uncertainty of detergent binding in the presence of high concentrations of sucrose or salt.

Guanidine hydrochloride, although capable of dissociating and completely unfolding most water-soluble proteins,[72,73] cannot be assumed to have the same effect on membrane proteins,[74] and unless independent methods have been used to verify the complete absence of structure, studies in guanidine hydrochloride should be regarded with suspicion.[44]

Once reliable estimates of molecular weights of constituent polypeptides are available, determination of stoichiometry can lead to an estimate of the minimum molecular weight of the complex. Comparison of this value with the measured molecular weight of the complex is a check on the self-consistency of the data, and also may indicate whether the protomer corresponding to the minimum molecular weight is capable of further self-association. Un-

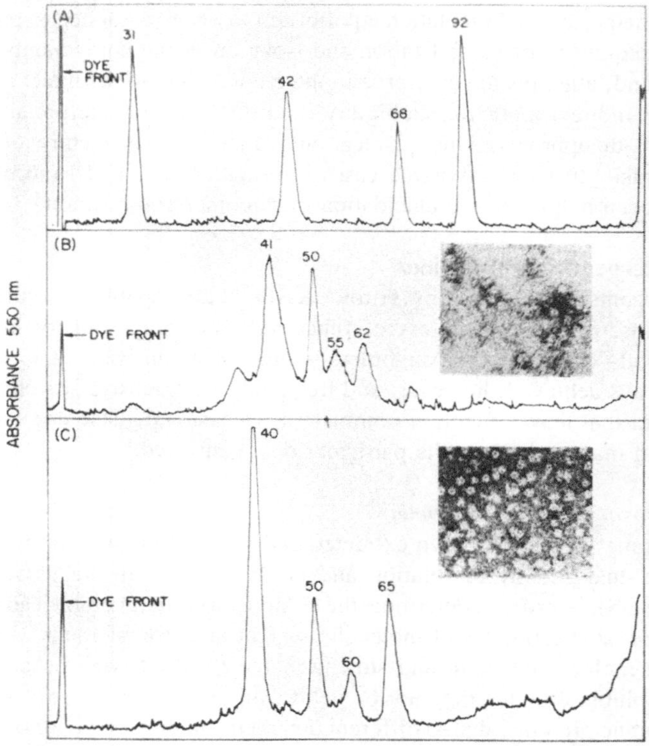

FIGURE 4. Gel electrophoresis scans of purified acetylcholine receptor from *E. electricus* (B) and *T. californica* (C). In (A), molecular weight standards are shown. These data reveal the composition of the protein, but by themselves cannot yield unambiguous stoichiometry due to the low staining intensity of the 55 and 60 kdalton subunits. Insets show electron micrographs of the receptors visualized by negative staining.

fortunately, for the reasons outlined above, estimates of stoichiometry are frequently in gross error.

One of the most careful examinations of stoichiometry is that of the acetylcholine receptor from *E. electricus*.[35] Conti-Tronconi et al.[35] overcame the problems presented by the poor staining of the γ-subunit through the use of simultaneous sequence analysis of N-terminal regions of the four subunits, establishing the stoichiometry unambiguously as 2:1:1:1 (Figure 4). Good agreement was obtained between the molecular weight of the complex and the sum of molecular weights of the subunits in the above stoichiometry.

Estimation of stoichiometry and of the molecular weight of a heteropolymeric oligomer may also be distorted by partial dissociation of one or more of the constituent polypeptides or by partial proteolysis. The ambiguity in the stoichiometry of the H^+-ATPase of mesophilic bacteria appears to have arisen largely from dissociation of some of the subunits.[75] The use of enzyme from thermophilic organisms overcame this problem leading to a stoichiometry of $\alpha_3\beta_3\gamma\delta\epsilon$.[75] Identical values were subsequently obtained with the ATPase of the mesophile *Escherichia coli* by use of radioactively labeled proteins.[76]

Even the use of SDS to study the subunits of membrane proteins may be fraught with problems, if proteins are not completely dissociated by the detergent.[44,77,78] The coat protein of the filamentous bacteriophage f1 remains dimeric in SDS solution.[78] The major sialoglycoprotein from erythrocyte membranes, glycophorin A, remains partially associated in SDS, even after boiling, to an extent dependent on the buffer composition.[77]

2. Molecular Weight of Oligomer

The molecular weight of the oligomer can be estimated by use of hydrodynamic techniques in nonionic detergent solution. Again, the molecular weight can be determined unambiguously by means of sedimentation equilibrium. Furthermore, the Stokes radius, a measure of particle size, can be determined from gel filtration,[71] or from the molecular weight with the aid of additional data from sedimentation velocity.[44,69] It is essential that these data be as accurate and precise as possible if meaningful conclusions are to be drawn: an uncertainty of 10% in molecular weight estimates would result in the inability to distinguish unequivocally between tetramer, pentamer, and hexamer.

In order to obtain a reliable estimate of molecular weight in detergent solutions, a measure of the effective partial specific volume of the protein is required.[79] This quantity includes the effects of the interactions between protein molecule and detergent (and any other bound ligand), and its use permits the determination of the anhydrous molecular weight of the protein, excluding bound ligands. The apparent partial specific volume, ϕ', may be calculated, given a knowledge of the binding and partial specific volumes of solvent components.[44,69] In the case where the only significant ligand is detergent:

$$(1 - \phi' \rho) = (1 - \bar{v}_P \rho) + \delta_D (1 - \bar{v}_D \rho) \tag{1}$$

where $\bar{v}_P$ is the partial specific volume of the pure protein, $\bar{v}_D$ is that of the detergent when bound to the protein, and δ_D is the mass of detergent bound per gram of protein. In the case of residual bound lipids or other ligands, additional terms are required.[69] It is reasonable to approximate $\bar{v}_P$ from the protein composition[80] and to assume that $\bar{v}_D$ is unaltered on binding.[78] Values of $\bar{v}_D$ for a wide range of detergents have been tabulated.[81] It is important in this type of analysis to work above the critical micelle concentration of the detergent to avoid possible abrupt changes in $\bar{v}_D$ at this point.[69] However, for most commonly used detergents, whose critical micelle concentration values are low,[82] this will not usually be a problem.

The limitation of Equation 1 in the study of membrane proteins in detergent is generally the quantity δ_D. For some common detergents for which radioactively labeled forms are commercially available, this quantity may be determined relatively easily. For other detergents, and when protein amounts are limited, the determination of δ_D may not be so straightforward. Reynolds and Tanford[83] have shown that if the solution density ρ is adjusted to equal the reciprocal of the detergent partial specific volume, the contribution of detergent in Equation 1 vanishes, regardless of the value of δ_D. In practice, density can be adjusted by the use of D_2O or $D_2{}^{18}O$. The use of sucrose or salts to adjust the solution density is to be avoided as preferential solvation of the protein by solvent components cannot be excluded, and would require additional, unknown terms in Equation 1. On the other hand, isotopically labeled water would be expected to exchange freely with H_2O.[83] Obviously, this technique is only applicable to detergents whose partial specific volume is close to 1, otherwise the required density is not accessible. Fortunately, many nonionic detergents as well as phospholipids satisfy this criterion.[81,83] The technique has been used successfully in a number of cases,[83] and in the case of the acetylcholine receptor[84] yielded results in very good agreement with the sum of the subunit molecular weights.[35,85]

Although the approach to membrane protein analysis using detergent extraction has been remarkably successful in recent years, it invites the fundamental objection that, by solubilizing the membrane, we have disrupted the very structure we wished to understand. In addition, the associations between membrane proteins may be intimately linked with lipid packing requirements in the membrane,[86,87] so that the presence of detergent may perturb integral protein interactions, leading to artifactual aggregation or dissociation and a loss of functional integrity.

B. Chemical Cross-Linking

The use of chemical cross-linking reagents with appropriate functional groups and spacer groups can detect spatial proximity of proteins within the intact membrane.[88] Typical of these reagents is dimethyl suberimidate in which the imidoester groups are reactive towards amino groups on the protein surface.[89] Loss of protein bands from the electrophoresis pattern, and the appearance of new bands with molecular weight equal to the sums of those of the component species, are taken as evidence for oligomers within the membrane.[90,91] This interpretation, however, carries considerable ambiguity. Diffusion of proteins within the membrane may lead to adventitious cross-linking of proteins that are not necessarily close neighbors.[25,92] This problem is aggravated by the fact that all of the copies of a given polypeptide float to the same depth in the bilayer, increasing the likelihood of cross-linking abundant proteins.[10,11] A further ambiguity of interpretation arises when several proteins of similar molecular weight are involved in cross-linking.

Some resolution of the ambiguity in determining near neighbors may be achieved by the use of cleavable cross-linking reagents such as dimethyl dithiobis(propionimidate).[93] These reagents again cross-link adjacent proteins via amino groups, but can be cleaved by thiol compounds to regenerate the protein components of the cross-linked complexes. In combination with two-dimensional electrophoresis techniques, such reagents are capable of high resolution in the determination of protein oligomers,[93,94] although the problem of adventitious cross-linking remains. Cross-linking with the aid of short-lived photoreactive cross-linking reagents may answer some of the objections to simple cross-linking studies.[25,92] If the lifetime of the photoinduced cross-linking reagent is sufficiently short compared with the time between random collisions of membrane proteins, adventitious cross-linking is minimized.

C. Electron Microscopy and Electron Diffraction

In favorable cases, where the proteins concerned form quasicrystalline arrays, electron microscopy can provide relatively direct information on the spatial organization of membrane proteins. Bacteriorhodopsin forms quasi-crystalline arrays in the purple membrane of *Halobacterium halobium* of sufficient regularity to allow the determination of the molecular structure to a resolution of 7 Å.[18] Reconstructed images suggested that the protein was arranged in trimers with rotational symmetry and that these trimers were further arranged in a hexagonal array.[18,95] After solubilization in nonionic detergent or reconstitution into lipid vesicles, however, bacteriorhodpsin can exist as an apparently functional monomer,[96] showing the importance of local protein concentration, and possibly of protein-lipid interactions, in determining oligomer formation in the intact functioning membrane.

The cyclic hexamer of the gap junction protein from rat hepatocytes can be visualized as a ring-like structure in negatively stained gap junction membranes (Figure 5).[97] Image filtration and enhancement can improve the resolution of these structures, and clearly reveals the sixfold rotational symmetry for the gap junction protein.[97,98]

Porin, the nonspecific transport protein of *E. coli*, can be visualized in arrays as an apparently trimeric unit, with a separate pore for each subunit of the trimer.[22,99,100] These trimers are, in turn, packed into arrays, the details of which depend on the lipid protein ratio.[22] Unlike the trimers of bacteriorhodopsin that appear to be only weakly associated, those of porin are stable enough to withstand dissociation by SDS.[101]

Even more complex organization has recently been revealed from Fourier image analysis electron microscopy of the photosynthetic membrane of *Rhodopseudomonas viridus*.[102] This study revealed a regular pattern comprising a large central structure protruding from both surfaces of the membrane and surrounded by six smaller bodies (Figure 6). These structures are thought to represent the central photosynthetic reaction center and surrounding light-harvesting chlorophyll-protein complexes.[102]

For some proteins it has proved possible to prepare two-dimensional crystals or quasi-

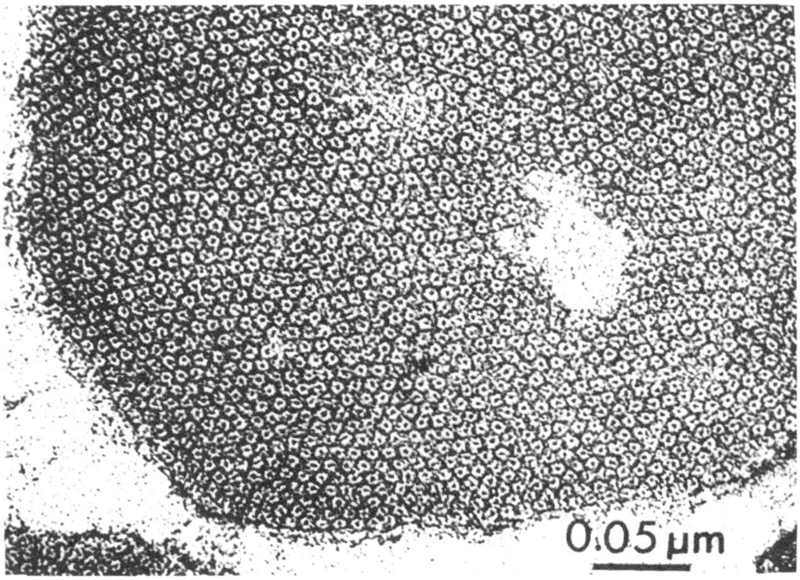

FIGURE 5. Electron micrograph of negatively stained isolated gap junctions, showing hexagonal packing of particles. (Reproduced from Hertzberg, E. L. and Gilula, N. B., *J. Biol. Chem.*, 254, 2138, 1979. With permission.)

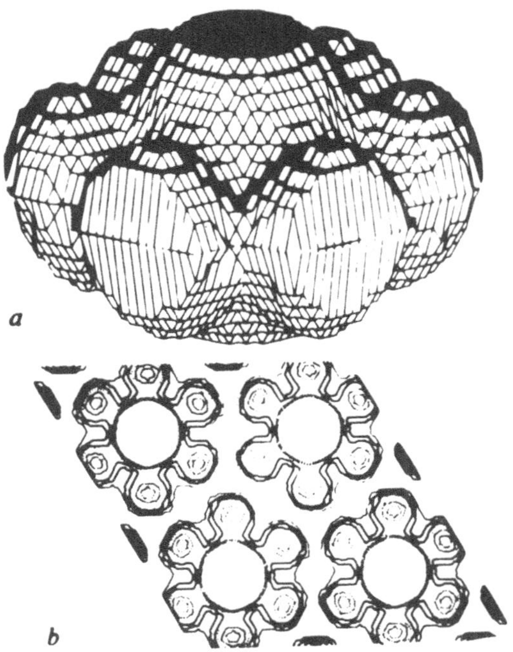

FIGURE 6. (a) Three-dimensional model of a photosynthetic reaction center of *R. viridus*, reconstructed from electron diffraction data obtained from negatively stained lattices. (b) Four unit cells, sectioned along a central plane parallel to the membrane sheet, and showing six "satellite" regions of mass density surrounding the central structure. (Reprinted from Miller, K. R., *Nature (London)*, 300, 53, 1982. With permission.)

crystalline arrays. Examination of two-dimensional crystals of cytochrome *c* oxidase has verified the dimeric nature of this enzyme, with its axis of twofold symmetry normal to the membrane plane.[103-105] Frequently, detergent-solubilized and purified proteins can be persuaded to crystallize,[106] allowing, in favorable cases, a comparison of the organization of the detergent-solubilized protein with that in the native membrane. Whereas mammalian rhodopsin crystallized from Tween® 80 retained a dimeric state and allowed resolution to 22 Å,[107] crystallization of cytochrome *c* oxidase from deoxycholate may have resulted in the monomer, indicating that deoxycholate had perturbed the native organization.[40,103,108]

Recently, it has been possible to produce three-dimensional crystals of integral membrane proteins, either by cocrystallizing with appropriate low molecular weight detergents (such as octyl glucoside) or with small amphiphiles such as heptane-1,2,3-triol.[106] Mitochondrial cytochrome *c* oxidase from beef heart[109] and cytochrome bc complex[110] have both recently been crystallized with only small residual traces of lipid. Furthermore, the photosynthetic reaction center of *R. viridus* has also been crystallized[111] and X-ray diffraction data have been collected to 3 Å.[111] It is to be hoped that high-resolution X-ray diffraction studies of such crystalline proteins may lead to a clearer picture of protein organization in these complex protein oligomers in the near future.

D. Spectroscopic Probes

Fluorescence techniques have frequently been used to determine structural relationships between membrane protein. Fluorescence energy transfer is capable of yielding information on the distance between appropriately labeled molecules in reconstituted membrane systems.[112] In the case of mammalian rhodopsin, fluorescence energy transfer suggests an interaction between rhodopsin molecules in lipid vesicles, although in detergent micelles the protein appears to be monomeric.[113]

Measurement of the decay of fluorescence polarization can also yield information on the rotational diffusion coefficient of membrane-bound proteins, which in turn is related to the effective cross-sectional area of the protein molecule in the bilayer.[114,115]

The dimeric state of band 3 protein, the anion channel of erythrocyte membranes, has been demonstrated in the intact membrane by Cherry,[24] who showed that the decay of fluorescence anisotropy of eosin-labeled band 3 protein was indistinguishable from that of the cross-linked dimer. Rotational diffusion of mammalian rhodopsin in retinal rod outer segment membranes has been determined by means of transient photodichroism[63] and by means of saturation transfer electron spin resonance.[62] The rotational behavior of rhodopsin in the native membrane was mimicked closely by the behavior of the protein in reconstituted lipid vesicles at comparable protein/lipid ratios,[49,50] consistent with transient formation of rhodopsin dimers in both environments.

The rotational diffusion of proteins in the lipid bilayer is a more sensitive measure of oligomer size than is lateral diffusion.[115,116] However, examination of fluorescence recovery after photobleaching[116,117] is capable of detecting gross clustering[116] or restricted lateral mobility through interactions with cytoskeletal networks.[118]

E. Neutron and X-Ray Scattering

Some estimate of the overall size and shape of a membrane protein may be obtained from neutron scattering[119-121] or low-angle X-ray scattering.[122-126] Although structural details are not resolvable by these techniques, it has proved possible to use the results as a bridge between the size and shape of protein particles in the membrane, and the size and shape of the protein in reconstituted vesicles or detergent micelles,[120] in which environment characterization may be made with less ambiguity.

Neutron scattering data have also provided an estimate of center-to-center spacing between protomers in the dimer of acetylcholine receptor.[37]

F. Radiation Inactivation

Some information concerning the size of biologically active oligomeric protein in a membrane may be obtained from a study of radiation inactivation.[127-131] These methods depend on the cross-section of the protein oligomer in the membrane[127,131] and rely upon the assumption that the active oligomer is sufficiently compact for significant energy transfer to occur within all parts of it.[130,131] This assumption, however, may not always be justified.[127] In the case of the acetylcholine receptor from *Torpedo,* the apparent molecular weight from radiation inactivation was found to be approximately 300,000, consistent with a single protomer.[130] Since many other studies have indicated that a substantial proportion of this protein exists as a dimer, linked through a disulfide bridge,[34] it would appear that the link between the protomers is insufficient for energy transfer between them.

Although this technique is, in principle, capable of detecting the oligomeric size of a *biologically active* species, and thus is potentially capable of elucidating the structure of transient oligomers produced in response to hormonal stimulation,[128,132] there are a number of interpretive difficulties inherent in the technique. First, samples must be dry to avoid generation of free radical hydroperoxides, and thus, the oligomeric structure of the enzyme may be altered during dehydration. Second, it is very difficult to quantitate the absolute dose, and empirical calibration with known proteins is usually relied upon.[130,131]

Nevertheless, as with the other techniques listed above, it may be possible to examine the state of the protein in the native membrane, at the very least as a check on the consistency of the data obtained with solubilized samples.

VI. INTERACTION WITH PERIPHERAL PROTEINS

The cytoplasm and the membrane interact in a variety of ways, and protein-protein interactions bridge the gap between the cytoplasm and the true membrane. The peripheral proteins are believed to interact principally with integral proteins of the membrane,[9] although interactions with lipid almost certainly contribute to the binding.[133] These proteins can generally be solubilized from the appropriate membrane by relatively mild, nondetergent, extraction procedures. About 40% of the mitochondrial inner membrane can be extracted in this way without disruption of the membrane.[6] The peripheral membrane proteins are capable of interacting, in turn, with other elements of the cytoplasm. The morphological determinant of cells, the cytoskeleton, may be attached at points to the membrane. In the case of the red blood cell, the cytoskeleton is very closely membrane limited,[134] and many of the interactions are now well understood.

The study of the interaction of peripheral proteins with membrane integral proteins is technically more amenable than that of interactions within the bilayer. Providing the peripheral protein can be isolated, its rebinding to the membrane-bound sites is a relatively straightforward problem exemplified by recent developments in the understanding of the red blood cell cytoskeleton.

A. The Red Cell Membrane Cytoskeleton

The red blood cell possesses the extraordinary ability to be flexible enough to pass through narrow capillaries and yet strong enough to resist fragmentation during its long and turbulent passage in the circulation. This ability is usually ascribed to the existence of a cytoskeleton underlying and attached to the membrane.[135-140] The cytoskeleton appears to be partly responsible for the biconcave disc shape of the red cell.[135] As well, it allows dissipation of lateral shear forces in the membrane, permitting the cell to undergo changes in shape in response to external stress, and yet links distant parts of the membrane, limiting bending of the bilayer, and preventing fragmentation.[135]

The major elements of the cytoskeleton are spectrin and actin, linked together to form a two-dimensional protein network (Figure 7).[140] Spectrin is comprised of two different poly-

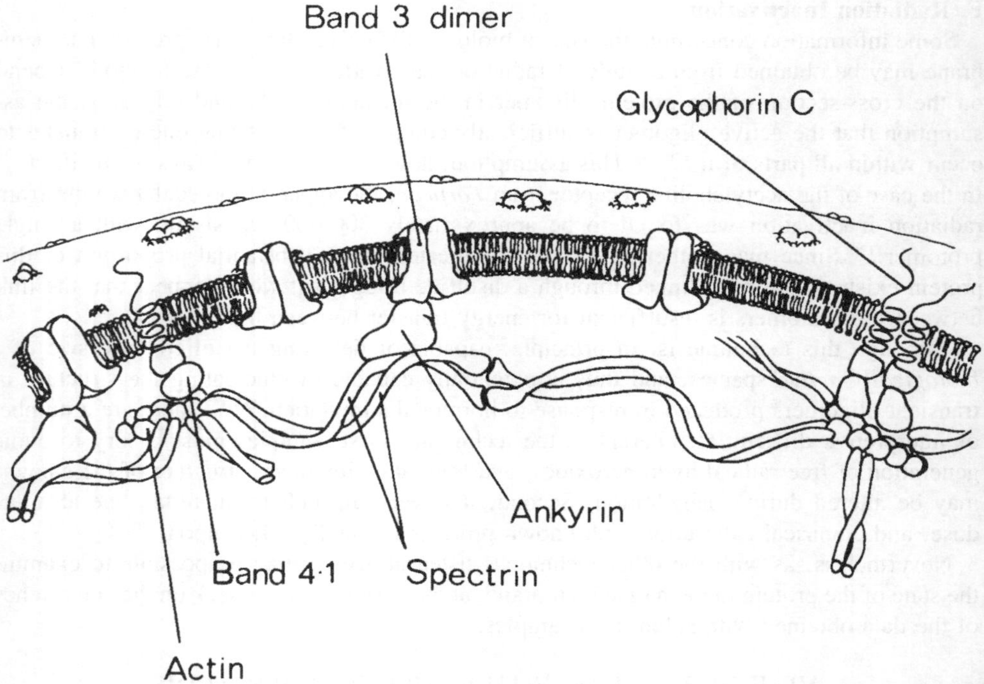

FIGURE 7. Arrangement of the proteins of the erythrocyte membrane cytoskeleton. Spectrin dimers, shown as paired flexible rods, associate in a head-to-head fashion to produce higher oligomers. These are further cross-linked into an extended network by interaction with actin clusters. The cytoskeleton is believed to be bound to the membrane through interactions mediated by ankyrin to band 3 protein and by band 4.1 protein to glycophorin C.

peptides, closely associated side by side to form a heterodimer.[141] The heterodimer is capable of self-associating further, to the tetramer,[141-143] and possibly higher association states.[144,145] Each heterodimer of spectrin possesses a single binding site for actin oligomers,[146] so that oligomers of spectrin beyond the tetramer are polyvalent towards actin and thus lead to extensively cross-linked networks.[146,147] The interaction between spectrin and actin is believed to be potentiated by component 4.1, another major component of the cytoskeleton.[148,149]

The cytoskeleton can be disrupted, and spectrin and actin solubilized by low ionic strength extraction.[150] However, purified spectrin can reassociate with a high-affinity binding site on spectrin-depleted membranes when the ionic strength is raised.[151] A thorough study of this binding revealed that spectrin is linked to the membrane via an intermediate link with component 2.1, now known as "ankyrin",[152] which in turn is bound to band 3 protein (the anion channel).[153] Although ankyrin is not solubilized at low ionic strength, it can be dissociated from the band 3 protein by means of incubation in 1 *M* NaCl,[154] and is, therefore, a peripheral protein. Treatment with 1 *M* salt also solubilizes band 4.1 protein, which has been purified, and is shown to bind to spectrin with high affinity at a site near the actin-binding site.[154,155]

While it is now clear that ankyrin links the cytoskeleton to the membrane via band 3 protein, the binding site for component 4.1 remains more elusive. There is growing evidence that one of the sialoglycoproteins, known variously as glycophorin C[156] or PAS-2 protein,[157] may be linked to the cytoskeleton, possibly via interaction with band 4.1,[157] and a high molecular weight complex comprising band 4.1, glycophorin C and glycophorin A, has recently been observed in two-dimensional electrophoresis.[169]

The anion channel, band 3 protein, appears to play a pivotal role in the structural organization of the red cell membrane proteins. In addition to its self-association,[23-26] red cell band 3 protein appears to interact closely with glycophorin, as demonstrated by the decrease in rotational diffusion of band 3 protein after cross-linking glycophorin.[158] Band 3 protein also possesses specific binding sites for the glycolytic enzymes aldolase,[159,160] glyceraldehyde-3-phosphate dehydrogenase[23] and phosphofructokinase,[161] for hemoglobin,[162] and for several peripheral membrane proteins including ankyrin and component 4.2.[23,153,154]

Aldolase and glyceraldehyde-3-phosphate dehydrogenase apparently bind to the same cytoplasmic domain of band 3 protein.[163] Although the binding of the glycolytic enzymes is diminished in physiological salt concentrations, a considerable proportion of these enzymes is believed to be membrane-bound in vivo.[164]

While only about 20% of the band 3 protein of the red blood cell is directly linked to the cytoskeleton through ankyrin,[153] the lateral mobility of most of the band 3 population is severely restricted by the presence of the cytoskeleton.[118] This effect may arise through the trapping of the band 3 protein in the interstices of the cytoskeleton network, and is discussed in detail in Chapter 3.

VII. PROSPECT

The developments discussed in the last section concerning the organization of proteins in the red blood cell membrane have largely taken place within the last 10 years. To some extent, advances in this area have been rapid due to the relative ease of preparing clean, intact membranes. Material from other sources is not always so readily obtained. Nevertheless, it is likely that understanding of the organization of other membrane systems will grow rapidly in the near future. In view of the increased complexity of many membranes in comparison with the red cell, these developments will be most interesting.

Understanding of the link between structure and function, however, is a more difficult question. In spite of the appeal of the model in which the symmetry axis of an oligomer may provide a transport pore, hard evidence is elusive. For example, there is good evidence of half-of-sites reactivity in the inhibition of cytochrome *c* oxidase by 1 mol arylazidocytochrome *c* per mole of dimeric oxidase,[165] but the functional significance of this finding is compromised by the functional capacity of monomeric cytochrome *c* oxidase.[47]

Although oligomer formation of bacteriorhodopsin in the purple patches of *H. halobium* membranes may be functionally important in terms of cooperativity effects in proton pumping,[166] there is some real doubt whether the purple patches are relevant in vivo.[167]

Stabilization of the enzyme against denaturation has been suggested as a functional role for oligomer formation in sarcoplasmic reticulum Ca^{2+}-ATPase,[67] but whether this is relevant in the native membrane rather than a peculiarity in detergent micelles has not been unambiguously demonstrated.

It is to be hoped that increased resolution of structure made possible by recent advances in the preparation of three-dimensional crystals[106,109-111] may lead to better understanding of both structure and function in some cases.

One of the most important unanswered questions in membrane biochemistry concerns the insertion of proteins into membranes and the segregation of different proteins to functionally appropriate locations. It appears likely that at least some of this behavior is directed or controlled by specific protein-protein interactions, and that detailed study of such interactions may help to answer some of the questions. There is cause for optimism in some of the insights already gained.[168]

REFERENCES

1. **Guidotti, G.,** Membrane proteins, *Annu. Rev. Biochem.*, 41, 731, 1972.
2. **Freedman, R. B.,** Membrane-bound enzymes, in *Membrane Structure*, Finean, J. B. and Michel, R. H., Eds., Elsevier/North-Holland, Amsterdam, 1981, 161.
3. **Gutierrez-Merino, C.,** On the rate of association and dissociation processes of intrinsic membrane proteins, *Biochim. Biophys. Acta*, 728, 179, 1983.
4. **Monod, J., Wyman, J., and Changeux, J. P.,** On the nature of allosteric transitions: a plausible model, *J. Mol. Biol.*, 12, 88, 1965.
5. **De Pierre, J. W. and Ernster, L.,** Enzyme topology of intracellular membranes, *Annu. Rev. Biochem.*, 46, 201, 1977.
6. **Sjöstrand, F. S.,** The structure of mitochondrial membranes: a new concept, *J. Ultrastruct. Res.*, 64, 217, 1979.
7. **Srere, P. A.,** Structure of mitochondrial innermembrane-matrix compartments, *Trends Biochem. Sci.*, 7, 375, 1982.
8. **Bretscher, M. S.,** A major protein which spans the human erythrocyte membrane, *J. Mol. Biol.*, 59, 351, 1971.
9. **Singer, S. J. and Nicolson, G. L.,** The fluid mosaic model of the structure of cell membranes, *Science*, 175, 720, 1972.
10. **Craig, W. S. and Kyte, J.,** Stoichiometry and molecular weight of the minimum asymmetric unit of canine renal $Na^+ + K^+$-ATPase, *J. Biol. Chem.*, 255, 6262, 1980.
11. **Kyte, J.,** Molecular considerations relevant to the mechanism of active transport, *Nature (London)*, 292, 201, 1981.
12. **McGavin, S.,** Axes of rotational symmetry in protein structures, *J. Theor. Biol.*, 26, 335, 1970.
13. **Klingenberg, M.,** Membrane protein oligomeric structure and transport function, *Nature (London)*, 290, 449, 1981.
14. **Klotz, I. M., Darnall, D. W., and Langerman, N. R.,** Quaternary structure of proteins, in *The Proteins*, Vol. 1, 3rd ed., Neurath, H. and Hill, R. L., Eds., Academic Press, New York, 1975, 293.
15. **Nakae, T., Ishii, J., and Tokunaga, M.,** Subunit structure of functional porin oligomers that form permeability channels in the outer membrane of *E. coli*, *J. Biol. Chem.*, 254, 1457, 1979.
16. **Mihara, K., Blobel, G., and Sato, R.,** *In vitro* synthesis and integration into mitochondria of porin, a major protein of the outer mitochondrial membrane of *S. cerevisiae*, *Proc. Natl. Acad. Sci. U.S.A.*, 79, 7102, 1982.
17. **Roos, N., Benz, R., and Brdiczka, D.,** Identification and characterization of the pore-forming protein in the outer membrane of rat liver mitochondria, *Biochim. Biophys. Acta*, 686, 204, 1982.
18. **Henderson, R. and Unwin, P. N. T.,** Three-dimensional model of purple membrane obtained by electron microscopy, *Nature (London)*, 257, 28, 1975.
19. **Heymann, E. and Mentlein, R.,** Cross-linking experiments for the elucidation of the quaternary structure of carboxyl esterase in the microsomal membrane, *Biochem. Biophys. Res. Commun.*, 95, 577, 1980.
20. **Wilson, I. A., Skehel, J. J., and Wiley, D. C.,** Structure of the haemagglutinin membrane glycoprotein of influenza virus at 3A resolution, *Nature (London)*, 289, 366, 1981.
21. **Varghese, J. N., Laver, W. G., and Colman, P. M.,** Structure of the influenza virus glycoprotein antigen neuraminidase at 2.9A resolution, *Nature (London)*, 303, 38, 1983.
22. **Dorset, D. L., Engel, A., Haner, M., Massalski, A., and Rosenbusch, J. P.,** Two-dimensional crystal packing of matrix porin, *J. Mol. Biol.*, 165, 701, 1983.
23. **Yu, J. and Steck, T. L.,** Associations of band 3, the predominant polypeptide of the human erythrocyte membrane, *J. Biol. Chem.*, 250, 9176, 1975.
24. **Cherry, R. J.,** Dimeric association of band 3 in the erythrocyte membrane demonstrated by protein diffusion measurements, *Nature (London)*, 277, 493, 1979.
25. **Mikkelsen, R. B. and Wallach, D. F. H.,** Photoactivated cross-linking of proteins within the erythrocyte membrane core, *J. Biol. Chem.*, 251, 7413, 1976.
26. **Pappert, G. and Schubert, D.,** The state of association of band 3 protein of the human erythrocyte membrane in solutions of nonionic detergents, *Biochim. Biophys. Acta*, 730, 32, 1983.
27. **Nakashima, H. and Makino, S.,** State of association of band 3 protein from bovine erythrocyte membrane in non-ionic detergent, *J. Biochem.*, 88, 933, 1980.
28. **Appell, K. C. and Low, P. S.,** Partial structural characterization of the cytoplasmic domain of the erythrocyte membrane protein, band 3, *J. Biol. Chem.*, 256, 11104, 1981.
29. **Reithmeier, R. A. F.,** Fragmentation of the band 3 polypeptide from human erythrocyte membranes, *J. Biol. Chem.*, 254, 3054, 1979.
30. **Klingenberg, M.,** The ADP-ATP carrier in mitochondrial membranes, in *The Enzymes of Biological Membranes: Membrane Transport*, Vol. 3, Martonosi, A. N., Ed., Plenum Press, New York, 1976, 383.

31. **Kyte, J.**, Structural studies of sodium and potassium ion-activated adenosine triphosphatase, *J. Biol. Chem.*, 250, 7443, 1975.
32. **Ottolenghi, P. and Jensen, J.**, The K^+-induced apparent heterogeneity of high-affinity binding sites in (Na^+ + K^+) ATPase can only be due to the oligomeric structure of the enzyme, *Biochim. Biophys. Acta*, 727, 89, 1983.
33. **Hebert, H., Jorgensen, P. L., Skriver, E., and Maunsbach, A. B.**, Crystallization patterns of membrane-bound (Na^+ + K^+) ATPase, *Biochim. Biophys. Acta*, 689, 571, 1982.
34. **Lindstrom, J., Merlie, J., and Yogeeswaran, G.**, Biochemical properties of acetylcholine receptor subunits from *Torpedo californica*, *Biochemistry*, 18, 4465, 1979.
35. **Conti-Tronconi, B. M., Hunkapiller, M. W., Lindstrom, J. M., and Raftery, M. A.**, Subunit structure of the acetylcholine receptor from *E. electricus*, *Proc. Natl. Acad. Sci. U.S.A.*, 79, 6489, 1982.
36. **Zingsheim, H. P., Neugebauer, D. C., Barrantes, F. J., and Frank, J.**, Membrane-bound acetylcholine receptor from *T. marmorata:* structural details, *Proc. Natl. Acad. Sci. U.S.A.*, 77, 952, 1980.
37. **Wise, D. S., Schoenborn, B. P., and Karlin, A.**, Structure of acetylcholine receptor dimer determined by neutron scattering and electron microscopy, *J. Biol. Chem.*, 256, 4124, 1981.
38. **Saraste, M.**, How complex is a respiratory complex?, *Trends Biochem. Sci.*, 8, 139, 1983.
39. **Capaldi, R. A.**, Arrangement of proteins in the mitochondrial inner membrane, *Biochim. Biophys. Acta*, 694, 291, 1982.
40. **Robinson, N. C. and Capaldi, R. A.**, Interaction of detergents with cytochrome oxidase, *Biochemistry*, 16, 375, 1977.
41. **Azzi, A.**, Cytochrome c oxidase: towards a clarification of its structure, interaction and mechanism, *Biochim. Biophys. Acta*, 594, 231, 1980.
42. **Wikstrom, M., Krab, K., and Saraste, M.**, *Cytochrome Oxidase — a Synthesis*, Academic Press, London, 1981.
43. **Saraste, M., Pentilla, T., and Wikstrom, M.**, Quaternary structure of bovine cytochrome oxidase, *Eur. J. Biochem.*, 115, 261, 1981.
44. **Tanford, C. and Reynolds, J. A.**, Characterization of membrane proteins in detergent solutions, *Biochim. Biophys. Acta*, 457, 133, 1976.
45. **Poyton, R. O. and Schatz, G.**, Cytochrome c oxidase from bakers' yeast, *J. Biol. Chem.*, 250, 752, 1975.
46. **Merle, P. and Kadenbach, B.**, The subunit composition of mammalian cytochrome c oxidase, *Eur. J. Biochem.*, 105, 499, 1980.
47. **Georgevich, G., Darley-Usmar, V. M., Malatesta, F., and Capaldi, R. A.**, Electron transfer in monomeric forms of beef and shark heart cytochrome c oxidase, *Biochemistry*, 22, 1317, 1983.
48. **Kawato, S., Sigel, E., Carafoli, E., and Cherry, R. J.**, Rotation of cytochrome oxidase in phospholipid vesicles, *J. Biol. Chem.*, 256, 7518, 1981.
49. **Kusumi, A., Sakaki, T., Yoshizawa, T., and Ohnishi, S.**, Protein-lipid interaction in rhodopsin recombinant membranes as studied by protein-rotational mobility and lipid alkyl chain flexibility measurements, *J. Biochem.*, 88, 1103, 1980.
50. **Kusumi, A. and Hyde, J. S.**, Spin-label saturation-transfer electron spin-resonance detection of transient association of rhodopsin in reconstituted membranes, *Biochemistry*, 21, 5978, 1982.
51. **Ackers, G. K. and Johnson, M. L.**, Linked functions in allosteric proteins, *J. Mol. Biol.*, 147, 559, 1981.
52. **Wyman, J.**, Linked functions and reciprocal effects in hemoglobin: a second look, *Adv. Protein Chem.*, 19, 223, 1964.
53. **Nichol, L. W. and Winzor, D. J.**, Binding equations and control effects, in *Protein Interaction Patterns*, Frieden, C. and Nichol, L. W., Eds., John Wiley & Sons, New York, 1981, 337.
54. **Cuatrecasas, P.**, Membrane receptors, *Annu. Rev. Biochem.*, 43, 169, 1974.
55. **Ross, E. M. and Gilman, A. G.**, Biochemical properties of hormone sensitive adenylate cyclase, *Annu. Rev. Biochem.*, 49, 533, 1980.
56. **Schlegel, W., Kempner, E. S., and Rodbell, M.**, Activation of adenylate cyclase in hepatic membranes involves interactions of the catalytic unit with multimeric complexes of regulatory proteins, *J. Biol. Chem.*, 254, 5168, 1979.
57. **Levitzki, A.**, The β-adrenergic receptor and its mode of coupling to adenylate cyclase, *CRC Crit. Rev. Biochem.*, 10, 81, 1981.
58. **Moss, J. and Vaughan, M.**, Activation of adenylate cyclase by choleragen, *Annu. Rev. Biochem.*, 48, 581, 1979.
59. **Martin, R. B., Kennedy, E. L., and Doberska, C. A.**, The effect of fluoride on the state of aggregation of adenylate cyclase in rat liver plasma membranes, *Biochem. J.*, 188, 137, 1980.
60. **Sardet, C., Tardieu, A., and Luzzati, V.**, Shape and size of bovine rhodopsin: a small-angle X-ray scattering study of a rhodopsin-detergent complex, *J. Mol. Biol.*, 105, 383, 1976.

61. **Fung, B. K. K., Hurley, J. B., and Stryer, L.,** Flow of information in the light-triggered cyclic nucleotide cascade of vision, *Proc. Natl. Acad. Sci. U.S.A.*, 78, 152, 1981.
62. **Baroin, A., Bienvenue, A., and Devaux, P. F.,** Spin-label studies of protein-protein interactions in retinal rod outer segment membranes, *Biochemistry*, 18, 1151, 1979.
63. **Cone, R. A.,** Rotational diffusion of rhodopsin in the visual receptor membrane, *Nature (London), New Biol.*, 236, 39, 1972.
64. **Lookman, T., Pink, D. A., Grundke, E. W., Zuckermann, M. J., and de Verteuil, F.,** Phase separation in lipid bilayers containing integral proteins-computer simulation studies, *Biochemistry*, 21, 5593, 1982.
65. **Helenius, A. and Simons, K.,** Solubilization of membranes by detergents, *Biochim. Biophys. Acta*, 415, 29, 1975.
66. **McCaslin, D. R. and Tanford, C.,** Different states of aggregation for unbleached and bleached rhodopsin after isolation in two different detergents, *Biochemistry*, 20, 5212, 1981.
67. **Moller, J. V., Lind, K. E., and Anderson, J. P.,** Enzyme kinetics and substrate stabilization of detergent - solubilized and membraneous (Ca^{2+} + Mg^{2+})-activated ATPase from sarcoplasmic reticulum: effect of protein-protein interactions, *J. Biol. Chem.*, 255, 1912, 1980.
68. **Brady, G. W., Fein, D. B., Harden, M. E., Spehr, R., and Meissner, G.,** A liquid diffraction analysis of sarcoplasmic reticulum, *Biophys. J.*, 34, 13, 1981.
69. **Tanford, C., Nozaki, Y., Reynolds, J. A., and Makino, S.,** Molecular characterization of proteins in detergent solutions, *Biochemistry*, 13, 2369, 1974.
70. **Martinez-Carrion, M., Sator, V., and Raftery, M. A.,** The molecular weight of an acetylcholine receptor isolated from *Torpedo californica*, *Biochem. Biophys. Res. Commun.*, 65, 129, 1975.
71. **Fish, W. W., Reynolds, J. A., and Tanford, C.,** Gel chromatography of proteins in denaturing solvents, *J. Biol. Chem.*, 245, 5166, 1970.
72. **Tanford, C., Kawahara, K., and Lapanje, S.,** Proteins as random coils. I. Intrinsic viscosity and sedimentation coefficients in concentrated guanidine hydrochloride, *J. Am. Chem. Soc.*, 89, 729, 1967.
73. **Fish, W. W., Mann, K. G., and Tanford, C.,** The estimation of polypeptide chain molecular weights by gel filtration in 6M guanidine hydrochloride, *J. Biol. Chem.*, 244, 4989, 1969.
74. **Juliano, R. L.,** The solubilization and fractionation of human erythrocyte membrane proteins, *Biochim. Biophys. Acta*, 266, 301, 1972.
75. **Yoshida, M., Soue, N., Hirata, H., Kagawa, Y., and Ui, N.,** Subunit structure of adenosine triphosphatase, *J. Biol. Chem.*, 254, 9525, 1979.
76. **Foster, D. L. and Fillingame, R. H.,** Stoichiometry of subunits in the H^+-ATPase complex of *E. coli*, *J. Biol. Chem.*, 257, 2009, 1982.
77. **Furthmayr, H. and Marchesi, V. T.,** Subunit structure of human erythrocyte glycophorin A, *Biochemistry*, 5, 1137, 1976.
78. **Makino, S., Woolford, J. L., Tanford, C., and Webster, R. E.,** Interaction of deoxycholate and of detergents with the coat protein of bacteriophage f1, *J. Biol. Chem.*, 250, 4327, 1975.
79. **Casassa, E. F. and Eisenberg, H.,** Thermodynamic analysis of multicomponent solutions, *Adv. Protein Chem.*, 19, 287, 1964.
80. **Cohn, E. J. and Edsall, J. T.,** *Proteins, Amino Acids and Peptides*, Reinhold, New York, 1943, 157.
81. **Steele, J. C. H., Tanford, C., and Reynolds, J. A.,** Determination of partial specific volumes for lipid-associated proteins, *Methods Enzymol.*, 48, 11, 1978.
82. **Helenius, A., McCaslin, D. R., Fries, E., and Tanford, C.,** Properties of detergents, *Methods Enzymol.*, 56, 734, 1979.
83. **Reynolds, J. A. and Tanford, C.,** Determining of molecular weight of the protein moiety in protein-detergent complexes without direct knowledge of detergent binding, *Proc. Natl. Acad. Sci. U.S.A.*, 73, 4467, 1976.
84. **Reynolds, J. A. and Karlin, A.,** Molecular weight in detergent solution of acetylcholine receptor from *Torpedo californica*, *Biochemistry*, 17, 2035, 1978.
85. **Raftery, M. A., Hunkapillar, M. W., Strader, C. D., and Hood, L. E.,** Acetylcholine receptor: complex of homologous subunits, *Science*, 208, 1454, 1980.
86. **Israelachvili, J. N.,** Refinement of the fluid mosaic model of membrane structure, *Biochim. Biophys. Acta*, 469, 221, 1977.
87. **Israelachvili, J. N., Marcelja, S., and Horn, R. G.,** Physical principles of membrane organization, *Q. Rev. Biophys.*, 13, 121, 1980.
88. **Peters, K. and Richards, F. M.,** Chemical cross-linking: reagents and problems in studies of membrane structure, *Annu. Rev. Biochem.*, 46, 523, 1977.
89. **Davies, G. E. and Stark, G. R.,** Use of dimethyl suberimidate, a cross-linking reagent, in studying the subunit structure of oligomeric proteins, *Proc. Natl. Acad. Sci. U.S.A.*, 66, 651, 1970.
90. **Steck, T. L.,** Cross-linking the major proteins of the isolated erythrocyte membrane, *J. Mol. Biol.*, 66, 295, 1972.

91. **Hucho, F., Bandini, G., and Suarez-Isla, B. A.,** The acetylcholine receptor as part of a protein complex in receptor-enriched membrane fragments from *Torpedo californica* electric tissue, *Eur. J. Biochem.*, 83, 335, 1978.
92. **Kiehm, D. J. and Ji, T. H.,** Photochemical cross-linking of cell membranes, *J. Biol. Chem.*, 252, 8524, 1977.
93. **Wang, K. and Richards, F. M.,** An approach to nearest-neighbour analysis of membrane proteins, *J. Biol. Chem.*, 249, 8005, 1974.
94. **Briggs, M. M. and Capaldi, R. A.,** Near-neighbour relationships of the subunits of cytochrome *c* oxidase, *Biochemistry*, 16, 73, 1977.
95. **Hayward, S. B. and Stroud, R. M.,** Projected structure of purple membrane determined to 3.7 A resolution by low temperature electron microscopy, *J. Mol. Biol.*, 151, 491, 1981.
96. **Heyn, M. P., Cherry, R. J., and Dencher, N. A.,** Lipid-protein interactions in bacteriorhodopsin-dimyristoyl phosphatidylcholine vesicles, *Biochemistry*, 20, 840, 1981.
97. **Hertzberg, E. L. and Gilula, N. B.,** Isolation and characterization of gap junctions from rat liver, *J. Biol. Chem.*, 254, 2138, 1979.
98. **Unwin, P. N. T. and Zampighi, G.,** Structure of the junction between communicating cells, *Nature (London)*, 283, 545, 1980.
99. **Steven, A. C., Ten Heggeler, B., Muller, R., Kistler, J., and Rosenbusch, J. P.,** Ultrastructure of a periodic protein layer in the outer membrane of *Escherichia coli*, *J. Cell Biol.*, 72, 292, 1977.
100. **Garavito, R. M., Jenkins, J., Jansonius, J. N., Karlsson, R., and Rosenbusch, J. P.,** X-ray diffraction analysis of matrix porin, an integral membrane protein from *Escherichia coli* membranes, *J. Mol. Biol.*, 164, 313, 1983.
101. **Rosenbusch, J. P.,** Characterization of the major envelope protein from *Escherichia coli*, *J. Biol. Chem.*, 249, 8019, 1974.
102. **Miller, K. R.,** Three-dimensional structure of a photosynthetic membrane, *Nature (London)*, 300, 53, 1982.
103. **Henderson, R., Capaldi, R. A., and Leigh, J. S.,** Arrangement of cytochrome oxidase molecules in two-dimensional vesicle crystals, *J. Mol. Biol.*, 112, 631, 1977.
104. **Frey, T. G., Chan, S. H. P., and Schatz, G.,** Structure and orientation of cytochrome oxidase in crystalline membranes, *J. Biol. Chem.*, 253, 4389, 1978.
105. **Fuller, S. D., Capaldi, R. A., and Henderson, R.,** Structure of cytochrome *c* oxidase in deoxycholate-derived two-dimensional crystals, *J. Mol. Biol.*, 134, 305, 1979.
106. **Michel, H.,** Crystallization of membrane proteins, *Trends Biochem. Sci.*, 8, 56, 1983.
107. **Corless, J. M., McCaslin, D. R., and Scott, B. L.,** Two-dimensional rhodopsin crystals from disk membranes of frog retinal rod outer segments, *Proc. Natl. Acad. Sci. U.S.A.*, 79, 1116, 1982.
108. **Seki, S., Hayashi, H., and Oda, T.,** Studies on cytochrome oxidase, *Arch. Biochem. Biophys.*, 138, 110, 1970.
109. **Ozawa, T., Tanaka, M., and Wakabayashi, T.,** Crystallization of mitochondrial cytochrome oxidase, *Proc. Natl. Acad. Sci. U.S.A.*, 79, 7175, 1982.
110. **Ozawa, T., Tanaka, M., and Shimomura, Y.,** Crystallization of cytochrome bc_1 complex, *Proc. Natl. Acad. Sci. U.S.A.*, 80, 921, 1983.
111. **Michel, H.,** Three-dimensional crystals of a membrane protein complex, *J. Mol. Biol.*, 158, 567, 1982.
112. **Stryer, L.,** Fluorescence energy transfer as a spectroscopic ruler, *Annu. Rev. Biochem.*, 47, 819, 1978.
113. **Borochov-Neori, H., Fortes, P. A. G., and Montal, M.,** Rhodopsin in reconstituted phospholipid vesicles. II. Rhodopsin-rhodopsin interactions detected by resonance energy transfer, *Biochemistry*, 22, 206, 1983.
114. **Cherry, R. J., Burkli, A., Busslinger, M., Schneider, G., and Parish, G. R.,** Rotational diffusion of band 3 proteins in the human erythrocyte membrane, *Nature (London)*, 263, 389, 1976.
115. **Saffman, P. G. and Delbruck, M.,** Brownian motion in biological membranes, *Proc. Natl. Acad. Sci. U.S.A.*, 72, 3111, 1975.
116. **Vaz, W. L. C., Criado, M., Madeira, V. M. C., Schoellmann, G., and Jovin, T. M.,** Size dependence of the translational diffusion of large integral membrane proteins in liquid crystalline phase lipid bilayers. A study using fluorescence recovery after photobleaching, *Biochemistry*, 21, 5608, 1982.
117. **Criado, M., Vaz, W. L. C., Barrantes, F. J., and Jovin, T. M.,** Translational diffusion of acetylcholine receptor (monomeric and dimeric forms) of *Torpedo marmorata* reconstituted into phospholipid bilayers studied by fluorescence recovery after photobleaching, *Biochemistry*, 21, 5750, 1982.
118. **Fowler, V. and Branton, D.,** Lateral mobility of human erythrocyte integral membrane proteins, *Nature (London)*, 268, 23, 1977.
119. **Osborne, H. B., Sardet, C., Michel-Villaz, M., and Chabre, M.,** Structural study of rhodopsin in detergent micelles by small angle neutron scattering, *J. Mol. Biol.*, 123, 177, 1978.
120. **Bartholdi, M., Barrantes, F. J., and Jovin, T. M.,** Rotational molecular dynamics of the membrane-bound acetylcholine receptor revealed by phosphorescence spectroscopy, *Eur. J. Biochem.*, 120, 389, 1981.

121. **Yeager, M., Schoenborn, B., Engelman, D., Moore, P., and Stryer, L.,** Neutron diffraction analysis of the structure of rod photoreceptor membranes in intact retinas, *J. Mol. Biol.*, 137, 315, 1980.
122. **Dratz, E. A. and Hargrave, P. A.,** The structure of rhodopsin and the rod outer segment disk membrane, *Trends Biochem. Sci.*, 8, 128, 1983.
123. **Dratz, E. A., Miljanich, G. P., Nemes, P. P., Gaw, J. E., and Schwartz, S.,** The structure of rhodopsin and its disposition in the rod outer segment disk membrane. *Photochem. Photobiol.*, 29, 661, 1979.
124. **Worthington, C. R.,** X-ray studies on membranes, in *The Enzymes of Biological Membranes*, Vol. 1, Martonosi, A. N., Ed., Plenum Press, New York, 1976, 1.
125. **Schwartz, S., Cain, J. E., Dratz, E. A., and Blasie, J. K.,** An analysis of lamellar X-ray diffraction from disordered membrane multilayers with application to data from retinal rod outer segments, *Biophys. J.*, 15, 1201, 1975.
126. **Le Maire, M., Moller, J. V., and Tardieu, A.,** Shape and thermodynamic parameters of a Ca^{2+}-dependent ATPase, *J. Mol. Biol.*, 150, 273, 1981.
127. **Kempner, E. S. and Schlegel, W.,** Size determination of enzymes by radiation inactivation, *Anal. Biochem.*, 92, 2, 1979.
129. **Harmon, J. T., Kahn, C. R., Kempner, E. S., and Schlegel, W.,** Characterization of the insulin receptor in its membrane environment by radiation inactivation, *J. Biol. Chem.*, 255, 3412, 1980.
129. **Vegh, K., Spiegler, P., Chamberlain, C., and Mommaerts, W. F. H. M.,** The molecular size of the Ca^{2+}-transport ATPase of sarcotubular vesicles estimated from radiation inactivation, *Biochim. Biophys. Acta*, 163, 266, 1968.
130. **Lo, M. M. S., Barnard, E. A., and Dolly, J. O.,** Size of acetylcholine receptors in the membrane. An improved version of the radiation inactivation method, *Biochemistry*, 21, 2210, 1982.
131. **Pollard, E. C., Guild, W. R., Hutchinson, F., and Setlow, R. B.,** The direct action of ionizing radiation on enzymes and antigens, *Prog. Biophys. Biophys. Chem.*, 5, 72, 1955.
132. **Houslay, M. D., Ellory, J. C., Smith, G. A., Hesketh, T. R., Stein, J. M., Warren, G. B., and Metcalfe, J. C.,** Exchange of partners in glucagon receptor-adenylate cyclase complexes, *Biochim. Biophys. Acta*, 467, 208, 1977.
133. **Marinetti, G. V. and Crain, R. C.,** Topology of amino-phospholipids in the red cell membrane, *J. Supramol. Struct.*, 8, 191, 1978.
134. **Ziparo, E., Lemay, A., and Marchesi, V. T.,** The distribution of spectrin along the membranes of normal and echinocytic human erythrocytes, *J. Cell Sci.*, 34, 91, 1978.
135. **Ralston, G. B.,** The structure of spectrin and the shape of the red blood cell, *Trends Biochem. Sci.*, 3, 195, 1978.
136. **Haest, C. W. M.,** Interactions between membrane skeleton proteins and the intrinsic domain of the erythrocyte membrane, *Biochim. Biophys. Acta*, 694, 331, 1982.
137. **Branton, D., Cohen, C. M., and Tyler, J. M.,** Interactions of cytoskeletal proteins on the human red blood cell membrane, *Cell*, 24, 24, 1981.
138. **Ralston, G. B.,** Spectrin and the red cell membrane cytoskeleton, in *Membranes and Transport*, Vol. 2, Martonosi, A. N., Ed., Plenum Press, New York, 1982, 415.
139. **Tyler, J. M., Cohen, C. M., and Branton, D.,** Molecular association of the erythrocyte cytoskeleton, in *Membranes and Transport*, Vol. 2, Martonosi, A. N., Ed., Plenum Press, New York, 1982, 409.
140. **Hainfeld, J. F. and Steck, T. L.,** The submembrane reticulum of the human erythrocyte: a scanning electron microscope study, *J. Supramol. Struct.*, 6, 301, 1977.
141. **Shotton, D. M., Burke, B. E., and Branton, D.,** The molecular structure of human erythrocyte spectrin, *J. Mol. Biol.*, 131, 303, 1979.
142. **Ralston, G. B.,** Physico-chemical characterization of the spectrin tetramer from bovine erythrocyte membranes, *Biochim. Biophys. Acta*, 455, 163, 1976.
143. **Ungewickell, E. and Gratzer, W. B.,** Self-association of human spectrin, *Eur. J. Biochem.*, 88, 379, 1978.
144. **Morrow, J. S., Haigh, W. B., and Marchesi, V. T.,** Spectrin oligomers: a structural feature of the erythrocyte membrane, *J. Supramol. Struct.*, 17, 275, 1981.
145. **Morris, M. and Ralston, G. B.,** Self-association of human spectrin, *Proc. Aust. Biochem. Soc.*, 15, 11, 1983.
146. **Cohen, C. M., Tyler, J. M., and Branton, D.,** Spectrin-actin associations studied by electron microscopy of shadowed preparations, *Cell*, 21, 875, 1980.
147. **Brenner, S. L. and Korn, E. D.,** Spectrin-actin interaction, *J. Biol. Chem.*, 254, 8620, 1979.
148. **Ungewickell, E., Bennett, P. M., Calvert, R., Ohanian, V., and Gratzer, W. B.,** *In vitro* formation of a complex between cytoskeletal proteins of the human erythrocyte, *Nature (London)*, 280, 811, 1979.
149. **Fowler, V. and Taylor, D. L.,** Spectrin plus band 4.1 cross-link actin, *J. Cell Biol.*, 85, 361, 1980.
150. **Marchesi, S. L., Steers, E., Marchesi, V. T., and Tillack, T. W.,** Physical and chemical properties of a protein isolated from red cell membranes, *Biochemistry*, 9, 50, 1970.

151. **Bennett, V. and Branton, D.,** Selective association of spectrin with the cytoplasmic surface of human erythrocyte membranes, *J. Biol. Chem.*, 252, 2753, 1977.
152. **Bennett, V. and Stenbuck, P. J.,** Identification and partial purification of ankyrin, the high affinity membrane attachment site for human erythrocyte spectrin, *J. Biol. Chem.*, 254, 2533, 1979.
153. **Bennett, V. and Stenbuck, P. J.,** Association between ankyrin and the cytoplasmic domain of band 3 isolated from the human erythrocyte membrane, *J. Biol. Chem.*, 255, 6424, 1980.
154. **Tyler, J. M., Hargreaves, W. R., and Branton, D.,** Purification of two spectrin-binding proteins: biochemical and electron microscopic evidence for site-specific reassociation between spectrin and bands 2.1 and 4.1, *Proc. Natl. Acad. Sci. U.S.A.*, 76, 5192, 1979.
155. **Tyler, J. M., Reinhardt, B. N., and Branton, D.,** Associations of erythrocyte membrane proteins, *J. Biol. Chem.*, 255, 7034, 1980.
156. **Furthmayr, H.,** Glycophorins A, B, and C: a family of sialoglycoproteins, *J. Supramol. Struct.*, 9, 79, 1978.
157. **Owens, J. W., Mueller, T. J., and Morrison, M.,** A minor sialoglycoprotein of the human erythrocyte membrane, *Arch. Biochem. Biophys.*, 204, 247,1980.
158. **Nigg, E. A., Bron, C., Girardet, M., and Cherry, R. J.,** Band 3-glycophorin A association in erythrocyte membranes demonstrated by combining protein diffusion measurements with antibody-induced cross-linking, *Biochemistry*, 19, 1887, 1980.
159. **Strapazon, E. and Steck, T. L.,** Binding of rabbit muscle aldolase to band 3, the predominant polypeptide of the human erythrocyte membrane, *Biochemistry*, 15, 1421, 1976.
160. **Strapazon, E. and Steck, T. L.,** Interaction of the aldolase and the membrane of human erythrocytes, *Biochemistry*, 16, 2966, 1977.
161. **Karadsheh, N. S. and Uyeda, K.,** Changes in allosteric properties of phosphofructokinase bound to erythrocyte membranes, *J. Biol. Chem.*, 252, 7418, 1977.
162. **Salhany, J. M., Cordes, K. A., and Gaines, E. D.,** Light-scattering measurements of hemoglobin binding to the erythrocyte membrane, *Biochemistry*, 19, 1447, 1980.
163. **Tsai, I.-H., Murthy, S. N. P., and Steck, T. L.,** Effect of red cell membrane binding on the catalytic activity of glyceraldehyde-3-phosphate dehydrogenase, *J. Biol. Chem.*, 257, 1438, 1982.
164. **Kliman, H. J. and Steck, T. L.,** Association of glyceraldehyde-3-phosphate dehydrogenase with the human red cell membrane, *J. Biol. Chem.*, 255, 6314, 1980.
165. **Bisson, R., Jacobs, B., and Capaldi, R. A.,** Binding of arylazidocytochrome *c* derivatives to beef heart cytochrome *c* oxidase: cross-linking in the high and low affinity binding sites, *Biochemistry*, 19, 4173, 1980.
166. **Hess, B., Kuschmitz, D., and Engelhard, M.,** Bacteriorhodopsin, in *Membranes and Transport*, Vol. 2, Martonosi, A. N., Ed., Plenum Press, New York, 1982, 309.
167. **Stoeckenius, W. and Bogomolni, R. A.,** Bacteriorhodopsin and related pigments of halobacteria, *Annu. Rev. Biochem.*, 51, 587, 1982.
168. **Meyer, D. I., Krause, E., and Dobberstein, B.,** Secretory protein translocation across membranes — the role of the "docking protein", *Nature (London)*, 297, 647, 1982.
169. **Elliott, C. and Ralston, G. B.,** Solubilisation of human erythrocyte band 4.1 protein in the non-ionic detergent Tween® 20, *Biochim. Biophys. Acta*, 775, 313, 1984.

151. Bennett, V. and Branton, D., Selective association of spectrin with the cytoplasmic surface of human erythrocyte membranes, J. Biol. Chem., 252, 2753, 1977.
152. Bennett, V. and Stenbuck, P. J., Identification and partial purification of ankyrin, the high affinity membrane attachment site for human erythrocyte spectrin, J. Biol. Chem., 254, [illegible], 1979.
153. Bennett, V. and Stenbuck, P. J., Association between ankyrin and the cytoplasmic domain of band 3 isolated from the human erythrocyte membrane, J. Biol. Chem., 255, 6424, 1980.
154. [illegible], J. M., [illegible], W. S., and [illegible], [illegible], [illegible] in Cell [illegible] and [illegible], [illegible] Acad. Sci., [illegible], 1980.
155. [illegible], J. M., [illegible], [illegible] and [illegible], [illegible], [illegible] Biol. Chem., 255, [illegible], 1980.
156. [illegible], H., [illegible] and [illegible], [illegible], 1978.
157. [illegible], J. W., [illegible], [illegible], and [illegible], [illegible], [illegible] Cell Res., [illegible], 1980.
158. [illegible], [illegible], [illegible], [illegible], [illegible], M., and [illegible], [illegible], [illegible], [illegible] 1980.
159. [illegible], [illegible] and [illegible], [illegible], [illegible] in the human [illegible], [illegible], 1976.
160. [illegible], [illegible] and [illegible], [illegible], Biochemistry, [illegible], 1980.
161. [illegible], N. S. and [illegible], [illegible], [illegible], J. Biol. Chem., 252, [illegible], 1977.
162. [illegible], J. M., [illegible], K. A., and [illegible], [illegible], [illegible] Biochemistry, 19, [illegible], 1980.
163. [illegible], [illegible], [illegible], and [illegible], [illegible], activity of glyceraldehyde-3-phosphate dehydrogenase, J. Biol. Chem., 25[illegible], 1980.
164. [illegible], [illegible] and [illegible], [illegible] human [illegible], J. Biol. Chem., 255, [illegible], 1980.
165. [illegible], [illegible], [illegible], and [illegible], [illegible], [illegible], 1981.
166. [illegible], [illegible], and [illegible], [illegible], [illegible], A., Ed., [illegible] Press, New York, 1982, [illegible].
167. [illegible], W. and [illegible], R. A., [illegible], [illegible], 1982.
168. Meyer, [illegible], [illegible], P., and [illegible], [illegible], the role of the [illegible], [illegible], 1980.
169. [illegible], C. and [illegible], G. B., [illegible], [illegible], 1980.

Chapter 3

LATERAL MOBILITY OF PROTEINS IN MEMBRANES

Reiner Peters

TABLE OF CONTENTS

I. INTRODUCTION

Cell function apparently requires that cell membranes have both dynamic and static properties. The insertion of newly synthetized proteins into the plasma membrane and their distribution on the cell surface or the uptake of material by pinocytosis and the subsequent recycling of membrane components are only two examples of dynamic events in membranes. The maintenance of size, shape, polarity, and permanent surface specializations, for instance, reflect membrane statics. This chapter will show that the duality of membrane dynamics is apparent in protein lateral mobility and makes it necessary to visualize cell membranes as fluid-solid composites. We assume that this architectural principle pertains to both major membrane coordinates: in-plane and perpendicular to the plane. The plasma membrane of many cells may feature on its cytoplasmic face a relatively stable and static protein network, the membrane skeleton. A more fluid lipid bilayer containing embedded proteins, attached to the membrane skeleton on its external face, is an element common to most cellular membranes. The membrane is completed, in many cases, by a layer of carbohydrate, e.g., the glycocalix of eukaryotes whose dynamic properties, however, have been little studied. The membrane bilayer — rather than a two-dimensional solution of integral proteins in a fluid lipid bilayer — probably is a patchwork made up from solid and fluid domains. Solid domains, mainly composed of protein oligomers, may alternate with more fluid lipo-protein bilayer regions. Phase segregation in the bilayer regions may contribute occasionally to the fluid-solid composite character.

In recent years many methods have been applied to the study of lateral mobility in membranes. There is, however, only one technique — fluorescence microphotolysis ("photobleaching") — that can be applied equally well to model membranes, isolated cell membranes, and living cells, thus facilitating the comparison and interpretation of data from different experimental systems. The chapter, therefore, focuses on results obtained by fluorescence microphotolysis. Rather than a complete catalog of experimental results, the chapter intends to provide a discussion of principles and general aspects and to illustrate them by selected examples. Therefore, sections on methodology, protein mobility in artificial membranes, the potential involvement of protein mobility in certain membrane functions, and the mechanisms by which protein mobility is restricted and possibly regulated in cell membranes have been included. For additional data and aspects the reviews by Webb,[1] Cherry,[2] Peters,[3] Edidin,[4] Jacobson and Wojcieszyn,[5] Schlessinger and Elson,[6] Webb et al.,[7] Vaz et al.,[8] Axelrod,[9] Koppel,[10] and Peters[11] may be consulted.

II. FLUORESCENCE MICROPHOTOLYSIS

Fluorescence microphotolysis is based on a property shared by many fluorescent chromophores. Exposed to high light intensities, the chromophores are photochemically modified and become nonfluorescent. The photochemistry of this process has not been studied systematically. One may guess, however, that minor modifications such as the removal of a double bond or the oxidation of a side chain can have an incisive effect on the fluorescence properties of a chromophore. The irreversible photolysis of fluorescent dyes has been known for a long time and, before the advent of sensitive electronic imaging techniques, was a major problem in fluorescence microscopy. We were confronted with the problem in 1973 when studying energy transfer in monolayers and erythrocyte membranes.[12] It then occurred to us that irreversible photolysis in combination with microfluorometry could serve as basis for diffusion measurements at the spatial resolution of the light microscope.[13]

Following its introduction, fluorescence microphotolysis has been considerably improved and extended.[14-26] Some steps in this development were the use of lasers as light source[14-16] and the control of laser intensities by acousto-optic modulation,[23] the application of periodic patterns as geometry of the illumination area,[17,25] the combination of fluorescence microphotolysis with total internal reflection,[19] the analysis of continuous microphotolysis,[20] and the measurement of rotational motion.[21] Fluorescence microphotolysis has been applied to a large range of topics: molecular motion of fluorescent lipid analogs and/or fluorescently labeled proteins in cell membranes,[13,27] in reconstituted bilayers,[28] in phospholipid monolayers at the air-water interface,[29] in free solution,[30] and in the cytoplasm of living cells.[31] Recently, fluorescence microphotolysis has been also used to measure transport *through* membranes in single cells and organelles.[26]

The measurement of molecular mobility in membranes by fluorescence microphotolysis typically proceeds as follows. First, the membrane is labeled fluorescently. In living cells labeling may be achieved, for instance, by incubating cells with a ligand such as a fluorescently labeled antibody or hormone which binds to a membrane protein with high affinity and specificity. After labeling, the specimen is positioned on the stage of a fluorescence microscope and a single cell is brought into focus. A small part of the cell surface, e.g., a circular spot of 1 μm radius, is illuminated by a microbeam of low intensity and fluorescence originating from the illuminated area is measured. The intensity of the microbeam is suddenly increased by several orders of magnitude. In a fraction of a second most chromophores which are present in the illuminated area become nonfluorescent. The initial low intensity of the microbeam is restored and fluorescence measured. Any transport of fluorescently labeled molecules from the surrounding membrane areas into the photolyzed area gives rise to a recovery of fluorescence. The time course of fluorescence recovery is used to derive values for the transport coefficients. In case of isotropic two-dimensional diffusion and a uniformly illuminated area of radius b the relation between fluorescence recovery r(t) and lateral diffusion coefficient D is simply given[16] by:

$$D = (0.88\ b^2)/(4\ t_{1/2}) \tag{1}$$

where $t_{1/2}$ is the time required for 50% fluorescence recovery. Fluorescence recovery r(t) is defined as $r(t) = [F(t) - F(0)]/[F(-) - F(0)]$ denoting fluorescence before, immediately after, at time t after, and very long after photolysis as F(−), F(0), F(t), and F(∞), respectively. If only a part of the labeled molecules is mobile, the magnitude R_m of the mobile fraction can be easily determined:

$$R_m = [F(\infty) - F(0)]/[F(-) - F(0)] \tag{2}$$

Equations (1) and (2) pertain to diffusive transport. In order to establish whether this case is given or whether other transport mechanisms (e.g., directed flow) prevail, more elaborate curve-fitting procedures have to be applied. A convenient and fast method in which the raw data are linearized by a reciprocal plot has been described recently.[24]

Instrumentation for fluorescence microphotolysis consists of a laser, a fluorescence microscope, a sensitive light-measuring device, a computer, and several optical components. In our set-up the line of an argon ion laser (Spectra Physics, model 167-07) is split into a weaker measuring and a stronger photolyzing beam. The photolyzing beam is blocked by an electromagnetic shutter which, for the purpose of photolysis, can be opened for intervals of 1/125 sec or longer. The beams are coaxially aligned and directed into the vertical illuminator of a fluorescence microscope (Zeiss, Universal® microscope). In front of the vertical illuminator, at an image plane of the microscope optics, a diaphragm is placed into the center of the laser beam. This has the effect that the diaphragm is imaged into the focal plane to yield a uniformly illuminated area. The method allows for easy change of size and geometry of the illuminated area. Another rather important feature of this type of illumination is its insensitivity to coaxial alignment of measuring and photolyzing beam which, with other methods, can be a major source of error. The microscope is equipped with a fast scanning stage (Zeiss) which allows to move the specimen in the focal plane in steps of 0.25 μm. The step rate is adjustable from 0/sec to a maximum of 10,000/sec. Fluorescence originating from the illuminated area is collected by the objective lens and imaged into a photometric attachment where it passes a field diaphragm. The field diaphragm is important for rejecting out-of-focus fluorescence and stray light. Fluorescence is projected on the cathode of a photomultiplier and than processed by single-photon counting equipment (ORTEC). The signal is transmitted to a computer (Digital Equipment, MINC 11/23), plotted in real time, and later evaluated under graphic visualization.

III. LATERAL MOBILITY OF PROTEINS IN RECONSTITUTED BILAYER MEMBRANES

Studies on reconstituted bilayers have largely contributed to the understanding of molecular motion in membranes and are the starting point for the interpretation of measurements pertaining to cell membranes. About a dozen different peptides and proteins have been incorporated into artificial bilayer membranes in order to study their lateral mobility, e.g., gramicidin S,[32] gramicidin C,[33] an apolipoprotein,[34] a phage coat protein,[35] cytochrome P-450,[36] glycophorin,[37] the band 3 protein and other proteins of the erythrocyte membrane,[38] bacteriorhodopsin,[39] acetylcholine receptor,[40] bovine rhodopsin,[41] and a Ca^{++}-activated adenosinetriphosphatase.[41] As in other reconstitution studies difficulties may arise from the circumstance that hydrophobic integral proteins have to be removed from their native environment in the cell membrane and inserted into artificial lipid bilayers without structural and functional modification. Furthermore, it is desirable to have control of the final concentration of protein in the bilayer and to obtain protein concentrations as high as in cell membranes. This problem has been satisfactorily solved in few cases, one of which is bacteriorhodopsin.[42-44]

Restricting the discussion for the moment to fluid membranes (i.e., to temperatures above the crystalline-liquid crystalline main phase transition temperature of the lipid) and to high lipid/protein ratios, there are, in particular, two conclusions of general implication: the lateral diffusion coefficient D of the protein is insensitive to protein molecular size, and the absolute D values are high and only slightly smaller than those of lipid probes. For instance, incorporated into bilayers of dimyristoylphosphatidylcholine (DMPC) the hydrophobic oligopeptide gramicidin S (molecular mass 1141 daltons) and the band 3 protein of the erythrocyte membrane (100,000 daltons) have diffusion coefficients of 1.8 μm^2/sec (24°C)[32] and 1.6 μm^2/sec (30°C),[38] respectively. The diffusion coefficient of lipid analogs under comparable conditions is close to 5.0 μm^2/sec.

Such a behavior had been predicted by the model calculations of Saffman and Delbrück.[45-48] These calculations are based on hydrodynamic principles and treat the lipid bilayer as fluid continuum. The protein molecule is modeled as cylinder of radius a and height h embedded in a fluid membrane of thickness h and viscosity η. The membrane is suspended in an aqueous phase of viscosity η_w. The exact solution of the problem has been recently obtained by Hughes et al.[47] and is rather complicated. However, for small values of η_w — more precisely for $\epsilon = (2\ \eta_w\ a)/(\eta\ h) < 0.1$ — the following simple relations hold for the lateral diffusion coefficient D and the rotational diffusion coefficient D_R:

$$D = \frac{k\,T}{4\,\pi\,\eta\,h}\left(\ln\frac{\eta\,h}{\eta_w\,a} - \gamma\right) \tag{3}$$

$$D_R = \frac{kT}{4\,\pi\,\eta\,h\,a^2} \tag{4}$$

where k is Boltzman's constant, γ is Euler's constant (0.5772), and T is absolute temperature. The ratio D/D_R is then given by:

$$\frac{D}{D_R} = a^2 \left(\ln \frac{kT}{4 \pi a^3 \eta_w D_R} - \gamma \right) \tag{5}$$

The Saffman-Delbrück equations have been recently tested[39] by studying the mobility of bacteriorhodopsin in bilayers of DMPC. Both the lateral and the rotational diffusion coefficient of bacteriorhodopsin were measured and it was thus possible to calculate a hydrodynamic molecular radius by application of Equation (5). For various values of temperature and lipid/protein ratio the calculated molecular radius ranged from 19.7 to 23.9 Å. These values reasonably compare with electron diffraction studies of crystalline bacteriorhodopsin samples[49] yielding a maximum radius of 17.5 Å. Furthermore, insertion of the molecular radius into Equation (3) permitted calculation of η, the viscosity of the membrane phase, which was found to vary between 1.1 and 3.5 P for molar lipid/protein ratios between 210 and 90 (T = 32°C). In a further step η_w, the viscosity of the aqueous phase, was varied between 0.76 and 9.54 cP by addition of sucrose which, in accordance with Equation (3), yielded an approximately twofold decrease of D. Altogether these studies suggest that the Saffman-Delbrück equations correctly describe diffusion of a cylindrical integral protein in a fluid lipid bilayer.

It is obvious, however, that the Saffman-Delbrück equations deal with a simple case and that their applicabity to more complex model membranes or to cell membranes is doubtful. Two parameters, in particular, are not accounted for: the effects of protein portions residing outside the bilayer and the effects of high protein concentration.

Integral membrane proteins frequently have large portions extending into the aqueous phase. In case of the band 3 protein, for instance, almost half of the molecular mass (a 45,000-dalton fragment) is located at the cytoplasmic membrane face outside the bilayer. It is probable that the extra-bilayer portions, by interacting with the membrane skeleton, with extracellular structures or with adjacent cells, have important functions. Certainly they can strongly modify protein lateral mobility.

The concentration of proteins in cell membranes is usually very high. Molar lipid/protein ratios are frequently close to 50. The dependence of the lateral distribution of membrane proteins on the lipid/protein ratio has been modeled by Monte Carlo calculations.[50] Qualitatively, three states in the arrangement and aggregation can be distinguished: at high lipid/protein ratio protein molecules are isolated and dispersed at random; with decreasing lipid/protein ratio the monomolecular lipid shells of the protein molecules (the annular lipid) come into touch; and, finally, at lipid/protein ratios as found in cell membranes, free lipid and annular lipid is partly lost and the protein molecules form a continuous network. The effect on lateral diffusion of protein aggregation has been explored by percolation theory.[51] Here, protein aggregates are modeled by impermeable patches. Long-range diffusion is relatively sensitive to the area fraction of impermeable patches. At a critical value of the area fraction, the percolation threshold, diffusion is completely blocked. These model calculations are only a first approximation to the complicated situation in cell membranes. They may give, however, at least qualitatively a correct impression. In the bacteriorhodopsin/DMPC system,[39] for instance, the diffusion coefficient of bacteriorhodopsin decreased from 3.4 to 0.15 μm^2/sec when the molar lipid/protein ratio was decreased from 210 to 30. A particularly large drop of D (about tenfold decrease) was observed between lipid/protein ratios of 90 and 30. This range of lipid/protein ratios corresponds to a transition from lipid shell contact to partially loss of free and annular lipid. It is important that lipid mobility is much less influenced by protein concentration. For the same reduction of lipid/protein ratio (90 to 30) the lipid probe diffusion coefficient dropped only from 2.5 to 0.73 μm^2/sec.

Upon transition of the lipid bilayer from a liquid-crystalline to a crystalline state protein lateral diffusion coefficients are reduced by two to three orders of magnitude. However, the temperature at which the protein becomes immobilized may substantially differ from the main phase transition temperature of the lipid. Glycophorin, incorporated into DMPC bilayers at a molar lipid/protein ratio of 4500:1, is immobilized at 15°C, i.e., about 9°C below the lipid phase transition temperature.[37] This has been attributed to a local fluidizing of the bilayer around protein molecules. In other systems this effect has not been seen as clearly. In the bacteriorhodopsin/DMPC system a different feature became apparent: the protein strongly influences energy and cooperativity of the lipid phase transition. At high protein concentrations the phase transition virtually disappears and Arrhenius plots become linear.[39]

IV. LATERAL MOBILITY OF PROTEINS IN CELL MEMBRANES

Speculation and theory largely stimulated interest in membrane dynamics. In 1968, for instance, Adam and Delbrück[52] considered the reaction of a ligand with a membrane-localized target. Situated initially in solution the ligand can take two pathways in order to hit the target: diffusion in solution or diffusion to the membrane, unspecific adsorption to the membrane, and diffusion on the membrane surface. Calculation of diffusion times showed that the latter pathway by "reduction of dimensionality" could shorten diffusion times and thus enhance reaction rates. This idea has been followed up[53-55] and applied to a number of biological problems, although experimental evidence is still ambiguous. Another example for the large impact of stimulating ideas is the "mobile receptor hypothesis"[56] which will be discussed below.

About 100 membrane proteins have been studied for their lateral mobility (data obtained up to 1981 are tabulated by Peters[3]). The diffusion coefficients of membrane proteins display an astonishing homogeneity: most of them are in the range of 0.01 to 0.05 μm^2/sec. There are only a few, but particularly interesting, exceptions, for example, rhodopsin having a very high mobility (D about 0.1 μm^2/sec) and the band 3 protein of the erythrocyte membrane with a low mobility (D about 0.001 μm^2/sec). The apparent homogeneity of protein lateral diffusion coefficients may be, however, misleading because the sample is biased: in most cases cultured cells have been studied. It may be reasonably expected that mobility in the membranes of fully differentiated cells *in situ* is more restricted than in cultured cells.

Lateral diffusion coefficients of 0.01 to 0.05 μm^2/sec which are typical for integral membrane proteins of cultured cells seem to be compatible with a fluid bilayer structure. It has been mentioned that the diffusion coefficient of bacteriorhodopsin in DMPC bilayers amounts to approximately 0.1 μm^2/sec at a molar lipid/protein ratio of 30. A heterogeneous lipid composition including high fractions of cholesterol may be sufficient to reduce protein diffusion coefficients to the values observed in membranes of cultured cells. It has been frequently argued that the difference between lipid probe and protein lateral diffusion coefficients is much larger in cellular than in reconstituted membranes and that this indicates restriction of protein diffusion by cytoskeletal elements. However, this argumentation extrapolates from reconstituted membranes with very low protein concentration to cell membranes with extremely high protein concentrations and does not take into account that an increase of protein density reduces lipid probe much less than protein diffusion coefficients. However, another feature of membrane protein dynamics is indeed incompatible with a fluid bilayer structure: membrane proteins regularly show immobile fractions. Their magnitude varies largely among different protein species but, on the average, may amount to approximately 0.3.

Thus, two categories of questions require particular attention: (1) is the lateral mobility of membrane proteins directly involved in certain membrane functions? Is lateral diffusion in these cases a rate-limiting and regulative parameter?; and (2) given a cellular requirement for relatively stable and permanent membrane structures, how is lateral mobility restricted?

In the following we will discuss these questions, taking hormone-stimulated adenylate cyclase as an example for the former and the erythrocyte membrane as example for the latter category. The potential role of lateral mobility in other membrane processes has been discussed in the references: electron transfer in the mitochondrial inner membrane,[57] visual transduction,[58] entrance of viruses into cells,[59] triggering of immunological responses,[60] and formation of the motoric endplate.[61]

A. The Potential Role of Lateral Mobility in Signal Transduction: Hormone-Sensitive Adenylate Cyclase

Hormone-sensitive adenylate cyclase is a molecular machinery which receives, transduces, amplifies, and translates the hormonal signal at the level of the plasma membrane (for review, see References 62 and 63). Its relevance is illuminated by the fact that it occurs in virtually all cells (with the exception of mammalian erythrocytes). Hormone-sensitive adenylate cyclase catalyzes the conversion of adenosine triphosphate (ATP) to cyclic 3′,5′-adenosine monophosphate (cAMP) which — as "second messenger" — carries the hormonal signal into the cell and, for instance, by activation of protein kinase, triggers off the hormonal response. Hormone-sensitive adenylate cyclase can be activated by many different hormones (adrenocorticotropic hormone, calcitonin, beta-adrenergic catecholamines, human chorionic gonadotropin, follicle-stimulating hormone, glucagon, histamine, luteinizing hormone, luteinizing hormone releasing factor, melanocyte-stimulating hormone, parathyroid hormone, prostaglandin E_1, serotonin, thyreotropin-releasing hormone, thyreoid-stimulating hormone, and vasopressin). However, individual cells usually respond to only one or a few different hormones. An exception is the rat epididymal fat cell which is sensitive to up to seven different hormones.

Hormone-sensitive adenylate cyclase consists of at least three physically independent entities all of which are integral membrane proteins: R, the receptor, C, the catalytic unit, and N, the guanine nucleotide binding regulatory protein. R, C, and N have been characterized and, in part, isolated. Chemical and biochemical studies are impeded by the circumstance that these proteins are rare. A number of 1000 receptors per cell, for instance, is normal and thus, in order to achieve a purification, a 100,000-fold enrichment is necessary. This was only possible when methods such as radioimmunoassay and affinity chromatography became available.

R contains the hormone-binding site on the extracellular membrane face. Apparently, receptors for different hormones are different proteins, each of which specifically binds only one kind of hormone. Among these the beta-adrenergic receptor has been studied most carefully (see Lefkowitz et al.[63]). In frog erythrocytes a single beta-adrenergic agonist binding peptide of 58,000 daltons has been found. But in avian erythrocytes two peptides of 40,000 and 50,000 daltons, respectively, and in mammalian cells two or three hormone-binding peptides of 62,000 to 65,000, 47,000 to 53,000, and 35,000 to 40,000 daltons, respectively, have been described. The purification of beta-adrenergic receptors from various sources has been achieved. It is, however, still unresolved why multiple proteins — each containing an appropriate hormone-binding site — exist. N has been isolated and purified.[64-68] It is an oligomer of 125,000 to 130,000 daltons with binding sites for guanine nucleotides, fluoride, choleratoxin, and other bacterial toxins. It is exposed at the internal membrane face. C — possibly because of its lability — has been less well characterized. Radiation inactivation analysis suggests a molecular mass of 190,000 daltons.

Adenylate cyclase can be only stimulated by hormone if R, N, and C all are present in a membrane-bound state. Current models assume[63] that the activation process can be described as a sequential process: the binding of hormone to receptor, the interaction of the hormone-receptor complex with N, the activation of N by exchange of guanosine diphosphate (GDP) against guanosine triphosphate (GTP), the interaction of activated N with C, and the activation of C. This scheme involves three coupled reaction cycles, one each for R, N, and C, respectively.

Qualitative considerations suggest lateral diffusion to play an important role in the adenylate cyclase system. Usually, only a small fraction of the cellular receptor pool needs to be activated by hormone in order to stimulate the maximal possible rate of cAMP production. Furthermore, in cells responding to many different hormones (the rat epididymal fat cell has been mentioned) the different receptors all compete for the same cyclase molecules. The "mobile receptor hypothesis",[56] therefore, assumes that R freely diffuses in the membrane and, upon binding of hormone, can activate many C molecules successively by collision. That R, N, and C are indeed mobile and can rapidly diffuse in the membrane plane over considerable distances was confirmed by fusion experiments.[69] In turkey erythrocytes the catalytic unit can be inactivated by heat treatment or by addition of *N*-ethylmaleimide. Friend erythroleukemia cells, on the other hand, contain an active catalytic unit but no receptors. Upon fusion of the two cell types the hormone-sensitive cyclase activity was restored in heterokaryons within a few minutes.

On the basis of the "mobile receptor hypothesis" many attempts have been made to quantitate the role of lateral diffusion for cyclase stimulation. Tolkovsky and Levitzki[70] analyzed the kinetics of cyclase activation and the binding of agonists and antagonists to the beta-adrenergic receptor of turkey erythrocytes under conditions in which N was permanently activated by sodium fluoride or nonhydrolyzable GTP analogs. The experimental results were compared with five different kinetic models but only one of them, the "collision-coupling mechanism" could fit the data sufficiently well. The collision coupling mechanism is based on the following scheme:

$$H + R \rightleftharpoons HR + C \xrightarrow{k} HRC \rightarrow HR + C^* \quad (6)$$

where H is hormone, C* is the active form of C, and k is the bimolecular rate constant of enzyme receptor collision. The collision-coupling mechanism predicts, for instance, that a reduction of the receptor number by an irreversible antagonist will reduce the rate constant of cyclase activation but will have no effect on the maximal cyclase activity. Inversely, a reduction of catalytic units will reduce the maximal cyclase activity but will not affect the rate constant of activation. These features were experimentally observed. A central aspect of the collision coupling mechanism is that k might be diffusion limited. Experimentally, this hypothesis was followed with a change in the "fluidity" of the lipid membrane phase. Houslay et al.[71,72] varied lipid composition and thermotropic properties of liver cell membranes by fusion with liposomes and measured the effect on glucagon-stimulated cyclase activity. The catalytic unit was sensitive to the lipid environment only when coupled to receptor.

It was concluded that phase transitions are restricted to the outer leaflet of the bilayer and, therefore, are sensed by R but not by (uncoupled) C. Hanski et al.,[73] studying the beta-adrenergic system in turkey erythrocytes, used *cis*-vaccenic acid to modify membrane "microviscosity" as monitored by steady-state fluorescence depolarization of diphenylhexatriene. It was found that k increased linearly 20-fold with decreasing microviscosity. On the basis of the collision coupling mechanism the lateral diffusion coefficient of R was estimated to be 0.004 to 0.09 $\mu m^2/sec$ at 25°C. Studies on the temperature dependence of cyclase activation[74] and on the mobility of intramembrane particles[75] also suggested that k is diffusion limited. Helmreich and colleagues[76] varied the lipid composition of Chang liver cell membranes by fusion with liposomes and measured the effect on the number of accessible receptors, cyclase activity, pyrene-dodecanoic mobility, and "phase transitions". "Fluidizing" of the membrane strongly decreased the number of accessible receptors which made it difficult to assess the role of lateral mobility for cyclase activation.

Measurements of long-range diffusion by fluorescence microphotolysis relating to the beta-adrenergic system have been performed only very recently.[77-79] In Chang liver cells[77,78] beta-adrenergic receptors were labeled with Alp-NBD, a fluorescent analog of the beta-antagonist alprenolol. The specificity of Alp-NBD was checked by fluorescence measurements on single cells showing that 60 to 75% of the Alp-NBD fluorescence could be displaced by propranolol and carazolol. Alp-NBD-labeled cells showed a patchy fluorescence. Patches were present even when labeling was performed for only 30 sec at 4°C. The specific fluorescence was to a large fraction (80%) immobile, whereas the mobile fraction (20%) had a diffusion coefficient of about 0.1 $\mu m^2/sec$. Interestingly, preincubation of the cells with high concentrations of the agonist (−)-isoproterenol decreased the patchy appearance and increased the mobile fraction. These results may well indicate that receptor mobility — in contradiction to widespread opinion — is not a necessary condition for adenylate cyclase stimulation. This conclusion is, however, not unequivocal because a substantial receptor fraction (20%) was found to be mobile and to have a rather large diffusion coefficient (0.1 $\mu m^2/sec$). Lipid probe mobility was studied[79] in membranes of turkey erythrocytes. The temperature dependence of the diffusion coefficient (Arrhenius plot) was linear when the outer leaflet of the bilayer had been labeled but displayed a discontinuity if both leaflets had been labeled. Thus, temperature-dependent changes seemed to be restricted to the inner leaflet of the bilayer. These changes, however, were not considered to indicate a phase separation process because fluorescence depolarization of *cis*- and *trans*-parinaric acid (which is sensitive to a phase separation) was temperature independent. *Cis*-vaccenic acid did not enhance lipid probe diffusion, a fact which apparently devaluates earlier conclusions.[73]

It is thus apparent that the role of lateral diffusion in the activation of hormone-sensitive adenylate cyclase is still unresolved. It seems that simpler experimental systems are needed in order to conduct unambigous experiments. Progress has been recently made in the reconstitution of adenylate cyclase in artificial bilayers.[80,81] This approach has confirmed[81] one of the essentials of the mobile receptor hypothesis, namely, that one molecule of the hormone-receptor complex can activate several molecules of the regulatory unit N.

B. Restriction of Lateral Mobility by Membrane Skeletons: The Erythrocyte Membrane

More than other cells the erythrocyte is subject to vigorous and rapidly changing mechanical stress. Shear forces in the bloodstream and the oxygenation-deoxygenation cycle in the capillaries of lung and peripheral tissue require a particularly high stability and flexibility of the erythrocyte surface. These parameters may have led to an evolutionary process in which a principle of cell membrane architecture was expressed in an extreme manner: the construction from two layers differing largely in dynamic and mechanical properties. A major component of the erythrocyte membrane skeleton is spectrin, a molecular species which is characterized by an elongated but highly variable shape, a composition of α- and β-subunits of about 240,000 and 220,000 daltons molecular mass, respectively, and an occurence as $(\alpha\beta)_2$ tetramer. Spectrin is presumably present in erythrocytes of all species.[82] The search for nonerythrocyte spectrin, unsuccessful in the beginning,[83] has seen a burst of confirming observations in the last 1 to 2 years (for review see Baines[84]). In particular, a protein referred to as fodrin[85] located at the cytoplasmic face of the nerve cell plasma membrane, and a protein from the intestinal brush border[86] have been characterized as spectrins. Components of the erythrocyte skeleton other than spectrin may be, also, of widespread occurrence.[87] Ankyrin, for instance, has been identified by immunological methods in many cells;[88] its colocalization with fodrin has also been shown.

The erythrocyte membrane can be physically separated into a membrane skeleton and a membrane skin (for review see References 89 to 91). Treatment of erythrocyte ghosts with Triton® X-100 removes the skin and leaves behind the skeleton which roughly maintains the shape of erythrocytes. The skeleton, on the other hand, can be dissociated and removed from the skin by low-ionic-strength treatment. The isolated skin is rather fragile and easily breaks up into small vesicles. The skeleton consists only of proteins. α- and β-subunits of spectrin, band 4.1 and band 5 (actin), are the main constituent occurring at a frequency of about 220,000, 230,000, and 510,000 copies per cell, respectively. The spectrin tetramers, by mediation of band 4.1 and actin, form a network at the cytoplasmic face of the membrane. The skeleton proteins are abundant enough to cover 50 to 70% of the membrane surface. Proteins of the lipo-protein bilayer (membrane skin) all are glycolysated at the portions extending into the outer medium. About 60 weight% of skin proteins are provided by copies of the band 3 protein which functions as anion channel and thus is involved in carbonate exchange. The band 3 protein is a single-chain polypeptide of 100,000 daltons molecular mass crossing the lipid bilayer several-fold. A 45,000-dalton fragment of the band 3 protein resides outside the bilayer at the cytoplasmic face. Other abundant components of the membrane skin are the glycophorins which are anchored in the bilayer only by short stretches of the peptide chain, while the bulk resides outside the bilayer on the external face. Skeleton and skin are coupled by specific protein-protein interactions. Ankyrin (bands 2.1 to 2.3) has a high-affinity binding site for spectrin and another one for the band 3 protein. The molar ratio of ankyrin to band 3 is 1:10. Band 4.1 has a high-affinity binding site for glycophorin C. Interactions between spectrin and lipids have been invoked as a means by which the pronounced asymmetry of the lipid bilayer is stabilized (the major fraction of phosphatidylethanolamine and -serine is found in the inner leaflet of the bilayer).

For a long time dynamic properties of erythrocyte membranes were only known from mechanical properties classifying the erythrocyte membrane as a tough, viscoelastic solid.[92] New aspects became apparent when Pinto da Silva,[93] by freeze-fracture electron microscopy, observed that intramembrane particles of erythrocyte membranes, normally dispersed at random on the fracture face, form aggregates under certain experimental conditions. Intramembrane particles are considered to be complexes of band 3 and glycophorin. A detailed analysis by Elgsaeter and Branton[94] established that the aggregability of intramembrane particles was correlated with the dissociation of spectrin from the membrane.

In addition to fluorescence microphotolysis fusion methods have been used to study the lateral mobility of proteins in the erythrocyte membrane. This method, introduced by Frye and Edidin,[95] was applied by Fowler and Branton[96] in the following manner: human erythrocytes were labeled fluorescently and fused with unlabeled erythrocytes by Sendai virus, the percentage of fused cell couples in which fluorescence had spread homogeneously over the whole surface was assessed visually by fluorescence microscopy, and from a plot of the fraction of uniformly fluorescent cells vs. time diffusion coefficients were estimated. This approach was placed on a more quantitative basis by measuring the spread of fluorescence in fused-cell couples in a microfluorimeter.[97] Polyethyleneglycol was preferred as fusing agent by some authors[97,98] because Sendai virus induces hemolysis and may affect membrane fluidity.

So far, only one protein of the erythrocyte membrane, the band 3 protein, has been studied for lateral mobility. In all experiments the band 3 protein was fluorescently labeled by direct reaction with dyes such as FITC,[13,96] dichlorotriazinylaminofluorescin (DTAF),[99] or eosinisothiocyanate (EITC).[100,101] The specificity of labeling was about 70% and, in most cases, band 3 and glycophorin were labeled together.

In the intact human erythrocyte membrane the mobility of band 3 is very small. The first study[13] — making no distinction between mobile and immobile fractions — determined the average diffusion coefficient to be 0.0003 μm^2/sec or smaller at 20 to 23°C. Fowler and Branton,[96] by the fusion method, estimated the minimum diffusion coefficient to be 0.0002 μm^2/sec in aged and 0.0006 μm^2/sec in fresh cells (T = 23°C). Schindler et al.[97] reported values of 0.002 μm^2/sec in ATP-depleted and 0.0092 μm^2/sec in fresh cells at 30°C; in addition to the mobile an immobile fraction was observed amounting to about 0.4. Golan and Veatch[101] measured a diffusion coefficient of 0.0050 ± 0.0022 μm^2/sec and a mobile fraction of 0.72 ± 0.07 in oxygen-depleted ghosts at 21°C.

Lipid mobility, in sharp contrast to protein mobility, is quite large in human erythrocyte membranes. Recent measurements,[102] placing particular emphasis on the elimination of potential radiation-induced artifacts, show the lipid probe diffusion coefficient in membranes of both intact erythrocytes and ghosts to be 0.8 μm^2/sec (25.1°C) which is in accordance with previous studies.[103,104]

The restriction of protein lateral mobility in the erythrocyte membrane is clearly related to the membrane skeleton. Measurements[105] of protein mobility in spectrin-deficient (spherocytic) erythrocyte variants of certain mouse strains support this notion. Lipid probe diffusion was found to be virtually identical in spectrin-deficient (S−) and normal (S+) membranes (D = 1.4 to 1.5 μm^2/sec, 23°C), whereas band 3 mobility was about 50-fold larger in S− than S+ membranes (0.25 and 0.0045 μm^2/sec, respectively). Mobile fractions were about 1.0 in S− and 0.6 in S+ membranes. A similarly large effect has been observed in human erythrocytes when spectrin was dissociated from the membrane by low ionic strength and elevated temperature.[101] In these studies a minimum mobility of D = 0.0043 μm^2/sec and R = 0.11 was reported for 46.0 m*M* sodium phosphate buffer, 21°C, whereas D increased to 0.065 and R to 0.83 in 13.3 m*M* buffer, 37°C.

Many attempts have been made to modify band 3 mobility by specific and gradual means, but the results are less impressive. Band 3 mobility is temperature dependent, decreasing about tenfold between 37 and 23°C; no mobility was detected at 0°C[96] (fusion method). ATP, possibly by phosphorylation of spectrin and loosening of skeleton-skin coupling, affects band 3 mobility. A twofold higher diffusion coefficient was reported for erythrocytes from fresh than from aged blood[96] (fusion method). Addition of an ATP-generating system to aged cells had a similar effect[96] (fusion method). Schindler et al.[97] also reported a fourfold increase of the diffusion coefficient when 12.5 m*M* ATP where added to ATP-depleted ghosts (fusion method). Cross-linking of spectrin into high molecular complexes by means of diamide[98] reduced the diffusion coefficient of band 3-glycophorin from 0.003 to 0.0018 μm^2/sec (fusion method). 2,3-Diphosphoglycerol (12.5 m*M*)[97] and triphosphoinositide[106] induced about a twofold increase of protein lateral diffusion coefficient. A tryptic, 72,000-dalton fragment of ankyrin added to ghost in relatively high concentration yielded about a 30% increase of band 3 mobility (fusion method).[107]

With these data a rough sketch of dynamics in the erythrocyte membrane can be drawn. Skeleton and skin are mainly coupled by specific protein-protein interaction, whereas lipid-protein interactions presumably can be neglected (lipid probe mobility is little restricted). The specific interactions between skeleton and skin (band 3-ankyrin-spectrin, glycophorin C-spectrin) are sufficient to link one tenth of the skin proteins directly to the skeleton. Skin proteins, however, do not occur in monomeric form. Dimers or tetramers have been described for band 3. More likely, however, the band 3 protein occurs in a temperature-dependent aggregation equilibrium.[108,109] Glycophorin may be part of these oligomeric complexes. The band 3-glycophorin complex has mobile and immobile fractions. The latter amounts to about 40% at 30°C. It is suggestive to identify the immobile fraction with that band 3 fraction linked directly or as part of a band 3 glycophorin complex to the skeleton by specific interaction. The mobile fraction is then represented by proteins trapped with their cytoplasmic fragments in the interstices of the skeletal meshwork. The diffusion coefficient of the mobile fraction is small because long-range movement is only possible if the mesh is rearranged, broken, or dissociated from the skin. Because several parameters (direct binding, aggregation, trapping) contribute to the restriction of lateral mobility, measures affecting only one parameter have only a small effect on lateral mobility. The complex behavior of the rotational diffusion of the band 3 protein[109-112] also indicates involvement of several parameters (ankyrin, temperature, cholesterol). Whether protein lateral mobility is related to mechanical properties of the membrane and whether these properties are regulated in vivo are still unknown.

V. CONCLUSION

Measurements of protein lateral mobility show that cell membranes have properties of both fluids and solids. Many membrane functions require a certain degree of protein mobility which is met by the measured diffusion coefficients. Whether lateral diffusion is centrally involved in certain membrane functions and, in some cases, constitutes a critical, rate-limiting, and regulating parameter is a question receiving much attention. In this chapter the question has been discussed taking hormone-sensitive adenylate cyclase as example. A review of the data shows that the question cannot yet be answered unequivocally. Considerable progress in reconstitution techniques and diffusion measurement methods makes it likely that the problem can be solved in the near future. Restriction of mobility in membranes is a common observation; most membrane proteins exhibit immobile (i.e., very slowly moving) fractions. In the case of the erythrocyte membrane the restriction of mobility can be related to a protein network, the membrane skeleton, which is located at the cytoplasmic face of the membrane and, by tight coupling to the lipo-protein bilayer, strongly influences mechanical and dynamic properties of the membrane. It seems that membrane skeletons are a widespread element of cellular membranes.

REFERENCES

1. **Webb, W. W.,** Lateral transport on membranes, in *Electrical Phenomenon at the Biological Membrane Level*, Roux, E., Ed., Elsevier, Amsterdam, 1977, 119.
2. **Cherry, R. J.,** Rotational and lateral diffusion of membrane proteins, *Biochim. Biophys. Acta*, 559, 289, 1979.
3. **Peters, R.,** Translational diffusion in the plasma membrane of single cells as studied by fluorescence microphotolysis, *Cell Biol. Int. Rep.*, 5, 733, 1981.
4. **Edidin, M.,** Molecular motions and membrane organization and function, in *Membrane Structure*, Finean, J. B. and Mitchell, R. H., Eds., Elsevier/North-Holland, New York, 1981, 37.
5. **Jacobson, K. and Wojcieszyn, J.,** On the factors determining the lateral mobility of cell surface components, *Comments Mol. Cell Biophys.*, 1, 189, 1981.
6. **Schlessinger, J. and Elson, E. L.,** Quantitative methods for studying the mobility and distribution of receptors on viable cells, in *Membrane Receptors: Methods for Purification and Characterization*, Jacobs, S. and Cuatrecasas, P., Eds., Chapman and Hall, London, 1981, 158.
7. **Webb, W. W., Barak, L. S., Tank, D. W., and Wu, E.-S.,** Molecular mobility on the cell surface, *Biochem. Soc. Symp.*, 46, 191, 1982.
8. **Vaz, W. L. C., Derzko, Z. I., and Jacobson, K. A.,** Photobleaching measurements of the lateral diffusion of lipids and proteins in artificial phospholipid bilayer membranes, in *Membrane Reconstitution*, Poste, E. and Nicolson, G. L., Eds., Elsevier, Amsterdam, 1982, 83.
9. **Axelrod, D.,** Lateral motion of membrane proteins and biological function, *J. Membrane Biol.*, 75, 1, 1983.
10. **Koppel, D. E.,** Fluorescence photobleaching as a probe of translational and rotational motion, in *Fast Methods in Physical Biochemistry and Cell Biology*, Sha'afi, R. I. and Fernandez, J. M., Eds., Elsevier, Amsterdam, 1983, 339.
11. **Peters, R.,** Fluorescence microphotolysis: diffusion measurements in single cells, *Naturwissenschaften*, 70, 294, 1983.
12. **Peters, R.,** The thickness of air-dried human erythrocyte membranes as determined by energy transfer, *Biochim. Biophys. Acta*, 330, 53, 1973.
13. **Peters, R., Peters, J., Tews, K. H., and Bähr, W.,** A microfluorimetric study of translational diffusion in erythrocyte membranes, *Biochim. Biophys. Acta*, 367, 282, 1974.
14. **Edidin, M., Zagyanski, Y., and Lardner, T. J.,** Measurement of membrane protein lateral diffusion in single cells, *Science*, 191, 466, 1976.
15. **Jacobson, K., Wu, E.-S., and Poste, G.,** Measurement of the translational mobility of concanavalin A in glycerol-saline solutions and on the cell surface, by fluorescence recovery after photobleaching, *Biochim. Biophys. Acta*, 433, 215, 1976.

16. **Axelrod, D., Koppel, D. E., Schlessinger, J., and Webb, W. W.,** Mobility measurement by analysis of fluorescence photobleaching recovery kinetics, *Biophys. J.*, 16, 1055, 1976.
17. **Smith, B. A. and McConnell, H. M.,** Determination of molecular motion in membranes using periodic pattern photobleaching, *Proc. Natl. Acad. Sci. U.S.A.*, 75, 2759, 1978.
18. **Koppel, D. E.,** Fluorescence redistribution after photobleaching. A new multipoint analysis of membrane translational dynamics, *Biophys. J.*, 28, 281, 1979.
19. **Thompson, N. L., Burghadt, T. P., and Axelrod, D.,** Measuring surface dynamics of biomolecules by total internal reflection fluorescence with photobleaching recovery or correlation spectroscopy, *Biophys. J.*, 33, 435, 1981.
20. **Peters, R., Brünger, A., and Schulten, K.,** Continuous fluorescence microphotolysis: a sensitive method for study of diffusion processes in single cells, *Proc. Natl. Acad. Sci. U.S.A.*, 78, 962, 1981.
21. **Smith, L. W., Weis, R. M., and McConnell, H. M.,** Measurement of rotational motion in membranes using fluorescence recovery after photobleaching, *Biophys. J.*, 36, 73, 1981.
22. **Smith, L. M., McConnell, H. M., Smith, B. A., and Parce, J. W.,** Pattern photobleaching of fluorescent lipid vesicles using polarized laser light, *Biophys. J.*, 33, 139, 1981.
23. **Garland, P.,** Fluorescence photobleaching recovery: control of laster intensities with an acousto-optic modulator, *Biophys. J.*, 33, 481, 1981.
24. **Yguerabide, J., Schmitt, J. A., and Yguerabide, E. E.,** Lateral mobility in membranes as detected by fluorescence recovery after photobleaching, *Biophys. J.*, 39, 69, 1982.
25. **Lanni, F. and Ware, B. R.,** Modulation detection of fluorescence photobleaching recovery, *Rev. Sci. Instrum.*, 53, 905, 1982.
26. **Peters, R.,** Nuclear envelope permeability measured by fluorescence microphotolysis of single liver cell nuclei, *J. Biol. Chem.*, 258, 11427, 1983.
27. **Schlessinger, J., Axelrod, D., Koppel, D. E., Webb, W. W., and Elson, E. L.,** Lateral transport of a lipid probe and labeled proteins on a cell membrane, *Science*, 195, 307, 1977.
28. **Wu, E.-S., Jacobson, K., and Papahadjopoulos, D.,** Lateral diffusion in phospholipid multilayers measured by fluorescence recovery after photobleaching, *Biochemistry*, 16, 3936, 1977.
29. **Peters, R. and Beck, K.,** Translational diffusion in phospholipid monolayers measured by fluorescence microphotolysis, *Proc. Natl. Acad. Sci. U.S.A.*, 80, 7183, 1983.
30. **Barisas, B. G. and Leuther, M. D.,** Fluorescence photobleaching recovery measurement of protein absolute diffusion constants, *Biophys. Chem.*, 10, 221, 1979.
31. **Wojcieszyn, J. W., Schlegel, R. A., Wu, E.-S., and Jacobson, K. A.,** Diffusion of injected macromolecules within the cytoplasm of living cells, *Proc. Natl. Acad. Sci. U.S.A.*, 78, 4407, 1981.
32. **Wu, E.-S., Jacobson, K., Szoka, F., and Portis, A.,** Lateral diffusion of a hydrophobic peptide, N-4-nitrobenzo-2-oxa-1,3-diazole gramicidin S, in phosopholipid multilayers, *Biochemistry*, 17, 5543, 1978.
33. **Tank, D. W., Wu, E.-S., Meers, P. R., and Webb, W. W.,** Lateral diffusion of gramicidin C in phospholipid multibilayers. Effects of cholesterol and high gramicidin concentrations, *Biophys. J.*, 40, 129, 1982.
34. **Vaz, W. L. C., Jacobson, K., Wu, E.-S., and Derzko, Z.,** Lateral mobility of an amphipathic apolipoprotein, Apo III, bound to phosphatidylcholine bilayers with and without cholesterol, *Proc. Natl. Acad. Sci. U.S.A.*, 76, 5645, 1979.
35. **Smith, L. M., Smith, B. A., and McConnell, H. M.,** Lateral diffusion of M13 coat protein in model membranes, *Biochemistry*, 18, 2256, 1979.
36. **Wu, E.-S. and Yang, C. S.,** Lateral diffusion of cytochrome P-450 in phospholipid multibilayers, *Fed. Prod., Fed. Am. Soc. Exp. Biol.*, 39, 1990, 1980.
37. **Vaz, W. L. C., Kapitza, H. E., Stümpel, J., Sackmann, E., and Jovin, T. M.,** Translational mobility of glycophorin in bilayer membranes of dimyristoylphosphatidylcholine, *Biochemistry*, 20, 1392, 1981.
38. **Chang, C.-H., Takeuchi, H., Ito, T., Machida, K., and Ohnishi, S.,** Lateral mobility of erythrocyte membrane proteins studied by the fluorescence photobleaching recovery technique, *J. Biochim.*, 90, 997, 1981.
39. **Peters, R. and Cherry, R. J.,** Lateral and rotational diffusional of backteriorhodopsin in lipid bilayers: experimental test of the Saffman-Delbrück equations, *Proc. Natl. Acad. Sci. U.S.A.*, 79, 4317, 1982.
40. **Criado, M., Vaz, W. L. C., Barrantes, F. J., and Jovin, Th. M.,** Translational diffusion of acetylcholine receptor (monomeric and dimeric forms) of *Torpedo marmorata* reconstituted into phospholipid bilayers studied by fluorescence recovery after photobleaching, *Biochemistry*, 21, 5750, 1982.
41. **Vaz, W. L. C., Criado, M., Madeira, V. M. C., Schoellmann, G., and Jovin, Th. M.,** Size dependence of the translational diffusion of large integral membrane proteins in liquid crystalline phase lipid bilayers. A study using fluorescence recovery after photobleaching, *Biochemistry*, 21, 5608, 1982.
42. **Cherry R. J., Müller, U., and Schneider, G.,** Rotational diffusion of bacteriorrhodsopin in lipid membranes, *FEBS Lett.*, 80, 465, 1977.

43. **Cherry, R. J., Müller, U., Henderson, R., and Heyn, M. P.,** Temperature-dependent aggregation of bacteriorhodopsin in dipalmitoyl- and dimyristoylphosphatidylcholine vesicles, *J. Mol. Biol.*, 121, 283, 1978.
44. **Cherry, R. J. and Godfrey, R. E.,** Anisotropic rotation of bacteriorhodopsin in lipid membranes, *Biophys. J.*, 36, 257, 1981.
45. **Saffman, P. J., and Delbrück, M.,** Brownian motion in biological membranes, *Proc. Natl. Acad. Sci. U.S.A.*, 72, 3111, 1975.
46. **Saffman, P. J.,** Brownian motion in thin sheets of viscous fluid, *J. Fluid Mech.*, 73, 593, 1976.
47. **Hughes, B. D., Pailthorpe, B. A., and White, L. R.,** The translational and rotational drag on a cylinder moving in a membrane, *J. Fluid Mech.*, 73, 593, 1976.
48. **Hughes, B. D., Pailthorpe, B. A., White, L. R., and Sawyer, W. H.,** Extraction of membrane microviscosity from translational and rotational diffusion coefficients, *Biophys. J.*, 37, 673, 1982.
49. **Henderson, R. and Unwin, P. N. T.,** Three-dimensional model of purple membrane obtained by electron microscopy, *Nature (London)*, 257, 28, 1975.
50. **Freire, E. and Snyder, B.,** Quantitative characterization of the lateral distribution of membrane proteins within the lipid bilayer, *Biophys. J.*, 37, 617, 1982.
51. **Saxton, M. J.,** Lateral diffusion in an archipelago. Effects of impermeable patches on diffusion in a cell membrane, *Biophys. J.*, 39, 165, 1982.
52. **Adam, G. and Delbrück, M.,** Reduction of dimensionality in biological diffusion processes, in *Structural Chemistry and Molecular Biology*, Rich, A. and Davidson, N., Eds., W. H. Freeman, San Francisco, 1968, 198.
53. **Richter, P. H. and Eigen, M.,** Diffusion controlled reaction rates in spheroidal geometry. Application to repressor-operator association and membrane bound enzymes, *Biophys. Chem.*, 2, 255, 1974.
54. **Berg, H. C. and Purcell, E. M.,** Physics of chemoreception, *Biophys. J.*, 20, 193, 1977.
55. **Wiegel, F. W. and De Lisi, C.,** Evaluation of reaction rate enhancement by reduction in dimensionality, *Am. J. Physiol.*, 12, R475, 1982.
56. **Cuatrecasas, P.,** Membrane receptors, *Ann. Rev. Biochem.*, 43, 169, 1974.
57. **Schneider, H. and Hackenbrock, Ch. R.,** The significance of protein and lipid mobility for catalytic activity in the mitochondrial membrane, in *Membranes and Transport*, Vol. 1, Plenum Press, New York, 1982, 431.
58. **Liebman, P. A. and Pugh, E. N., Jr.,** Control of rod disk membrane phosphodiesterase and a model for visual transduction, *Curr. Top. Membrane Transp.*, 15, 157, 1981.
59. **Simons, K., Garoff, H., and Helenius, A.,** *Sci. Am.*, 58, 1982.
60. **Weis, R., Balakrishnan, K., Smith, B. A., and McConnell, H. M.,** Stimulation of fluorescence in a small contact region between rat basophil leukemia cells and planar lipid membrane targets by coherent evanescent radiation, *J. Biol. Chem.*, 257, 6440, 1982.
61. **Stya, M. and Axelrod, D.,** Diffusely distributed acetylcholine receptors can participate in cluster formation on cultured rat myotubes, *Proc. Natl. Acad. Sci. U.S.A.*, 80, 449, 1983.
62. **Ross, E. M. and Gilman, A. G.,** Biochemical properties of hormone-sensitive adenylate cyclase, *Annu. Rev. Biochem.*, 49, 533, 1980.
63. **Lefkowitz, R. J., Stadel, J. M., and Caron, M. G.,** Adenylate cyclase-coupled beta-adrenergic receptors: structure and mechanisms of activation and desensitization, *Ann. Rev. Biochem.*, 52, 159, 1983.
64. **Pfeuffer, Th. and Helmreich, E. J. M.,** Activation of pigeon erythrocyte membrane adenylate cyclase by guanylnucleotide analogues and separation of a nucleotide binding protein, *J. Biol. Chem.*, 250, 867, 1975.
65. **Pfeuffer, Th.,** GTP-binding proteins in membranes and the control of adenylate cyclase activity, *J. Biol. Chem.*, 252, 7224, 1977.
66. **Naya-Vigne, J., Johnson, G. L., Bourne, H. R., and Coffino, Ph.,** Complementation analysis of hormone-sensitive cyclase, *Nature (London)*, 272, 720, 1978.
67. **Pfeuffer, Th.,** Guanine nucleotide-controlled interactions between components of adenylate cyclase, *FEBS Lett.*, 101, 85, 1979.
68. **Cassel, D. and Selinger, Z.,** Mechanism of adenylate cyclase activation by cholera toxin: inhibition of GTP hydrolysis at the regulatory site, *Proc. Natl. Acad. Sci. U.S.A.*, 74, 3307, 1977.
69. **Schramm, M.,** Hybridization and implantation of membrane components by fusion, in *Membranes and Transport*, Vol. 2, Martonosi, A. N., Ed., Plenum Press, New York, 1982, 555.
70. **Tolkovsky, A. M. and Levitzki, A.,** Mode of coupling between the β-adrenergic receptor and adenylate cyclase in turkey erythrocytes, *Biochemistry*, 17, 3795, 1978.
71. **Houslay, M. D., Metcalfe, J. C., Warren, G. B., Hesketh, T. R., and Smith, G. A.,** The glucagon receptor of rat liver plasma membrane can couple to adenylate cyclase without activating it, *Biochim. Biophys. Acta*, 436, 489, 1976.
72. **Houslay, M. D., Hesketh, T. R., Smith, G. A., Warren, G. B., and Metcalf, J. C.,** The lipid environment of the glucagon receptor regulates adenylate cyclase activity, *Biochim. Biophys. Acta*, 436, 495, 1976.

73. **Hanski, E., Rimon, G., and Levitzky, A.,** Adenylate cyclase activation by the β-adrenergic receptors as diffusion-controlled process, *Biochemistry,* 18, 846, 1979.
74. **Rimon, G., Hanski, E., and Levitzki, A.,** Temperature dependence of receptor, adenosine receptor, and sodium fluoride stimulated cyclase from turkey erythrocytes, *Biochemistry,* 19, 4451, 1980.
75. **Atlas, D., Volsky, D. J., and Levitzki, A.,** Lateral mobility of β-receptors involved in adenylate cyclase activation, *Biochim. Biophys. Acta,* 597, 64, 1980.
76. **Bakardjieva, A., Galla, H.-J., and Helmreich, E. J. M.,** Modulation of the β-receptor cyclase interactions in cultured change liver cells of phospholipid enrichment, *Biochemistry,* 18, 3016, 1979.
77. **Bakardjieva, A., Peters, R., Hekman, M., Hornig, H., Burgermeister, W., and Helmreich, E. J. M.,** Catecholamine-stimulated adenylate cyclase; an associating-dissociating system, in *Metabolic Interconversion of Enzymes,* Holzer, H., Ed., Springer-Verlag, Berlin, 1981, 378.
78. **Henis, Y. I., Hekman, M., Elson, E. L., and Helmreich, E. J. M.,** Lateral motion of receptors in membranes of cultured liver cells, *Proc. Natl. Acad. Sci. U.S.A.,* 79, 2907, 1982.
79. **Henis, Y. I., Rimon, G., and Felder, S.,** Lateral mobility of phospholipids in turkey erythrocytes. Implications for adenylate cyclase activation, *J. Cell Biol.,* 257, 1407, 1982.
80. **Citri, Y. and Schramm, M.,** Resolution, reconstitution and kinetics of the primary action of a hormone receptor, *Nature (London),* 287, 297, 1980.
81. **Pedersen, S. E. and Ross, E. M.,** Functional reconstitution of β-adrenergic receptors and the stimulatory GTP-binding protein of adenylate cyclase, *Proc. Natl. Acad. Sci. U.S.A.,* 79, 7228, 1982.
82. **Pinder, J., Phethean, J., and Gratzer, W. B.,** Spectrin in primitive erythrocytes, *FEBS Lett.,* 92, 278, 1978.
83. **Hiller, G. and Weber, K.,** Spectrin is absent in various tissue culture cells, *Nature (London),* 266, 181, 1977.
84. **Baines, A. J.,** The spread of spectrin, *Nature (London),* 301, 377, 1983.
85. **Levine, J. and Willard, M.,** Fodrin: axonally transported polypeptides associated with the internal periphery of many cells, *J. Cell Biol.,* 90, 631, 1981.
86. **Glenney, J. R., Glenney, Ph., Osborne, M., and Weber, K.,** *Cell,* 28, 843, 1982.
87. **Cohen, C. M., Foley, S. F., and Korsgen, C.,** A protein immunologically related to erythrocyte bound 4.1 is found on stress fibres of non-erythroid cells, *Nature (London),* 299, 648, 1982.
88. **Bennett, V.,** Immunoreactive forms of human erythrocyte ankyrin are present in diverse cells and tissues, *Nature (London),* 281, 597, 1979.
89. **Branton, D., Cohen, C. M., and Tyler, J.,** Interaction of cytoskeletal proteins on the human erythrocyte membrane, *Cell,* 24, 24, 1981.
90. **Sheetz, M. P.,** Membrane skeletal dynamics: role in modulation of red cell deformability, mobility of transmembrane proteins, and shape, *Sem. Hematol.,* 20, 175, 1983.
91. **Haest, C. W. M.,** Interactions between membrane skeleton proteins and the intrinsic domain of the erythrocyte membrane, *Biochim. Biophys. Acta,* 694, 331, 1982.
92. **Rand, R. P.,** Mechanical properties of the red cell membrane. II. Viscoelastic breakdown of the membrane, *Biophys. J.,* 4, 303, 1964.
93. **Pinto de Silva, P.,** Translational mobility of the membrane intercalated particles of human erythrocyte ghosts, *J. Cell Biol.,* 53, 777, 1972.
94. **Elgsaeter, A. and Branton, D.,** Intramembrane particle aggregation in erythrocyte ghosts. I. The effects of protein removal, *J. Cell Biol.,* 63, 1018, 1974.
95. **Frye, L. D. and Edidin, M.,** The rapid intermixing of all surface antigens after formation of mouse-human heterokaryons, *J. Cell Sci.,* 7, 319, 1969.
96. **Fowler, V. and Branton, D.,** Lateral mobility of human erythrocyte integral membrane proteins, *Nature (London),* 268, 23, 1977.
97. **Schindler, M., Koppel, D. E., and Sheetz, M. P.,** Modulation of membrane protein lateral mobility by polyphosphates and polyamines, *Proc. Natl. Acad. Sci. U.S.A.,* 77, 1457, 1980.
98. **Smith, D. K. and Palek, J.,** Modulation of lateral mobility of band 3 in the red cell membrane by oxidative cross-linking of spectrin, *Nature (London),* 297, 424, 1982.
99. **Sheetz, M. P., Schindler, M., and Koppel, D. E.,** Lateral mobility of integral membrane proteins is increased in spherocytic erythrocytes, *Nature (London),* 285, 510, 1980.
100. **Cherry, R. J., Burkli, A., Busslinger, M., and Schneider, G.,** Rotational diffusion of band 3 proteins in the human erythrocyte membrane, *Nature (London),* 263, 389, 1976.
101. **Golan, D. E. and Veatch, W.,** Lateral mobility of band 3 in the human erythrocyte membrane studied by fluorescence photobleaching recovery: evidence for control by cytoskeletal interactions, *Proc. Natl. Acad. Sci. U.S.A.,* 77, 2537, 1980.
102. **Bloom, J. A. and Webb, W. W.,** Lipid diffusibility in the intact erythrocyte membrane, *Biophys. J.,* 42, 295, 1983.
103. **Kapitza, H.-G. and Sackmann, E.,** Local measurement of lateral motion in erythrocyte membranes by photobleaching technique, *Biochim. Biophys. Acta,* 595, 55, 1980.

104. **Thompson, N. L. and Axelrod, D.,** Reduced lateral mobility of a fluorescent lipid probe in cholesterol-depleted erythrocyte membranes, *Biochim. Biophys. A*, 597, 155, 1980.
105. **Koppel, D. E., Sheetz, M. P., and Schindler, M.,** Matrix control of protein diffusion in biological membranes, *Proc. Natl. Acad. Sci. U.S.A.*, 78, 3576, 1981.
106. **Sheetz, M. P., Febbroriello, P., and Koppel, D. E.,** Triphosphoinositide increases glycoprotein lateral mobility in erythrocyte membranes, *Nature (London)*, 296, 91, 1982.
107. **Fowler, V. and Bennett, V.,** Association of spectrin with its membrane attachment site restricts lateral mobility of human erythrocyte integral membrane proteins, *J. Supramol. Struct.*, 8, 215, 1978.
108. **Dorst, H.-J. and Schubert, D.,** Self-association of band 3 protein from human erythrocyte membranes in aqueous solutions, *Hoppe-Seyler's Z. Physiol. Chem.*, 360, 1605, 1979.
109. **Nigg, E. A. and Cherry, R. J.,** Influence of temperature and cholesterol on the rotational diffusion of band 3 in the human erythrocyte membrane, *Biochemistry*, 18, 3457, 1979.
110. **Austin, R. H., Chan, S. S., and Jovin, Th. M.,** Rotational diffusion of cell surface components by time-resolved phosphorescence anisotropy, *Proc. Natl. Acad. Sci. U.S.A.*, 76, 5650, 1979.
111. **Nigg, E. A. and Cherry, R. J.,** Anchorage of a band 3 population at the erythrocyte cytoplasmic membrane surface: protein rotational diffusion measurements, *Proc. Natl. Acad. Sci. U.S.A.*, 77, 4702, 1980.
112. **Mühlebach, T. and Cherry, R. J.,** Influence of cholesterol on the rotation and self-association of band 3 in the human erythrocyte membrane, *Biochemistry*, 21, 4225, 1982.

Chapter 4

LATERAL DIFFUSION OF LIPIDS

Charles G. Wade

TABLE OF CONTENTS

I. INTRODUCTION

Many fundamental processes in cells require motions of membrane components. Lateral diffusion (D) is the displacement of molecules in a direction parallel to the plane of the membrane (Figure 1). Except for very special cases, the diffusion of phospholipids (and most other amphiphilic molecules) across the membrane (perpendicular to the plane) has a very low probability because of the large free energy change required to move the head group through the hydrophobic region. Consequently, the diffusion of phospholipids in membranes is visualized amost exclusively as a two-dimensional problem; the molecules undergo diffusion within the plane of the membrane while maintaining their orientation with respect to the membrane boundaries.

A principal stimulus for the interest in lateral motion of membrane components came from the work of Frye and Edidin[1] in 1970, who inferred lateral processes from cell fusion data. The first measurements of lipid lateral diffusion were done independently and almost simultaneously by Kornberg and McConnell[2,3] and by Sackmann and Träuble[4] slightly more than a decade ago. Using ESR spin labels but different analytical approaches, the two groups estimated $D \approx 10^{-8}$ cm^2/sec for phospholipids, a number which has withstood the test of time as well as numerous other experiments and has become the basis of comparison of lipid diffusion in membranes. The importance of lipid diffusion in biological processes is well illustrated by the breadth of topics involving lipid dynamics in these three volumes alone.

II. SCOPE

This review covers lipid lateral diffusion, but the close relationship of this topic to other membrane diffusion processes dictates that these other topics must be briefly covered as well. Protein rotation and diffusion in membranes is an extremely important topic which has been the subject of two recent general reviews[5,6] and two specific reviews.[7,8] Lipid diffusion has been reviewed in one of the general reviews.[6] The rather extensive literature on diffusion in lipid/water model membrane systems will only be summarized, in part, because general agreement seems to exist on the fundamental results in that field.

III. MEMBRANE MODELS

The bilayer structural model of the membrane, shown qualitatively in Figure 1, is so commonly used that we will not give it extensive discussion. The review by Sjöstrand[9] in this series gives a basic treatment. It is worth noting, however, that since the protein mass in a membrane may exceed that of the lipid mass, the structure is considerably different from that sometimes described as "proteins floating a sea of lipids." It is more correctly an interactive system composed of proteins and lipids; much data exist to indicate that the proteins at biological concentrations in membranes have a significant effect on lipid behavior.

IV. SIMPLE DIFFUSION MODELS

The diffusion coefficient has units of length²/time (e.g., cm^2/sec) because of the fundamental time-dependent diffusion equation which in one dimension has the form given in Equation 1:

$$\frac{\partial C(x,t)}{\partial t} = D \frac{\partial^2 C(x,t)}{\partial x^2} \qquad (1)$$

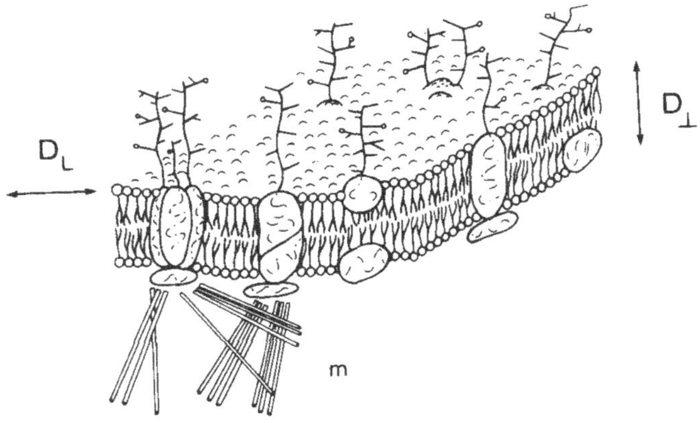

FIGURE 1. Bilayer membrane model. Microtubules, m, affect protein mobility by providing an interface to the cytoskeleton. D_L is lateral diffusion (referred to as D in text) and $D_\perp$ is transverse or "flip flop" diffusion.

Table 1
ILLUSTRATIVE TWO-DIMENSIONAL RANDOM WALK VALUES

Diffusion coefficient (cm^2/sec)	Example	Root mean square displacement (μm) In 1 msec	In 1 sec
10^{-5}	H_2O	2	60.
10^{-7}	Lipid	0.2	6
10^{-8}	Lipid	0.06	2.0
10^{-11}	Protein	0.002	0.06

In this expression, C(x,t) is the concentration of molecules at position x at time t. Solutions to a variety of diffusion equations are commonly derived in standard texts.[10]

It is instructive to begin with a simple model. Assume that a particular lipid molecule has a label which allows it to be identified but which will not alter its diffusion rate with respect to the other lipids. Assume further that the special lipid occupies a position in a planar, two-dimensional membrane structure and that its motion is recorded as it undergoes random walk diffusion. It is appropriate to calculate the average root mean square displacement, $\langle r^2 \rangle^{1/2}$, of the molecule from its initial position as a function of time. For the general case in the simplest type of isotropic, two-dimensional diffusion, this relationship is described by Equation (2):

$$\langle r^2 \rangle = 4Dt \qquad (2)$$

If the diffusion coefficient were similar to that of water, 10^{-5} cm^2/sec, an average molecule would be displaced about 64 μm (6400 Å) in 1 sec. A typical phospholipid, with a diffusion coefficient of 10^{-8} cm^2/sec, would be displaced on the average 2 μm (200Å) in 1 sec. Table 1 tabulates a variety of other diffusion data.

We can improve our simple model by making it heterogeneous and, thus, somewhat closer to a biological membrane in structure by incorporating proteins and regions of gel phase into the lipid structure. Now the membrane has regions which are relatively impenetrable to lipids, and there may be bounded regions from which lipids cannot escape. The value resulting from a lipid diffusion measurement in this model is more difficult to interpret. On a short time scale, a measurement will reflect the real lipid D. But on a long time scale, some lipids will reach barriers and be reflected, while others may diffuse along channels between the barriers. In general, large displacements will be less likely in systems with barriers than in systems without barriers. Restricted diffusion is present in almost all cell membranes and is a significant factor in many biological processes.

V. MATHEMATICAL MODELS OF DIFFUSION

On a molecular level, membranes pose a particular problem in the development of mathematical models for dynamics, including diffusion. A fundamental, though not widely appreciated, difficulty in membranes is that the motion of most molecules, especially that of lipids, is highly anisotropic. Therefore, many of the theories developed for isotropic systems are not necessarily valid. Analysis of molecular tumbling provides an example of the complexities anisotropy can introduce. In an isotropic fluid, tumbling is defined with reference to rotational viscosity, a single parameter derived from relationships of translational diffusion and fluid viscosity. In anisotropic systems such as liquid crystals, a quite different approach is needed.[11] The dynamics are correctly described by a second rank, symmetric tensor with zero trace, so that five independent components, each with the dimensions of viscosity, are necessary to completely describe the dynamics. It is not possible to ascribe viscosity, a comforting feature of isotropic fluids, to a single parameter; rather, the five components must be specified to define the motion of a molecule, such as a phospholipid, in an ordered environment.

Diffusion is somewhat less fraught with such problems, but theoretical difficulties are present. A rather nagging problem arises at a basic level in two-dimensional (as opposed to three-dimensional) diffusion mathematics: a solution to a two-dimensional diffusion equation is by no means assured. Molecular dynamics simulations on hard sphere and hard disk model liquids[12,13] indicate that the integral fundamental to the calculation of D diverges over long times. It has been argued, however,[14] that this difficulty does not arise in most biological membranes because they are, in fact, three-dimensional systems due to the interaction with water; further arguments are given by Saxton.[14]

In the next section, we will review models of membrane diffusion, but it is wise to keep in mind that intuition gained from experience with isotropic fluids may be in error and that mathematical modeling of membrane dynamics is still in its infancy.

It is useful to classify existing theories as continuum, where the membrane is considered to be a continuum, or as heterogeneous, where restricted diffusion is somehow recognized.

A. Continuum

The Saffman-Delbruck[15,16] theory, one of the more effective treatments of protein diffusion in membranes, models the protein as a cylinder which diffuses into a lipid bilayer continuum of viscosity η. Expressions for lateral and rotational diffusion coefficients have been derived,[17,18] and experiments[7,19] demonstrate that this model accurately describes protein dynamics in well-characterized dilute solutions of proteins in lipids. (For additional details, see Chapter 3.)

The assumptions on which this theory is derived preclude its application to diffusion of lipids, and efforts to apply it to experimental data on lipid diffusion have noted inconsistencies.[20] These findings are analogous to those resulting from applications of hydrodynamic theories to isotropic fluids.[21] It is well established[21] that continuum theories are semiquantitative when the diffusant is much larger than the solvent molecules, but are quite poor in the limit where the diffusant size approaches that of the solvent molecules.

B. Heterogeneous

1. Free Volume Model

Free volume theories[22,23] define the liquid as a lattice and permit only a fraction of the lattice sites to be available for diffusion. Dynamics are described in terms of a jump frequency to available free lattice sites. Galla et al.[8] have adapted this approach to describe the effects of cholesterol on lipid diffusion. They obtained a temperature-dependent expression for the jump frequency which adequately explained the experimental data. Their approach required the estimation of phenomenological constants. MacCarthy and Kozak[24] derived an alternative free volume approach which avoids the phenomenological constants, giving rather an expression for D in terms of the lattice spacing, the approximate close-packed area per molecule, and the average free volume (or area in two dimensions) per molecule. Semiquantitative agreement is obtained for data of D of pyrenedodecanoic acid in oleic acid monolayers at 25°C[25] and for pyrene lecthin in phospholipid/cholesterol vesicles at 50°C.[8] Improvements in this theory might be anticipated it the intermolecular potential were expanded to include steric interactions.

2. Polymer Diffusion Analogs

Schindler et al.[26] have noted the similarities between diffusion in network polymers and diffusion in reconstituted membranes. Similar properties include (1) diffusion rate and temperature dependence being a particular function of size and (2) diffusion rate often affected by presence of small molecules (which presumably disrupt the structure). Their approach, which builds on concepts first introduced by Lieb and Stein,[27] uses a relatively simple physical model. Random thermal motions of polymer chains can be thought of as forming holes into which the diffusant can move. The rate of diffusion is then controlled by the appearance of holes and by the probability distribution of hole sizes. Since diffusion occurs only when a sufficiently large hole appears nearby, a size and temperature dependence of diffusion is incorporated into the model. Specific interactions between proteins and other membrane components can lead to differential D changes. A significant prediction of the model is that rotational diffusion and translational diffusion are not necessarily related; a protein molecule could undergo rotational motions while being restricted with resepct to lateral diffusion. This means that membrane changes could affect D without significantly altering rotational diffusion. This work has prompted an exchange in the literature[28,29] on the question of the necessity for new membrane models.

The work of Pace and Chan[30] adapts polymer diffusion techniques to a model which assumes that diffusion is dominated by the dynamics of the head group interactions. The rate-determining step is taken to be the separation of two adjacent head groups by an amount which will just accomodate the diffusing molecule. Head group displacements thus reflect the probability of diffusion of a lipid. The jump frequency, and hence D, are assumed to have an Arrhenius form, and reaction rate theory[31] is used to approximate the viscosity. The model fits D in one chain and two chain membranes to within a factor of 2 or 3, and also provides a basis for an explanation of the effects of cholesterol on lipid D.

3. Percolation Theory Models

Saxton[14] has adapted to diffusion in heterogeneous membranes two well-established theories from the physics of electrical conductivity: effective medium theory and percolation theory. The heterogeneity is assumed to arise from the presence of proteins (impenetrable by lipids) and of gel phase regions in the lipid structure. Tenets central to Saxton's model are that membrane diffusion depends on the area fraction, X, of fluid lipid and on the relative permeability, r, of the two phases. That is,

$$r = D_s/D_f \tag{3}$$

where D_s is diffusion in the solid phase and D_f is in the fluid phase. In general, effective medium theory is valid when $D_s > 0$ and $r > 0$. For lipid mixtures in which lateral phase separations are present, experimental data indicate that r ranges from 0.01 to 0.6; thus, effective medium theory is expected to be valid. When proteins are present, $D_s = 0$, so percolation theory obtains. For reasons discussed further by Saxton,[14] the solutions used are constructed from combinations of the two theories. A set of universal solutions is obtained from which comparisons can be made with experimental data on X, r, and D. Few systems have been studied in sufficient detail to check the results, but Saxton[14] notes predictions of the model for diffusion-controlled reactions.

Slater and Caille[32] have applied percolation theory to the mixture dynamics of cholesterol/phospholipid systems using a lattice model with three site components: (1) pure, liquid crystalline phase lipid, (2) cholesterol-rich aggregates with cholesterol/phospholipid ratios of 1:3.5, and (3) pure, gel phase lipid. If $T < T_c$, the phospholipids in the aggregate are assumed to be in a less frozen state than in (3); if $T > T_c$, the phospholipids are assumed to be in a less fluid state than in (1). The model provides a quite satisfactory description of the thermodynamics of the mixture and also predicts phase transitions both above and below T_c. Experimental evidence indicating the possible presence of the two phase separations is surveyed by Slater and Caille.[32] The lattice is assumed to be static over the times of diffusion. Thus, for a molecule to diffuse over the lattice, it must "percolate" across regions. Diffusion between sites (1) is called easy, between sites (2) is intermediate (depends on T_c), and between sites (3) is hard. By relating cholesterol concentration to the size of the domains (>260 molecules), diffusion across the surface can be calculated. For $T > T_c$, applications of the model are in general agreement with most measurements. Below T_c, the model predicts D would be slow until the concentration of cholesterol was sufficiently high that the domains (2) were of a sufficient abundance to touch and form a large patch; then D suddenly increases. Evidence[33] and the model indicate that this occurs at approximately 20 mol% cholesterol.

4. Cholesterol/Phospholipid Models

Fluorescence recovery photobleaching experiments on plasma membranes from mouse fibroblasts and on binary mixtures of dimyristoylphosphatidylcholine (DMPC) and cholesterol[33] indicate anisotropic lipid diffusion may occur in banded regions. Bands or "ripples" appear within which diffusion is fast; the boundaries of the bands are areas of greatly reduced diffusion. Work on the mixtures indicates that the system is formed of "solid" ridges, principally DMPC, and "fluid" valleys composed of 20 mol% cholesterol in lipid. Owicki and McConnell[34] have derived a solution for photobleaching experiments in which the ridges are assumed to have a $D \approx 10^{-10}$ and the valleys are assumed to have a $D \approx 10^{-8}$. Diffusion within the valley is assumed unimpeded. Solutions to the equations are presented which are especially useful for fluorescence recovery experiments with grid bleaching; these solutions allow the extraction of the average lipid D. This model is only applicable for D when phase separations are present. A somewhat more general treatment using the percolation model[32] gives qualitative agreement with the results obtained above and below the phase transition temperatures.

5. Inclusion of Intermolecular Potential

All of the models discussed above omit the effects of intermolecular potential. Weaver[35] has presented a general formalism based on the Smoluchowski equation, and he derives an expression for D (Equation 4) which includes the potential energy function, V(r):

$$D = D_0 \times F(V[r]) \tag{4}$$

D_0 is the free value of diffusion, for example, the Saffman-Delbruck model. F(V[r]) is an average of V(r) over one lattice spacing. F can be evaluated for specific intermolecular potentials. For a sinusoidal potential of amplitude V_0, Weaver derives Equation (5):

$$D = (D_0/T)\exp(-V_0/kT) \quad (5)$$

In Equation (5), k is Boltzmann's constant. This approach indicates several general conclusions: (1) the measured D is always less than or equal to D_0; (2) the temperature dependence of D is the product of several factors, so the observed dependence may not follow that predicted by D_0; (3) D is not especially dependent on molecular length. Weaver applies a variety of V(r) forms[35] and shows the relationship of the resulting D expressions to models derived by others,[34,36] including the ridge/valley model of Owicki and McConnell[34] discussed above.

This theory has yet to be extensively applied to experimental data, but its general format is amenable to the incorporation of potential functions into diffusion models.

C. Planar Vs. Nonplanar

The limitations of applying planar diffusion theories to membranes have been investigated.[37-39] Aizenbud and Gershon[37] have modeled the membrane as a "wavy" surface, utilizing a potential which creates an egg carton appearance. Cone-shaped peaks on the surface simulate microvilli on the cell surface. No analytic solutions could be obtained, but numerical solutions were done for time intervals typical of fluorescence recovery half-lives, and the boundary conditions of microvilli size and bleached area were selected to be typical for those experiments. Illumination areas encompassed several microvilli. As the height of the microvilli increases, the area illuminated increases drastically and the boundary becomes mathematically complicated. The solutions depend rather critically on the ratio of the area illuminated to the perimeter of the illumination. For typical membranes containing microvilli 1 or 2 μm high, an analysis assuming a planar membrane will give a calculated D which is not less than half the actual D. That is, Equation (6)

$$0.5\,(\text{actual D}) < \text{calculated D} < \text{actual D} \quad (6)$$

is a general result. *Absolute* differences of this size (a factor of 2) are difficult to observe with fluorescence recovery photobleaching (see below). Data of Dragsten et al.[38] indicated no difference in diffusion of lipid analogs in lymphocytes with or without microvilli. Diffusion of lipids was found to be the same on the "smooth" regions (as measured by a scanning electron microscope) as on the microvillous regions of mouse egg surfaces.[39] These results are consistent with the theory if the "smooth" regions have small microvilli with dimensions below the electron microscope resolution. The conclusions of this model[37] are valid only for the time scales and dimensions relative to fluorescence recovery experiments. A barrier to the extension of this work to other experiments is the significant amount of computer time required to generate the solutions.

VI. METHODS OF MEASUREMENT

A. Magnetic Resonance Methods

1. Electron Spin Resonance

ESR utilizes resonance from a paramagnetic moiety, a spin label, present in a synthesized probe molecule. Spectral analysis, especially of line widths and line shapes, provides information not only on lateral diffusion but on other dynamics, especially rotation. Benga[40] has a general review in this volume.

The most common spin label is the nitroxide group (Figure 2); Heisenberg exchange effects broaden the line widths in ESR, and a model-dependent mathematical analysis yields

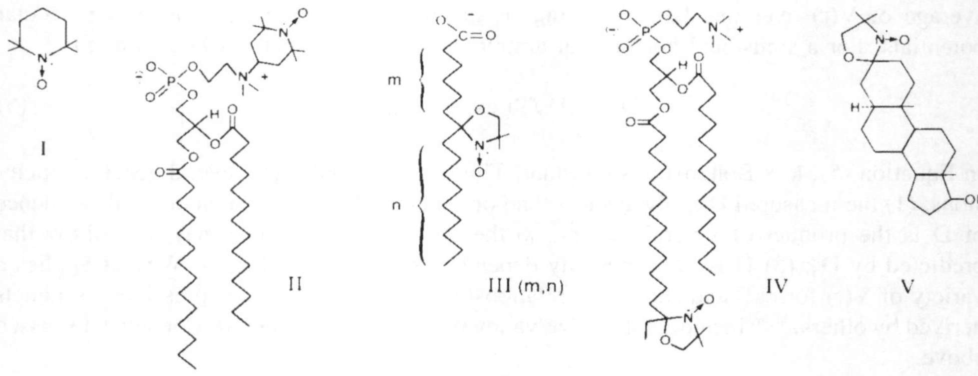

FIGURE 2. Molecular structures of spin labels. (I) TEMPO, 2,2,6,6-tetramethylpiperidine-1-oxyl; (II) head labeled phospholipid; (III [m,n]) fatty acid spin label, for example, a steric acid label 16-SASL: 2-(14-carboxy-tetradecyl)-2-ethyl-4,4-dimethyl-3-oxazolindinyl oxide; (IV) tail labeled phospholipid; (V) 3 nitroxide androstane, 17β-hydroxy-4′,4′-dimethylspiro[α5-androstane-3,2′ oxazolidin]-3′oxyl.

the diffusion coefficient. Devaux and McConnell[3] injected a high concentration of head group labeled phospholipid (II in Figure 2) into a small area of an oriented multibilayer phospholipid membrane, then studied the ESR signal for several hours. Träuble and Sackmann[4,41] used androstane-derived spin labels (V in Figure 2) in *Escherichia coli* membranes; their static method models spin exchange as a bimolecular process. The analysis relates the exchange broadening frequency, W_{ex}, to the area per lipid in the membrane, to a critical distance for the onset of exchange, and to the label jump frequency in the lipid lattice. The method studies exchange as a function of the label/lipid molar ratio, C, then uses Equation (7). Values of C range from 0.02 to 0.2.

$$W_{ex} \propto DC/(1 + C) \tag{7}$$

As noted in Section I, these experimental results were fundamental to development of the field.

Popp and Hyde[42] have used electron-electron double resonance (ELDOR) to obtain D for a stearic acid label (III[m,n] in Figure 2) in phospholipid vesicles. ELDOR results yield the product of the Heisenberg exchange rate times the electron spin lattice relaxation time, T_{1e}. Saturation recovery techniques were used to measure T_{1e}, and a model-dependent analysis of the exchange rate (essentially the bimolecular collision rate in this system) gave the diffusion coefficient. Their analysis uses the same model and membrane structural parameters as did Träuble and Sackmann.[4,41] An advantage of this method is that spin label/probe ratios as small as 0.06 could be used, less than those frequently required for nonELDOR experiments.

2. *NMR*

Pulsed gradient NMR,[43,44] NMR relaxation, and NMR lineshape methods have been used to measure diffusion in biological systems. NMR offers the advantage that special probes are not needed.

a. Pulsed Gradient NMR

The use of a Carr-Purcell pulse sequence with continuous[45] and pulsed[43] gradients has been a staple of diffusion measurements in liquids for 30 years. Until recently, these methods have been applicable only to isotropic fluids since relaxation due to dipolar effects prevented the observation of spin echoes in solids or in relatively rigid phases such as glasses and membranes. Two new methods have provided a means to lift dipolar broadening in membranes and permit the use of pulsed gradient NMR methods to study diffusion. One method is to use oriented multibilayers;[46,47] special properties of the dipolar Hamiltonian in such systems result in a minimization of dipolar interactions when the lipid chains make an angle of 54° 44′ with the static magnetic field, $\vec{H}_0$. When these conditions are satisfied, a spin echo can be observed, and diffusion can be measured.[48-50] The technique has now been applied to studies of many membrane systems with no significant areas of disagreement.[51-56] The experimental techniques and difficulties have been described,[54,57,58] and Stejskal[44] has reviewed methods to detect bound or restricted diffusion. Hrovat and Wade[59] demonstrated that monitoring the echo response to mismatched gradients provides an absolute determination of the size of the gradient, thus avoiding the use of calibration procedures with known standards such as glycerine. Oriented sample pulsed gradient NMR has the advantage that the analysis for D is the least model dependent of all methods of D determination. Most experimental errors (vibrations, pulse timing instabilities, etc.) tend to make the experimental D larger than the actual D. Where the resolution permits, Fourier transform methods[52,60] can be used to measure relative diffusion coefficients of different components. An obvious disadvantage of this technique is the requirement for oriented macroscopic samples; this has limited the method to model membranes.

Dipolar interactions can also be removed by sophisticated NMR pulse techniques,[61-64] and recent work which combines these with pulsed gradient NMR has demonstrated that diffusion may now be measured in unoriented membranes. Silva-Crawford et al.[64] removed the proton dipolar interactions with a proton multipulse (REV-8)[63] sequence, then superposed a Carr Purcell spin echo pulsed gradient experiment to measure D in phospholipid/water multibilayers. The results agree with NMR results from oriented membranes. This method can also be used below the gel-liquid crystalline transition; a value of $D \approx 1.6 \times 10^{-10}$ was obtained[64] at 25°C for DMPC in a 15% w/w D_2O membrane. This method is applicable to a wide variety of systems, but it requires a rather sophisticated spectrometer.

b. Relaxation and Lineshape NMR

Nuclear spin relaxation is dominated by the modulation of nuclear dipole-dipole interactions by molecular motions, including diffusion. Extraction of D requires a model-dependent analysis and often some serious assumptions about the relative contributions of various mechanisms. Proton relaxation studies[65-67] give values of D generally equal to[65-67] or slightly less than[68] pulsed gradient NMR results. Fisher and James[69] have used rotating frame relaxation to estimate $D = 4 \times 10^{-9}$ cm²/sec in dipalmitoylphosphatidylcholine (DPPC) at 45°C. P31 NMR line shapes are dependent on motion, and phospholipid diffusion has been deduced in several papers.[9,70,71] In sonicated vesicles, the line shapes depend rather critically on the vesicle tumbling rate and on the molecular motion in the vesicle wall. Cullis and co-workers[71] have presented an analysis. Experimental agreement[72] was quite good in dioleoylphosphatidylcholine (DOPC) vesicles for which vesicular tumbling rates were carefully varied.

Deuterium NMR splittings which arise in an anisotropic environment afford an excellent measure of molecular dynamics. Since deuterium has little dipolar interaction with protons, its spectrum is dominated by intramolecular effects. Models which reproduce deuterium line shapes and splittings in membranes include the effects of diffusion.[73] The results have been reviewed recently by Davis.[74]

B. Optical Methods

1. Fluorescence Recovery after Photobleaching (FRP)

FRP methods dominate the literature of molecular motion. This method has been the

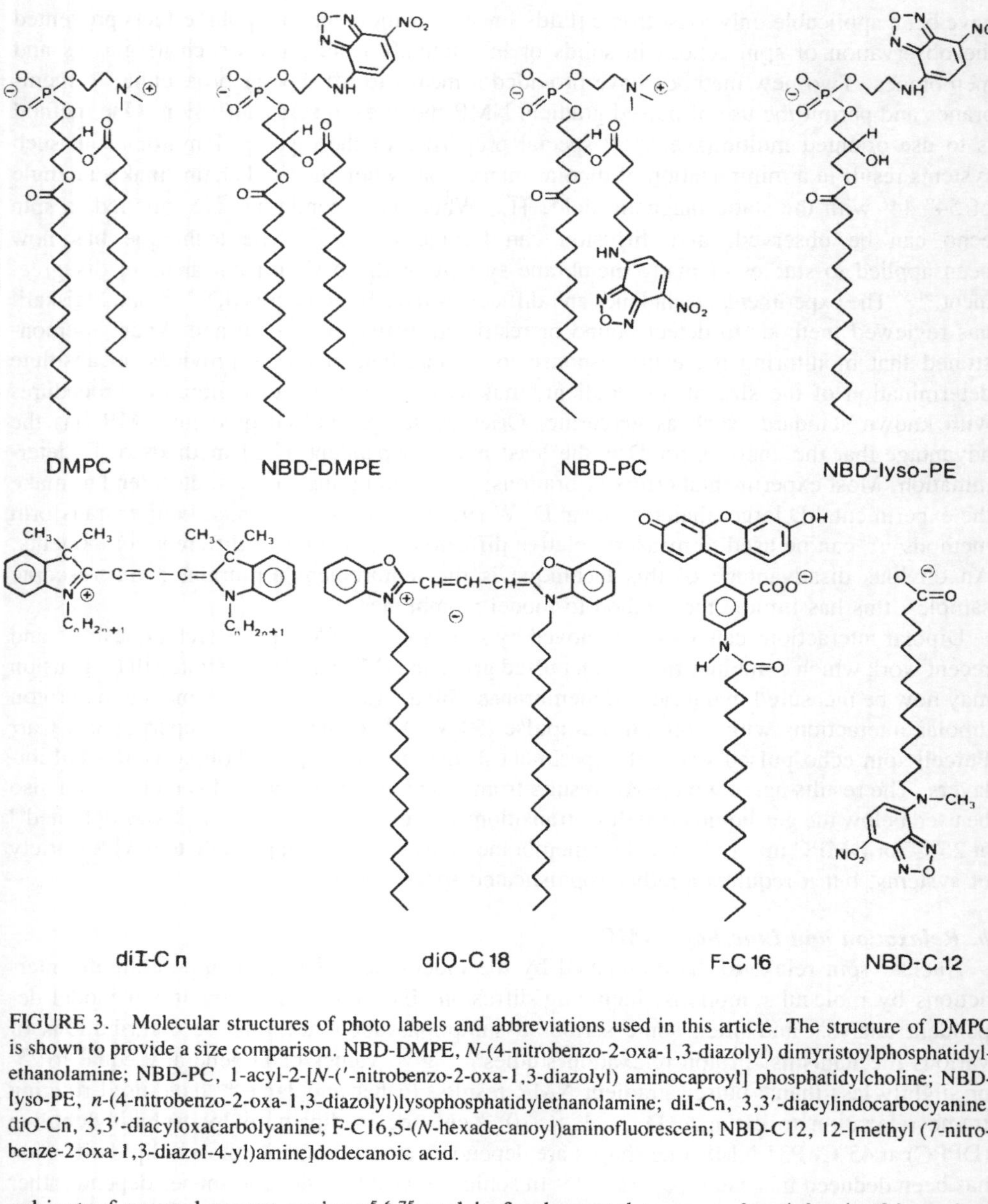

FIGURE 3. Molecular structures of photo labels and abbreviations used in this article. The structure of DMPC is shown to provide a size comparison. NBD-DMPE, *N*-(4-nitrobenzo-2-oxa-1,3-diazolyl) dimyristoylphosphatidylethanolamine; NBD-PC, 1-acyl-2-[*N*-('-nitrobenzo-2-oxa-1,3-diazolyl) aminocaproyl] phosphatidylcholine; NBD-lyso-PE, *n*-(4-nitrobenzo-2-oxa-1,3-diazolyl)lysophosphatidylethanolamine; diI-Cn, 3,3′-diacylindocarbocyanine; diO-Cn, 3,3′-diacyloxacarbolyanine; F-C16,5-(*N*-hexadecanoyl)aminofluorescein; NBD-C12, 12-[methyl (7-nitrobenze-2-oxa-1,3-diazol-4-yl)amine]dodecanoic acid.

subject of several recent reviews[5,6,75] and is fundamental to several articles in this series. Vaz et al.[6] have included lipid diffusion in their recent review. The method involves labeling a surface with a probe molecule which contains a fluorophore (Figure 3). A small circular region or a small grid[79] is photochemically bleached with a strong pulse of laser excitation transmitted through a microscope. Typically, the bleached area covers a few microns. As diffusion occurs, the fluorescence intensity of the area, monitored with low level excitation from the laser, increases. An analysis of the recovery yields the diffusion coefficient of the probe. Usually this involves measurement of the half-life of the recovery. Theoretical analyses[34,77] provide a method of separating results of diffusion from results of uniform flow. FRP can be utilized in living systems and can be used to measure diffusion coefficients as small as 10^{-12} cm^2/sec. Thus, it is one of the few methods which can measure D below the phase transition in membranes.[34,76] The precision of the value of D depends on the determination of the dimensions of the bleached spot, a quantity which is usually diffraction limited. This introduces errors in the range of ±15 to ±30% when comparing data from various sources,[78] but relative data on one apparatus may be more reproducible.

Other factors may influence FRP results. Fluorescence recovery is almost never 100%; for most labeled protein probes in natural membranes it is never more than 80% and may be nearer to 50%. The difference is usually attributed to the presence of immobile proteins in the bleached area. Bretscher[79] has expressed concern about the lack of definitive understanding about the low rate of recovery, but some evidence now exists (see Webb et al.[75] for a discussion) to support the position that cytoskeletal effects on proteins are responsible for the reduced recovery. Another factor in FRP studies is the possible effects of the excitation on the system, in particular, cross-linking or photodamage. A significant body of research exists on this topic; in general, the effects appear to be minimal. Discussions and evaluations are in recent reviews.[5,6,75,80] Recovery rates for lipid analogs are usually quite high, but a practical problem in studies of cells may be internalization of the probe. If some of the probes move to interior membranes, there is often no convenient way to distinguish the response of these from those in the cell membrane.

2. *Other Optical Methods*

Excimer fluorescence from fluorophores such as pyrene is diffusion controlled. A two-dimensional diffusion model may be used to extract diffusion coefficients from the bimolecular kinetics.[81,82] Excimer emission can also be used to measure transverse diffusion.[83,84]

Formation of an excited complex (exciplex) between unlike molecules is sensitive to polarity, molality, and diffusion rates. Kano et al.,[85] used exciplex formation between aniline derivatives and either pyrene or pyrene decanoic acid to measure diffusion rates of these small molecules in liposomes of DPPC above and below the phase transition.

Flash photolysis studies of triplet-triplet annihilation can yield diffusion coefficients for the fluorophores, which usually contain aromatic moieties.[86]

VII. AN OVERVIEW OF METHODS AND RESULTS

All of the above techniques, with the exception of NMR, measure the behavior of probe molecules, the use of which immediately raises two questions. One of these is whether probe diffusion is an accurate monitor of solvent diffusion. In most cases, probes with structures similar to that of the host exhibit similar diffusion characteristics.[5,6,87] For lipid diffusion, head group interactions are important, so an amphipathic probe is important in order to accurately monitor lipid diffusion. A second question is whether the probe affects the system, so that diffusion is altered by the presence of the probe. Answers to this question are harder to obtain. For cells, metabolic tests must be included.

As discussed below, there is general agreement on lipid diffusion values in model membranes. A diffusion coefficient of 10^{-8} cm^2/sec is equal to that expected from theory for a fluid with a viscosity of about 1 P, roughly that estimated for bilayers. Extrapolation of this behavior to protein diffusion, however, leads to a major discrepancy. Theory[4-7,19,20,75] predicts a protein D about one quarter of that of a lipid; measured values, however, are often 40 to 100 times less than that of lipids. This difference is attributed to cytoskeletal effects on the proteins (Figure 1) and to protein-protein binding.[4-7,19,20,75]

The time involved in measuring D may be crucial to the interpretation (see Section IV). Restricted diffusion becomes more evident as the time involved increases. Table 2 provides a comparison of methods in which a characteristic time is involved.

VIII. APPLICATIONS

A. Model Membranes

1. *Fluid (Liquid Crystalline) Phase*

Data on lipid diffusion in model membranes are voluminous. A recent, thorough survey[6]

Table 2
CHARACTERISTIC TIMES IN COMMON DIFFUSION MEASUREMENT METHODS

Method	Characteristic time (sec)
ESR static concentration	$\sim 10^{-7}$
Triplet kinetics	10^{-3} to 10^{-6}
Pulsed gradient NMR	≤ 0.05
FRP	~1—1,000
ESR concentration gradient	~4,000—10,000

Table 3
TYPICAL FRP LATERAL DIFFUSION DATA[a] AT 38°C

Probe	DMPC[b]	Egg-PC[c]	Brain-PS[d]
di0-C18[e]	9.2 ± 0.4	8.9 ± 0.3	7.8 ± 0.1
di0-C16	9.0 ± 0.9	—	9.5 ± 0.2
diI-C18	11.9 ± 0.3	10.9 ± 0.9	9.8 ± 1.4
diI-C12	11.3 ± 0.6	—	10.1 ± 0.6
NBD-PE	9.3 ± 2.0	10.1 ± 0.4	—
NBD-lyso PE	7.9 ± 1.6	11.8 ± 1.2	—
NBD-C12	18.5 ± 2.4	11.5 ± 0.8	—

Note: Data are D × 10^8 cm²/sec.

[a] From Derzko and Jacobson.[87]
[b] Dimyristoylphosphatidylcholine.
[c] Egg phosphatidylcholine.
[d] Brain phosphatidylserine.
[e] See Figure 3.

lists more than 10 studies using nonFRP methods and more than 15 studies using FRP methods; a wide variety of phospholipids were subjects of the studies. This section will provide only a general summary; the reader is referred to Vaz et al.[6] for a more detailed presentation.

In the fluid phase, results in recent years have tended to converge. The diffusion coefficient of most probe molecules is largely independent of probe chain length and of the type of phospholipid. Values of D fall between 10^{-8} and 2×10^{-7} cm²/sec with an activation energy of 4 to 8 kcal/mol. For example, Vaz et al.[6] plot results from four laboratories on diffusion of NBD-PE (Figure 3) and two structurally similar analogs in three lipids over temperature ranges of 10 to 50°C. All of the data, without regard to phospholipid type or transition temperature, can be fitted in an Arrhenius plot with a correlation coefficient of 0.9 (see Figure 4 in Reference 6). Given that calibration errors from lab to lab may be significant, the agreement is astonishing. An example of data of FRP studies with different probes is given in Table 3, taken from some of the extensive work of Derzko and Jacobson.[87] The probes, with the exception of NBD-C12, show very similar behavior in a quite varied selection of phospholipids. NBD-C12 represents another common observation: small probes often show anomalous behavior in certain lipids, and they tend to have larger energies of activation ($\approx$ 13 to 15 kcal/mol). Presumably, this behavior represents a diffusion mechanism different from that of lipids, perhaps dissociation or solubilization in the water layer.

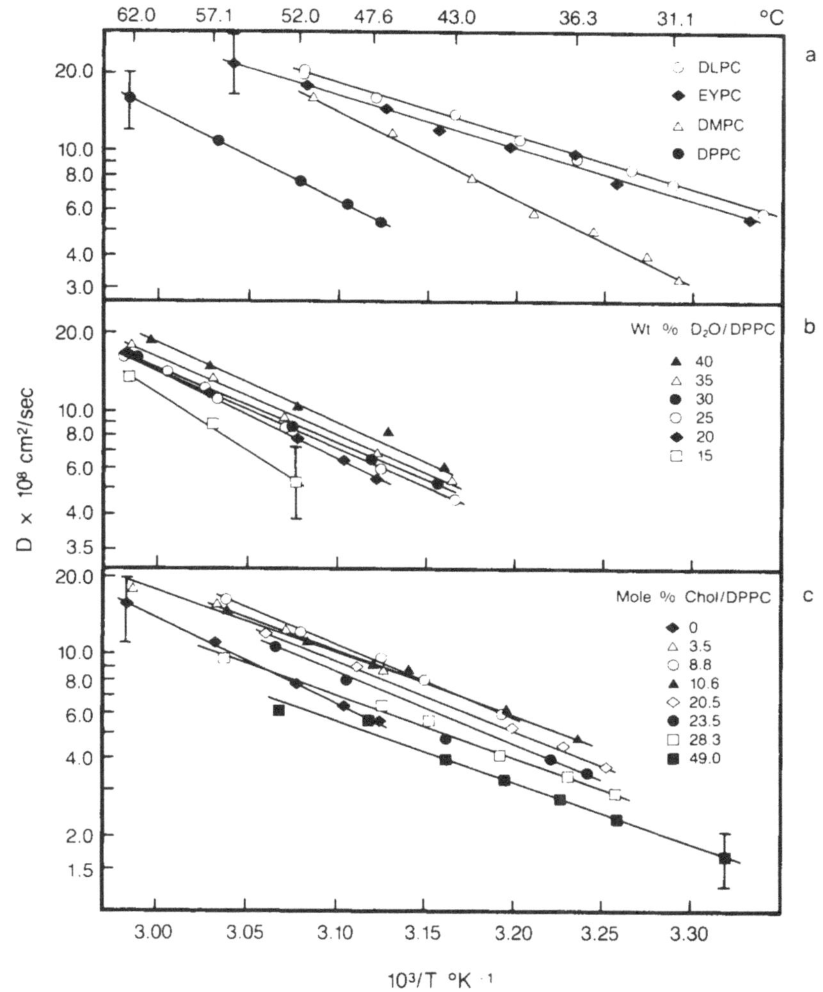

FIGURE 4. Oriented membrane pulsed gradient NMR data on lipid diffusion, on hydration effects, and on cholesterol effects. All data are from Wade and Kuo.[52] DLPC, dilauryl-phosphatidylcholine; EYPC, egg yolk phosphatidylcholines; DMPC, dimyristoylphosphatidylcholine; DPPC, dipalmitoylphosphatidylcholine; Chol, cholesterol.

Pulsed gradient NMR methods[51-56] do show a dependence of D on the type of phospholipid, the amount of hydration, and the amount of cholesterol. Figure 4 shows representative data. For saturated lipids at the same hydration and temperature, D decreases monotonically as the chain length increases. The lipid D is definitely hydration dependent over 15 to 40% D_2O w/w, varying in DPPC over this range, for example, by a factor of 2. Hydration effects on D are significant in FRP, also. McCown et al.,[90] using NBD-PE in egg phosphatidylcholine (egg-Pc), found D increased by a factor of 8 as hydration changed from 5 to 100 wt%.

2. Gel Phase

The high degree of lipid packing and chain ordering in this phase produces an almost solid system; consequently, D is much slower than in the fluid phase. Typical values[32,75,87,91,92] are 10^{-10} to 5×10^{-12} cm²/sec with a large activation energy (≈ 30 to 40 kcal/mol). Smith and McConnell[76] obtained D ≈ 2×10^{-10} for NBD-egg-PE (phosphatidylethanolamine) in the gel phase of DMPC at 23°C, in good agreement with an earlier estimate of Wu et al.[91]

An extensive study by Derzko and Jacobson[87] of 13 FRP probes in DMPC and brain bovine phosphatidylserine (PS) at temperatures from 3 to 20°C showed two-component diffusion in most cases. The slower component, with values from 1.7×10^{-10} to 2×10^{-12}, presumably represents diffusion in the pure gel state. The faster component, by factors of 10 to 100, could represent diffusion along defects or diffusion in small fluid regions around an impurity.

Klaussner and Wolf[93] have shown that diI-Cn (Figure 3) may selectively partition into the gel or into the fluid region of phospholipids. A critical factor is the chain length of the probe relative to the acyl chain length on the phospholipid. When the probe chains are longer than the lipid chains, the probe preferentially dissolves in the gel phase. The findings are significant for FRP studies of diffusion in natural membranes (see below).

Pulsed gradient NMR accompanied by a multipulse technique to remove dipolar broadening has yielded[64] a $D = 2 \times 10^{-10}$ in unoriented DPPC liposomes at 25°C. Results in small (~300 Å) unilamellar vesicles tend to give faster gel phase diffusion results than those for liposomes. P31 NMR line widths[70] and triplet-triplet annihilation techniques[86] on these small vesicles yield 1×10^{-9} and 1.6×10^{-8}, respectively. The results may be due to the time scales of the methods used or to the fact that small vesicles have a high curvature and different packing from liposomes.[6] The Slater-Caille percolation model[14] predicts (see Section V.B.3) that D values may depend on the measurement time scale since for shorter times fewer domains would be sampled over the lattice.

3. *Effects of Cholesterol*

Extensive studies have been conducted on the effect of cholesterol on lipid membranes. Cholesterol tends to have a condensing effect on lipids above the phase transition and a fluidizing effect (alteration in lateral packing organization) below the phase transition. A frequent observation is that a wide range of effects occur for cholesterol contents from 5 to 20 mol%. Recently, evidence[33,34,76,91,92] has emerged that at concentrations below 20% cholesterol, a lateral phase separation exists: one phase has about 20 mol% cholesterol while the other contains pure lipid. Below the phase transition, the phase containing cholesterol is more fluid. Freeze fracture electron microscopy[33,92] indicates that at temperatures below the pure phase transition, a banded pattern of the two phases appears. Above 20 mol% cholesterol, the banded pattern disappears. The gel phase (DPPC) bands, about 1.3 μm wide, act as barriers to lipid diffusion in the more fluid cholesterol/lipid phase. Lipid diffusion in solutions containing more than 20% cholesterol is similar to that occurring in pure lipid systems. FRP results[34] for D as a function of cholesterol concentration show a substantial change in D when the cholesterol concentration is around 18 to 20 mol%. Above T_c, diffusion decreases by a factor of 2 or 3. Below T_c, diffusion increases by a factor of 10. The Owicki-McConnell theory[34] qualitatively describes the observations below T_c, and the Slater-Caille[32] percolation model provides a qualitative description for the behavior above and below T_c (see Section V.B.4).

Oriented membrane pulsed gradient NMR studies of lipid/cholesterol mixtures above T_c give results which agree with FRP data at high (>20 mol%) cholesterol concentrations but which differ from FRP data at lower concentrations. In DPPC, Kuo and Wade[52] found an increase (up to twofold) in lipid D as the cholesterol concentration increased from 0 to 10 mol%; as the cholesterol concentration increased further, diffusion smoothly decreased. Lindblom et al.[55] used a similar technique on unsaturated lipids: egg yolk lecithin, dioleoyllecithin, and palmitoyloleoyllecithin. Their results indicated only a slight increase in lipid D as cholesterol increased to 20 mol%. They did find, from deuterium NMR and sodium-23 NMR, a significant effect on the ordering of the choline head group as the cholesterol concentration increased. In both these studies, the water concentration was held constant at 20% w/w. The differences between the two NMR studies may reflect a difference in diffusive behavior of saturated[52] and unsaturated[53] lipids.

Table 4
PROBE DIFFUSION IN CELLS UNDERGOING FERTILIZATION. DIFFUSION COEFFICIENTS IN D × 10^9 cm^2/sec

	Sea urchin eggs[a]			Mouse eggs[b]	
Probe	Unfertilized	Fertilized	Effect of fertilization	Unfertilized	Fertilized
diI-C10	3.4	2.1	↓	—	—
diI-C12	1.9	3.9	↑	—	—
diI-C14	4.5	5.5	↑	3.8	3.4
diI-C16	5.9	3.1	↓	6.3	7.7
diI-C18	Does not label		—	4.7	4.3
diI-C20	—	—	—	7.1	5.9
diI-C22	—	—	—	6.6	4.7
NBD-PC	2.0	4.8	↑	—	—

[a] Data from Wolf et al.[95]
[b] Data from Wolf et al.[96]

Alecio et al.[94] have used NBD-labeled cholesterol to measure cholesterol diffusion in DMPC liposomes and have compared the results with measurements on NBD-labeled PE on the same liposomes. Results obtained with the two probes both above and below the phase transition were identical to within experimental error. They also measured diffusion as a function of cholesterol content in the liposomes. Significant differences between diffusion coefficients of the two probes were observed only at cholesterol concentrations between 5 and 20 mol% at temperatures below the major phase transition. Within that limited range, the cholesterol-labeled probe diffused twice as fast as the lipid-labeled probe. Above 20% cholesterol, the diffusion rates were again equivalent. These results are consistent with data which indicate that a lateral phase separation occurs in cholesterol/lipid systems when the cholesterol concentration is below 20 mol%.

B. Natural Membranes and Living Cells

FRP studies of diffusion in living cells have appeared in the last 5 years and will undoubtedly increase in importance.

Klaussner and Wolf[93] have added an important facet to the field. They have shown that diI-Cn (Figure 3) preferentially bind into different lipid phases as a function of the acyl chain length. In membranes which contain both fluid and gel phases, the diI-Cn's with n slightly greater than the mean chain length of the lipids shows a high preference for the gel phase and thus a slower diffusion rate. Potentially then, probes can be preferentially solubilized in microenvironments in the cell. This can be utilized for selective studies; alternatively, if the selective solubility occurs but is undetected, data analysis is suspect. It should be noted that in fluid phase model membrane systems, as noted above, diI-Cn diffusion is approximately independent of chain length. Wolf and co-workers[95,96] demonstrated these effects in FRP studies of plasma membrane order changes which accompany fertilization. By using a series of probes, diI-Cn, with varying n, they obtained quite different results for two systems. In sea urchin eggs,[95] D shows a relative minimum at n = 12 for unfertilized eggs, but a relative maximum at n = 14 for fertilized eggs (Table 4). The dyes appeared inhomogeneously distributed when viewed under a fluorescent microscope. The percentage recovery was rather small (36 to 77%) for lipid analogs; it was always smaller in fertilized eggs than in unfertilized eggs. For mouse eggs,[96] unfertilized and fertilized eggs showed similar trends: maxima at n = 16 and 20 (Table 4). Percentage recovery was unaffected by fertilization for any chain length. The effect of fertilization on D is chain length-dependent but to a lesser extent than for sea urchin eggs. Little change upon fertilization is observed

Table 5
DIFFUSION OF LIPID ANALOGS IN LIVING CELLS[97]

Probe	Cell	D × 10^9 cm^2/sec
diI-C18	3T3	8.96 ± 0.8
diI-C18	SV3T3	8.68 ± 1.32
Fl-GM1	3T3	4.97 ± 0.7
Fl-GM1	SV3T3	4.24 ± 1.25
CONA	3T3	0.129 ± 0.086
CONA	SV3T3	0.126 ± 0.083

for n = 14 and 16; a slight increase is observed for n = 16 and slight decreases are observed for n = 20 and 22. One possible explanation for the changes is the existence of gel and fluid phases, the relative amounts of which change during fertilization; an alternative explanation is chain length-dependent binding of probes to proteins. The authors propose that, whatever the specific effect, a reordering occurs in the sea urchin eggs upon fertilization; furthermore, the order change is probably due to membrane lipid introduced by extensive cortical reaction. Differing results for mouse eggs can be explained by the reduced cortical reactions in these compared to sea urchin eggs.

FRP has been used to monitor diffusion alterations which may accompany known surface membrane phenotype changes in cell transformation. Eldridge et al.[97] used lipid analogs and larger photolabeled molecules to compare changes in cell lines of 3T3 mouse fibroblasts and Simian virus transformed SV3T3 cells. These two derived lines showed the common characteristics which distinguish normal and transformed cells, including increased agglutinability and an enhanced redistribution of lectin receptors in SV3T3 relative to 3T3. Evidence exists that receptor mobility is faster in transformed than in normal lines. This study applied FRP to cells in petri dishes at 37°C utilizing as probes diI-C18, a fluorescent labeled ganglioside derivative (*Fl-GM1*), and a rhodamine-labeled Conconavalin A (*CONA*). Diffusion of larger molecules was studied using rhodamine-labeled stearoyldextran molecules and antimouse cell surface antigens. When viewed under a microscope, all labels showed uniform distributions over the cells, but diI-C18 showed evidence of internalization. The mobility of the smaller labels (diI-C18 and Fl-GM1) showed no differences within experimental error over the cell lines (Table 5). Diffusion of the larger molecules, however, was 50% slower in SV3T3, and the mobile fraction of CONA receptors was lower on SV3T3 cells. All previous spectroscopic methods, with the exception fluorescence polarization, have indicated that membranes of transformed cells have fluidity comparable to or lower than that in normal cells. Receptor mobility and fluidity changes have often been invoked to explain the increased agglutinability of transformed cells over normal cells. Eldridge et al.[97] found that lipid fluidity is not the sole determinant of protein mobility. Another finding was that control of the surface receptors functions independently of membrane fluidity and functions differently for different types of receptors. A third was that changes in agglutinability resulting from transformation cannot be simply interpreted in terms of changes in membrane fluidity and receptor mobility.

Searls and Edidin[98] measured diI-C16 diffusion in plasma membranes of two lines of teratocarcinoma-derived endothermal cells and five lines of embryonal-derived cell lines. Fluorescence recovery was 75 to 88%. Fastest diffusion (8 × 10^{-9} cm^2/sec) was found in embryonal carcinoma cell lines. Endothermal derivatives of the same tumors averaged 5 × 10^{-9} cm^2/sec. Treatment of the carcinoma cells with retinoic acid, a process which mimics the earliest events of differentiation, produced a slowing of diffusion, the final values being near those of the endothermal cells. The change was paralleled, however, by a two- to threefold increase in free and esterified cholesterol contents of the plasma membrane. These changes may be sufficient to account for the changed values of D. Other factors which may be involved in the observation are changes in surface microvilli and possible internalization of the probes.[97]

Changes in diffusion of lipid and protein analogs during differentiation and during the cell growth cycle in C1300 mouse neuroblastoma cells have been studied by de Laat et al.[99,100] Lipid diffusion was monitored with diI-C18 and with a fluorescein-labeled analog of ganglioside GM1 (*F-GM1*). Membrane proteins were labeled with rhodamine-labeled rabbit antibodies against mouse E14 cells. Topical heterogeneity was observed in differentiating cells.[99] Measurements were made after 24 hr of differentiation using as the FRP circle an area small compared to the area of extending neurites. Control experiments were done on a nonextending surface location in the perikaryon region. Upon differentiation, the diffusion of F-GM1 increased from 5.7 ± 0.4 to $7.1 \pm 0.3 \times 10^{-9}$ cm^2/sec (24 hr) and to $9.9 \pm 0.5 \times 10^{-9}$ cm^2/sec (48 hr). Labeled protein diffusion doubled to 3.5×10^{-10} cm^2/sec after 24 hr. The lipid diffusion results show an optical heterogeneity in the plasma membrane of differentiating neuroblast cells; a probable explanation is a higher concentration of fluid domains in the neurites. The relative increase in protein diffusion relative to lipid diffusion after 24 hr may be significant. It is consistent with many other observations that lipid and protein diffusion are subject to different constraints. Work from the same laboratory[100] with synchronized neuroblastoma cells shows that lipid diffusion and protein diffusion change with the cell cycle. The probes were incubated with the cells for 10 to 30 min at 37°C and measurements were done with washed cells at 20 to 24°C. Diffusion of diI-C18 and F-GM1 were qualitatively (though not quantitatively) similar: a minimum during mitosis ($\approx 2 \times 10^{-9}$ cm^2/sec); a two- to threefold increase during G1; maintenance of this high level during S; a decrease during G2 to the mitosis level. The lipid results showed a strong correlation with results of fluorescence polarization measurements of molecular rotational mobility. Diffusion of the protein analog showed some similarities with the lipid observations during the initial portions of the cell cycle: a minimum during M ($\approx 3 \times 10^{-10}$ cm^2/sec) and a threefold increase during G1. Past this point in the cycle, however, protein diffusion deviates from the lipid observations. Late in G1, protein diffusion begins a steady decline which continues through S and G2 to M. The authors suggest that the increase in protein diffusion at the beginning of G1 result from the increase in lipid mobility; constraints which arise in G1, S, and G2 are responsible for the deviations from lipid behavior. Overall, the cell cycle-dependent changes might reflect membrane changes which are pertinent for growth control. The work demonstrates that FRP can be used to define changes occurring during cell growth. Boonstra et al.[101] have experimentally altered the lipid diffusion in these cells and correlated FRP diffusion measurements with changes in electrical membrane properties and with changes in cation transport. Short-term supplementation of serum-free chemically defined growth medium with oleic, linoleic, or stearic acid resulted in incorporation of these fatty acids into the cells. The unsaturated fatty acids were found mainly in the phospholipid (45%) and triacylglyceride (40%) fractions. Stearic acid uptake was relatively low compared to that of the unsaturated acids. It was present mainly as free fatty acid (50%) and as an incorporant in the phospholipids (30%). FRP measurements with diI-C18 at 39°C showed significant differences in effects of saturated and unsaturated acids. When oleic and linoleic were incorporated, D increased approximately 20% compared to the controls; this increase was observed 1.5 hr after supplementation. However, no effect was observed with stearic acid. These observations are positively correlated with rotational viscosity values obtained with fluorescence polarization methods on the same cells. A variety of other physical and chemical measurements showed effects for both oleic and linoleic acid incorporation but no effects for stearic acid incorporation. Addition of unsaturated acids was accompanied by a marked depolarization of the membrane potential, while the addition of stearic gave no change. The permeability of sodium ions increased, and that of potassium ions decreased upon addition of unsaturated acids but showed no change upon addition of stearic acid. The

Table 6
DIFFUSION OF diI-C16 IN HUMAN FIBROBLAST MEMBRANES[102]

	D × 10⁸ cm²/sec			E$_a$ below 25°C (kcal/mol)
	5°C	20°C	37°C	
Living cell PM[a]	0.35	0.8	1.6	9.0
Cell bottom surface PM[a]	0.39	1.05	2.1	10.7
PML[b]	0.51	2.5	6.3	17.0
WCL[c]	0.91	3.8	8.5	11.7

[a] PM = plasma membrane.
[b] PML = plasma membrane lipid extract.
[c] WCL = whole cell lipid extract.

lower value of incorporation of stearic acid into the phospholipids is probably not the source of the differences in behavior, since differences could be observed for equivalent small amounts of unsaturated acids. The results suggest that cell cycle-dependent modulations found in the permeability and dynamics of the plasma membrane of neuroblastoma 2A cells could arise from modulation in the degree of phospholipid unsaturation.

Jacobson and co-workers[102] have used FRP of diI-C18 (Figure 3) to compare lipid diffusion in membranes of living human fibroblasts with diffusion in inanimate membranes of the same cells. Measurements were made on the plasma membrane of the living cell, on surface-attached bottom surface ghosts, on multibilayers reconstituted from plasma membrane lipids (PML), and on multibilayers reconstituted from whole cell lipids (WCL) (Table 6). Diffusion in the living cell plasma membrane and in the bottom ghosts was the same, ranging from 3.5×10^{-9} at 5°C to 2×10^{-8} at 37°C with a transition at 25°C. Both reconstituted lipid systems showed a transition at 10°C and a break in the diffusion curve at 25°C. Above 25°C, all systems showed the same temperature dependence with an activation energy of 5 to 7 kcal/mol. WCL diffusion was the fastest at all temperatures (Table 6). Diffusion in the PML multibilayers was faster than that in the living and surface cell membrane; the difference increased with temperature, reaching a factor of 4 at 37°C. This difference is most likely due to the presence in the living cell plasma membranes of (glyco-)proteins and of cytoskeletal interactions, neither of which is present in PML membranes. That diffusion is slower in PML than in WCL membranes is thought to be due to compositional differences. The major difference found, among several pursued, is the cholesterol to total phospholipid molar ratio, which is 0.57 in PML compared to 0.41 in WCL. Although only minor modifications of the fluid mosaic model would be necessary to explain the lipid diffusion results, major changes would have to be invoked to explain the protein diffusion.

It is known that treatment of the S3G strain of HeLa cells with dexamethasone inhibits cholesterol synthesis and creates decreased cholesterol/protein ratios in the plasma membrane. Boullier et al.[103] have used FRP of diI-C18 on control cells and on dexamethasone-treated cells to monitor effects of lipid D which accompany these changes. Diffusion measurements were made 0, 2, 6, 12, and 24 hr after the addition of dexamethasone. At 37°C, the D was approximately 4.5×10^{-9} in both samples; the untreated control cells maintained this value for 24 hr. Diffusion in the treated cells began to increase 6 hr after treatment and reached 6.4×10^{-9} after 24 hr. This increase was closely correlated with a decrease in the cholesterol/protein ratio from a value of 19 ± 0.02 to a value of 16.4 ± 1.5 μg/mg. Rotational mobility (fluorescence polarization) also increased in the treated cells. The temperature dependence of D from 4 to 37°C, measured 24 hr after treatment, showed that the treated sample had a smaller activation energy. Although the treated cells showed faster D (relative to the controls) at 37°C, they showed roughly equivalent D at 25°C and slower D at 4°C. The changes in D with changes in cholesterol content were similar to those observed in model systems,[32-34] in endothermal cells[98] (above), and in cholesterol-depeleted erythrocyte membranes[89] (below).

Lipid diffusion measurements using FRP in lymphocyte and erythrocyte membranes gives results similar to those obtained in plasma membranes. Dragsten et al.[38] found that dil-C18 diffusion in mouse spleen lymphocytes had a value of $1.7 \pm 0.3 \times 10^{-8}$ cm²/sec at ambient temperatures with nearly complete recovery. No differences were observed between T- and B-cell subpopulations. Little or no effect was observed when capping was introduced by treatment with appropriate antigens. D in capped regions was slowed only slightly, although recovery was reduced to 75%. Treatment of the cells with sodium azide (poisons oxidative metabolism), colchicine (disrupts microtubules), and cytochalasin B (interferes with F-actin structures) produced no discernible effects on diffusion of either dil-C18 or labeled proteins. These results for lipids are similar to those observed in cell membrane.[104,105] The protein findings, however, differ from earlier results[105] which found that D of labeled proteins in the plasma membrane of L-6 (rat) myoblasts was unaffected by azide and colchicine but was reduced by a factor of 10 upon treatment with cytochalasin B. This disagreement in cytocholasin B results has not been explained. As noted in Section V.C above, Dragsten et al.[38] also provided evidence that lipid diffusion FRP results are insensitive to the presence of microvilli on the membrane surface. Human peripheral blood lymphocytes have an extensive microvilluous structure which can be altered if the cells are flattened. Flattening can be achieved by letting the lymphocytes adhere to an IgG-coated surface. After flattening, the cell surface is smooth and devoid of microvilli, at least to the resolution of an electron microscope. FRP measurements using dil-C18 in rounded (with microvilli) and in flattened (without microvilli) cells showed no differences in lipid diffusion. Values of $D \approx 9 \times 10^{-9}$ cm²/sec were obtained, somewhat smaller than those found in mouse lymphocytes. Thompson and Axelrod[89] also studied the diffusion of dil-C18 in human erythrocytes but found even slower diffusion, values of 2×10^{-9} being typical at 40°C. This work studied diffusion as a function of temperature (−3 to +40°C) in control and in cholesterol-depleted erythrocytes. Depletion of cholesterol content by 30 to 50% was accomplished by contact with DPPC sonicated dispersions at 37°C. The maximum depletion corresponds to a reduction in cholesterol mole fraction from 0.5 to 0.3. To reduce the scatter of day-to-day variations, all data were multiplicatively corrected to make the 30°C values in each set of control cell data agree with the average of the 30°C data for all control cell experiments. Recovery was greater than 90% in all experiments. At higher temperatures, diffusion was the same in both types of samples. At lower temperatures, diffusion was slower in the cholesterol-depleted cells by a factor of 1.6. The effects of cholesterol in this concentration range seem to parallel results in model membranes (see Section VIII.A.3 above). Lipid diffusion at the higher temperatures is much slower than that found in other studies. Sources of this difference may lie in the heterogeneous composition of erythrocytes, in chemical compositional changes which may accompany cholesterol depletion, and in incorporation of phospholipid into the cell membrane.

Excimer studies of diffusion of pyrenedecanoic acid and of pyrene-PC in erythrocyte ghosts and in lipids extracted from erythrocytes showed much faster diffusion at 35°C[106] than is usually found with FRP.[38,89] Excimer results give 1.6×10^{-7} for pyrene-PC in ghosts. Lateral diffusion in the reconstituted total lipid extract was a factor of two slower. Addition of cholesterol (20 mol%) to the extract decreased the jump time (and, hence, D) by 50%. Results in the intact ghosts may reflect the asymmetric distribution of cholesterol since diffusion in a lipid extract from the outer membrane gave diffusion similar to that of the intact ghosts. Pyrene-PC excimer fluorescence has been used to estimate transverse phospholipid exchange (''flip-flop'' diffusion) in ghosts.[106] The half-life of this process is 20 to 30 hr at 25°C and 100 min at 37°C. Transbilayer diffusion of lipids can be increased in vesicles by the presence of certain proteins.[107] Van Deenen and co-workers[107] have shown

Table 7
LATERAL DIFFUSION COEFFICIENTS OF LIPID PROBES IN A6 CELL MEMBRANES AT 20°C[108,109]

Probe	$10^9 \times D$ cm²/sec Apical	Basal	Pass tight junction
dil-C14	9.7 ± 3.6	11.9 ± 2.3	Yes
dil-C16	7.9 ± 3.4	10.2 ± 5.6	Yes
F-C12	13.3 ± 4.6	11.2 ± 3.2	Yes
F-C16	4.2 ± 1.8	—	No
NBD-PC	2.9 ± 1.9	4.2 ± 1.7	No

that glycophorin and partially purified band 3 proteins increase PC transbilayer exchange but other proteins (e.g., cytochrome *c* oxidase) do not.

The flip-flop diffusion process may be important in biological processes. Dragsten et al.[108,109] have applied FRP of a variety of lipid analog photolabels to monitor diffusion and molecular asymmetry in the plasma membrane of epithelial cells. Their work indicates that the tight junction may be a barrier to lateral diffusion for molecules which cannot easily undergo transverse diffusion from the outer to the inner membrane surface. The cell lines studied, A6 (derived from toad kidney), LLC-PK (from pig kidney), and MDCK (from dog kidney), grow in culture with the apical (mucosal) membrane uppermost. Dragsten et al.[108] developed a method to detach intact A6 cells from the dish and fold them over to expose the basolateral (serosal) surface. Thus, they could selectively label either of the two membranes. By utilizing certain growth characteristics and careful microscope depth focus, they could also selectively bleach and observe either membrane. If the entire cell were bleached, no replenishment occurred, indicating that there was no transfer of probes from neighboring cells. Consequently, diffusion within each membrane and between membranes could be studied, and the role of the tight membrane could be probed. Several probes were used (Figure 3): diI-Cn (n = 14, 16, 18, 20), F-Cn (n = 12, 16), and NBD-PC. Diffusion was studied for cells at confluence, and the results were independent of which surface was initially labeled. Diffusion of two of the probes (F-C16 and NBD-PC) was blocked by the tight junction; both stayed in the membrane into which they were placed. All the other probes diffused past the tight junction at 20°C (Table 7). Two points were evident in the study. The penetrating probes have D values similar to those found in other cell membranes at 20°C; the nonpenetrating probes have D values generally half as fast as the penetrating probes. The implication is that F-C16 and NBD-PC interact differently with the cell surface. The differential barrier posed by the tight junction could be removed by subjecting the cells to chemicals known to disrupt the tight junction filament network. Especially striking was the contrast between F-C12 and F-C16. F-C12 rapidly penetrates the tight junction until roughly 30% of the fluorescence is on the apical membrane, regardless of which membrane is initially labeled. Furthermore, in contrast to F-C16, the behavior of F-C12 is temperature dependent (Table 7). If the cells were cooled to 10°C or lower, labeled, then maintained at that temperature, F-C12 was restricted by the tight junction although it still diffused within the membrane into which it was initially placed. The results in Table 7 were uncorrelated with differences in aqueous solute transport (through the tight junction network) measured in the different cell lines, with head group charges, or with the partition coefficient of the probe between the aqueous medium and the membrane. There was a strong correlation between tight junction effects and the transverse diffusion of the probes measured in membranes of vesicles of DOPC and in membranes of murine YAC cells. Addition of probes to solutions of either the vesicles or the YAC cells results in a labeling of the outer half of the

Table 8
FLUORESCENCE QUENCHING FOR FLIP FLOP RATES[108,109]

Quenching of F-12 and F-C16 Fluorescein by Antifluorescein Antibody

Probe	Labeling temp. (°C)	Measurement temp. (°C)	Quenching (%)
F-C16	20	20	71 ± 6
F-C16	20	5	76 ± 6
F-C12	5	5	72 ± 9
F-C12	20	20	56 ± 15
F-C12	20	5	32 ± 12

Quenching of diI-C16 Fluorescence by Iodide

Sample	Iodide Quenching (%)	Quenching after Triton® (%)
diI-C16 added to formed vesicles	17 ± 4	44 ± 7
Vesicles formed from diI-C16 plus lipid	15 ± 3	50 ± 5

membrane. Those probes which do not easily undergo flip-flop transfer are susceptible to nearly 100% fluorescence quenching by agents acting at the outer membrane/water interface. The initial location of the probes could be controlled in the vesicles; incorporation of the probes prior to vesicle preparation resulted in roughly equal concentrations of the label on the inner and outer surfaces. Antifluorescein antibody was used to quench the labels in the YAC solutions and iodide was used for the vesicle systems. The results in Table 8 show that the probes which undergo transverse diffusion are not restricted by the tight junction. The results indicate, furthermore, that while F-C12 undergoes flip-flop diffusion at 20°C, it does not do so at 5°C. F-C16, on the other hand, does not appreciably flip-flop at any of the temperatures. Thus, the difference in behavior of F-C12 and F-C16 can be qualitatively explained by the temperature-dependence results. Only 17% of the diI-C16 can be quenched by aqueous soluble quenchers, an indication of a relatively-efficient flip-flop process. Treatment of the vesicles with Triton® X-100, which disrupts the vesicular structure to one of a micellar form, allows more complete quenching since the labels have increased exposure. Overall, the results[108] show that epithelial cells can segregate some lipophilic components to create and maintain an asymmetric chemical distribution over the surface of the cell.

IX. CONCLUDING OBSERVATIONS

Studies of lipid lateral diffusion are rapidly evolving from a stance in which the rates themselves were of interest to a stance in which the measurements are being used to solve other questions regarding the cell membrane. The overwhelming bulk of the studies on living systems have been done with FRP. It is anticipated that as the selectivity, either in solubility or in binding, of the FRP probes improves, this field will undergo even more applications to membrane and cellular dynamics. The use of selective quenching to determine structural information, applied in a few cases discussed in Section VIII, is another area which could be exploited in FRP studies. This technique has been used with good success in fluorescence studies of biological systems to yield both dynamic and static structural information.

ACKNOWLEDGMENTS

I wish to acknowledge many useful discussions about membrane dynamics with Ed Samulski, Barton Smith, Mingjien Chien, Roberta Matthews, An-Li Kuo, Mirko Hrovat, Dick Rhyne, Jim Bartholomew, and Sue Hawkes. For this review, special thanks are due Kim Wade for her assistance in information retrieval, Mona Wan for her efforts in obtaining many references, and Chris Wade for his hours at a word processor preparing this manuscript.

REFERENCES

1. **Frye, L. D. and Edidin, M.,** The rapid intermixing of cell surface antigens after formation of mouse-human heterokaryons, *J. Cell. Sci.*, 7, 319, 1970.
2. **Kornberg, R. D. and McConnell, H. M.,** Lateral diffusion of phospholipids in a vesicle membrane, *Proc. Natl. Acad. Sci. U.S.A.*, 68, 2564, 1971.
3. **Devaux, P. and McConnell, H. M.,** Lateral diffusion in spin labeled phosphatidylcholine multilayers, *J. Am. Chem. Soc.*, 94, 4475, 1970.
4. **Träuble, H. and Sackmann, E.,** Studies on the crystalline-liquid cyrstalline phase transition of lipid model membranes. III, *J. Am. Chem. Soc.*, 94, 4499, 1972.
5. **Cherry, R. I.,** Rotational and lateral diffusion of membrane proteins, *Biochim. Biophys. Acta*, 559, 289, 1979.
6. **Vaz, W., Derko, Z. I., and Jacobson, K. A.,** Photobleaching measurements of the lateral diffusion of lipids and proteins in artificial phospholipid bilayer membranes, *Cell Surface Rev.*, 8, 83, 1982.
7. **Peters, R.,** Lateral mobility of proteins in membranes, in *Structure and Properties of Cell Membranes*, Vol. 1, Benga, Gh., Ed., CRC Press, Boca Raton, Fla., chap. 2.
8. **Galla, H., Hartman, W., Thielen, V., and Sackmann, E.,** On two-dimensional passive random walk in lipid bilayers and biological membranes, *J. Membr. Biol.*, 48, 215, 1979.
9. **Sjöstrand, F. S.,** Models of molecular architecture of cell membranes, in *Structure and Properties of Cell Membranes*, Vol. 1, Benga, Gh., Ed., CRC Press, Boca Raton, Fla., 1984, chap. 1.
10. **Margenau, H. and Murphy, S. M.,** *The Mathematics of Physics and Chemistry*, Van Nostrand, New York, 1943.
11. **deGennes, P. G.,** *The Physics of Liquid Crystals*, Oxford University Press, London, 1970.
12. **Alder, B. J. and Wainwright, T.,** Velocity autocorrelations for hard spheres, *Phys. Rev. Lett.*, 18, 988, 1967.
13. **Pusey, P. N.,** Study of brownian motion by intensity fluctuation spectroscopy, *Philos. Trans. R. Soc. London, Ser. A.*, 293, 429, 1979.
14. **Saxton, M. J.,** Lateral diffusion in an archipelago. Effects of impermeable patches on diffusion in a cell membrane, *Biophys. J.*, 39, 165, 1982.
15. **Saffman, P. G. and Delbruck, M.,** Brownian motion in biological membranes, *Proc. Natl. Acad. Sci. U.S.A.*, 72, 3111, 1975.
16. **Saffman, P. G.,** Brownian motion in thin sheets of viscous fluid, *J. Fluid Mech.*, 73, 593, 1976.
17. **Singer, S. and Nicholson, S.,** The fluid mosaic model of the structure of cell membranes, *Science*, 175, 720, 1972.
18. **Hughes, B. D., Pailthorpe, B. A., White, L. R., and Sawyer, W. H.,** Extraction of membrane microviscosity from translational and rotational diffusion coefficients, *Biophys. J.*, 37, 673, 1982.
19. **Peters, R. and Cherry, R.,** Lateral and rotational diffusion of bacteriorhodopsin in lipid bilayers: experimental test of the Saffman-Delbrueck equations, *Proc. Natl. Acad. Sci. U.S.A.*, 79, 4317, 1982.
20. **Vaz, W. and Hallman, D.,** Experimental evidence against the applicability of the Saffman-Delbrueck model to the translational diffusion of lipids in phosphatidylcholine bilayer membranes, *FEBS Lett.*, 152, 287, 1983.
21. **Birks, J. B.,** *The Photophysics of Aromatic Molecules*, Wiley-Interscience, New York, 1970.
22. **Cohen, M. and Turnbull, D.,** Molecular transport in liquids and gases, *J. Chem. Phys.*, 31, 1164, 1959.
23. **Turnbull, D. and Cohen, M.,** On the free volume model of the liquid-glass transition, *J. Chem. Phys.*, 52, 3038, 1970.
24. **MacCarthy, J. E. and Kozak, J. J.,** Lateral diffusion in fluid systems, *J. Chem. Phys.*, 77, 2214, 1982.

25. **Loughran, T., Hatlee, M., Patterson, L., and Kozak, J.,** Monomer-eximer dynamics in spread monolayers. I. Lateral diffusion of pyrene dodecanoic acid at the air-water interface, *J. Chem. Phys.*, 72, 5791, 1980.
26. **Schindler, M., Osborn, M., and Koppel, D.,** Lateral mobility in reconstituted membranes — comparisons with diffusions in polymers, *Nature (London)*, 283, 346, 1980.
27. **Lieb, W. R. and Stein, W. D.,** *Curr. Top. Membr. Transp.*, 2, 1971.
28. **Jahnig, F.,** No need for a new membrane model, *Nature (London)*, 289, 694, 1981.
29. **Koppel, D. E., Sheetz, M. P., and Schindler, M.,** Reply from D. E. Koppel, M. J. Osborn, and M. Schindler, *Nature (London)*, 289, 696, 1981.
30. **Pace, R. J. and Chan, S. I.,** Molecular motions in lipid bilayers. III. Lateral and transverse diffusion in bilayers, *J. Chem. Phys.*, 76, 424, 1982.
31. **Glasstone, S., Laidler, K. J., and Eyring, H.,** *The Theory of Rate Processes*, McGraw-Hill, New York, 1941, 481.
32. **Slater, G. and Caille, M.,** A percolation model for lateral diffusion in cholesterol-phospholipid mixtures, *Biochim. Biophys. Acta*, 686, 249, 1982.
33. **Copeland, B. and McConnell, H. M.,** The rippled structure in bilayer membranes of phosphatidylcholine and binary mixtures of phosphatidylcholine and cholesterol, *Biochim. Biophys. Acta*, 599, 95, 1980.
34. **Owicki, J. and McConnell, H. M.,** Lateral diffusion in inhomogeneous membranes. Model membranes containing cholesterol, *Biophys. J.*, 30, 383, 1980.
35. **Weaver, D. L.,** Note on the interpretation of lateral diffusion coefficients, *Biophys. J.*, 38, 311, 1982.
36. **Koppel, D. E., Sheetz, M. P., and Schindler, M.,** Matrix control of protein diffusion in biological membranes, *Proc. Natl. Acad. Sci. U.S.A.*, 78, 3576, 1981.
37. **Aizenbud, B. M. and Gershon, N. D.,** Diffusion of molecules on biological membranes of nonplanar form. A theoretical study, *Biophys. J.*, 38, 287, 1982.
38. **Dragsten, P., Henkart, P., Blumenthal, R., Weinstein, J., and Schlessinger, J.,** Lateral diffusion of surface immunoglobin, Thy-1 antigen, and a lipid probe in lymphocyte plasma membrane, *Proc. Natl. Acad. Sci. U.S.A.*, 76, 516, 1979.
39. **Wolf, D., Handyside, A., and Edidin, M.,** Changes in the organization of the mouse egg plasma membrane upon fertilization and first cleavage: indications from the lateral diffusion rates of fluorescent lipid analogs, *Biophys. J.*, 38, 295, 1982.
40. **Benga, Gh.,** Spin labeling, in *Biochemical Research Techniques*, Wrigglesworth, J., Ed., John Wiley & Sons, London, 1983, 79.
41. **Sackman, E., Träuble, H., Galla, H., and Overath, P.,** Lateral diffusion, protein mobility, and phase transitions in *Escherichia coli* membranes. A spin label study, *Biochemistry*, 12, 5360, 1973.
42. **Popp, C. and Hyde, J. S.,** Electron-electron double resonance and saturation-recovery studies of nitroxide electron and nuclear spin-lattice relaxation times and Heisenberg exchange rates: lateral diffusion in dimyristoylphosphatidylcholine, *Proc. Natl. Acad. Sci. U.S.A.*, 79, 2559, 1982.
43. **Stejskal, E. O. and Tanner, J. E.,** Spin diffusion measurements: spin echoes in the presence of a time dependent field gradient, *J. Chem. Phys.*, 42, 288, 1965.
44. **Stejskal, E.,** Spin-echo measurement of self-diffusion in colloidal systems, *Adv. Mol. Relax. Proc.*, 3, 27, 1972.
45. **Carr, H. Y. and Purcell, E. M.,** Effects of diffusion on free precession in NMR experiments, *Phys. Rev.*, 94, 630, 1954.
46. **DeVries, J. J. and Berendsen, H. J. C.,** NMR measurements on a macroscopically ordered smectic liquid crystalline phase, *Nature (London)*, 221, 1139, 1969.
47. **Hemminja, M. and Berendsen, H. J. C.,** Magnetic resonance in ordered lecithin-cholesterol multilayers, *J. Magn. Res.*, 8, 133, 1972.
48. **Shimshick, E. J. and McConnell, H. M.,** Lateral phase separations in binary mixtures of cholesterol and phospholipids, *Biochim. Biophys. Res. Commun.*, 53, 446, 1973.
49. **Samulski, E. T., Smith, B., and Wade, C. G.,** NMR free induction decay and spin echoes in oriented model membrane bilayers, *Chem. Phys. Lett.*, 20, 167, 1973.
50. **Roeder, S., Burnell, E., Wade, C. G., and Kuo, A.-L.,** Determination of the lateral diffuson coefficient of potassium oleate in the lamellar phase, *J. Chem. Phys.*, 64, 4807, 1976.
51. **Tiddy, G., Hayter, J., Hecht, A., and White, J. W.,** NMR studies of water self-diffusion in the lamellar phase, *Ber. Bunsen. Phys. Chem.*, 78, 961, 1974.
52. **Wade, C. G. and Kuo, A.-L.,** Lipid lateral diffusion by pulsed nuclear magnetic resonance, *Biochemistry*, 18, 2300, 1979.
53. **Lindblom, G., Wennerstrom, H., Arvidson, G., and Lindman, B.,** Lecithin translational diffusion studied by pulsed nuclear magnetic resonance, *Biophys. J.*, 16, 128, 1976.
54. **Lindblom, G. and Wennerstrom, H.,** Amphiphile diffusion in model membrane systems studied by pulsed NMR, *Biophys. Chem.*, 6, 167, 1977.

55. **Lindblom, G., Johannson, L., and Arvidson, G.,** Effect of cholesterol in membranes. Pulsed nuclear magnetic resonance measurements of lipid lateral diffusion, *Biochemistry*, 20, 2204, 1981.
56. **Wennerstrom, H. and Lindblom, G.,** Biological and model membranes studied by nuclear magnetic resonance of spin one half nuclei, *Q. Rev. Biophys.*, 10, 67, 1977.
57. **Hrovat, M. and Wade, C. G.,** NMR pulsed-gradient diffusion measurements. I. Spin-echo stability and gradient calibration, *J. Magn. Res.*, 44, 62, 1981.
58. **Hrovat, M. and Wade, C. G.,** NMR pulsed gradient diffusion measurements. II. Residual gradients and lineshape distortions, *J. Magn. Res.*, 45, 67, 1981.
59. **Hrovat, M. and Wade, C. G.,** Absolute measurements of diffusion coefficients by pulsed nuclear magnetic resonance, *J. Chem. Phys.*, 73, 2509, 1980.
60. **James, T. L. and MacDonald, G.,** Measurement of the self-diffusion coefficients of each component in a complex system using pulsed gradient FT NMR, *J. Magn. Res.*, 11, 58, 1973.
61. **Waugh, J. S., Huber, L., and Haeberlen, U.,** Approach to high-resolution NMR in solids, *Phys. Rev. Lett.*, 20, 180, 1968.
62. **Mansfield, P., Orchard, M. J., Stalker, D. C., and Richards, K. H. B.,** Symmetrized multipulse nuclear-magnetic-resonance experiments in solids: measurement of the chemical-shift shielding tensor in some compounds, *Phys. Rev. B*, 7, 90, 1973.
63. **Rhim, W.-K., Elleman, D. D., and Vaughan, R. W.,** Analysis of multiple pulse NMR in solids, *J. Chem. Phys.*, 59, 3740, 1973.
64. **Silva-Crawford, M., Gerstein, B. C., Kuo, A.-L., and Wade, C. G.,** Measurement of diffusion in rigid bilayer systems by combined multiple pulse and pulse gradient NMR, *J. Am. Chem. Soc.*, 102, 3728, 1980.
65. **Lee, A., Birdsall, N., and Metcalfe, J.,** Measurement of fast lateral diffusion of lipids in vesicles and in biological membranes by 1H-nuclear magnetic resonance, *Biochemistry*, 12, 1650, 1973.
66. **Bloom, M., Burnell, E., MacKay, A., Nichol, C., Valic, M., and Weeks, G.,** Fatty acyl chain order in lecithin model membranes determined from proton magnetic resonance, *Biochemistry*, 17, 5758, 1978.
67. **Kroon, P. A., Kainosho, M., and Chan, S. I.,** Proton magnetic resonance of lipid bilayer membranes. Experimental determination of inter- and intramolecular nuclear relaxation in sonicated phosphatidylcholine bilayer vesicles, *Biochim. Biophys. Acta*, 433, 282, 1976.
68. **Brulet, P. and McConnell, H. M.,** Magnetic resonance spectra of membranes, *Proc. Natl. Acad. Sci. U.S.A.*, 72, 1451, 1975.
69. **Fisher, R. W. and James, T.,** Lateral diffusion of the phospholipid molecule in dipalmitoylphosphatidylcholine bilayers. An investigation using nuclear spin-lattice relaxation time in the rotating frame, *Biochemistry*, 17, 1177, 1978.
70. **Cullis, P. R.,** Lateral diffusion rates of phosphatidylcholine in vesicle membranes: effects of cholesterol and hydrocarbon phase transitions, *FEBS Lett.*, 70, 223, 1976.
71. **Burnell, E., Cullis, P. R., and DeKruijff, B.,** Effects of tumbling and lateral diffusion on phosphatidylcholine model membrane phosphorous-31-NMR lineshapes, *Biochim. Biophys. Acta*, 603, 63, 1980.
72. **Campbell, R., Meirovitch, E., and Freed, J.,** Slow-motional NMR line shapes for very anisotropic rotational diffusion. Phosphorus-31 NMR of phospholipids, *J. Phys. Chem.*, 83, 525, 1979.
73. **MacKay, A., Burnell, E., Nichol, C., Weeks, G., Bloom, M., and Valic, M.,** Effect of viscosity on the width of the methylene proton magnetic resonance line in sonicated phospholipid bilayer vesicles, *FEBS Lett.*, 88, 97, 1978.
74. **Davis, J. H., Maraviglia, B., Weeks, G., and Gadin, D. V.,** Bilayer rigidity of the erythrocyte membrane deuterium NMR of a perdeuterated palmitic acid probe, *Biochim. Biophys. Acta*, 550, 362, 1979.
75. **Webb, W. W., Barak, L. S., Tank, D. W., and Wu, E. S.,** Molecular mobility on the cell surface, *Biochem. Soc. Symp.*, 46, 191, 1981.
76. **Smith, B. A. and McConnell, H. M.,** Determination of molecular motion in membranes using periodic pattern photobleaching, *Proc. Natl. Acad. Sci. U.S.A.*, 75, 2759, 1978.
77. **Axelrod, D., Koppel, D., Shlessinger, J., Elson, E., and Webb, W.,** Mobility measurement by analysis of fluorescence photobleaching recovery kinetics, *Biophys. J.*, 16, 1055, 1976.
78. **Schneider, M. B. and Webb, W. W.,** Measurement of submicron laser beam radii, *Appl. Opt.*, 20, 1382, 1981.
79. **Bretscher, M. S.,** Lateral diffusion in eukaryotic cell membranes, *Trends Biochem. Sci.*, 5, 6, 1980.
80. **Wolf, D., Edidin, M., and Dragsten, P.,** Effect of bleaching light of measurements of lateral diffusion in cell membranes by the fluorescence photobleaching recovery methods, *Proc. Natl. Acad. Sci. U.S.A.*, 77, 2043, 1980.
81. **Galla, H. and Sackmann, E.,** Lateral diffusion in the hydrophobic region of membranes: use of pyrene excimers as optical probes, *Biochim. Biophys. Acta*, 329, 103, 1974.
82. **Vanderkooi, J. and Callis, J.,** Pyrene. A probe of lateral diffusion in the hydrophobic region of membranes, *Biochemistry*, 13, 4000, 1974.
83. **Roseman, M. A. and Thompson, T. E.,** Mechanism of the spontaneous transfer of phospholipids between bilayers, *Biochemistry*, 19, 439, 1980.

84. **Galla, H.-J., Theilen, U., and Hartmann, H.,** Tranversal mobility in bilayer membrane vesicles: use of pyrene lecithin as optical probe, *Chem. Phys. Lipids*, 23, 239, 1979.
85. **Kano, K., Kawazumi, H., Ogawa, T., and Sunamoto, J.,** Fluorescence quenching in lipsomal membranes. Exciplex as a probe for investigating artificial lipid membrane properties, *J. Chem. Phys.*, 85, 2204, 1981.
86. **Razi-Naqvi, K., Behr, J.-P., and Chapman, D.,** Methods for probing lateral diffusion of membrane components: triplet-triplet annihilation and triplet-triplet energy transfer, *Chem. Phys. Lett.*, 26, 440, 1974.
87. **Derzko, Z. and Jacobson, K.,** Comparative lateral diffusion of fluorescent lipid analogues in phospholipid multibilayers, *Biochemistry*, 19, 6050, 1980.
88. **Barisas, B. C.,** Criticality of beam alignment in fluorescence photobleaching recovery experiments, *Biophys. J.*, 29, 545, 1980.
89. **Thompson, N. L. and Axelrod, D.,** Reduced lateral mobility of a fluorescent lipid probe in cholesterol depleted erythrocyte membrane, *Biochim. Biophys. Acta*, 597, 155, 1980.
90. **McCown, J. T., Evans, E., Diehl, S. E., and Wiles, H. C.,** Degree of hydration and lateral diffusion in phospholipid multibilayers, *Biochemistry*, 20, 2134, 1981.
91. **Wu, E.-S., Jacobson, K., and Papahadjopoulos, D.,** Lateral diffusion in phospholipid multibilayers measured by fluorescence recovery after photobleaching, *Biochemistry*, 16, 3936, 1977.
92. **Rubenstein, J. R., Smith, B. A., and McConnell, H. M.,** Lateral diffusion in binary mixtures of cholesterol and phosphatidylcholines, *Proc. Natl. Acad. Sci. U.S.A.*, 76, 15, 1979.
93. **Klaussner, R. D. and Wolf, D. E.,** Selectivity of fluorescent lipid analogs for lipid domains, *Biochemistry*, 19, 6199, 1980.
94. **Alecio, M. R., Golan, D. E., Veatch, W. R., and Rando, R.,** Use of a fluorescent cholesterol derivative to measure lateral mobility of cholesterol in membranes, *Proc. Natl. Acad. Sci. U.S.A.*, 79, 5157, 1982.
95. **Wolf, D. E., Kinsey, W., Lennarz, W., and Edidin, M.,** Changes in the organization of the sea urchin egg plasma membrane upon fertilization: indications from the lateral diffusion rates of lipid soluble fluorescent dyes, *Dev. Biol.*, 81, 133, 1981.
96. **Wolf, D. E., Edidin, M., and Handyside, A.,** Changes in the organization of the mouse egg plasma membrane upon fertilization and first cleavage: indications from the lateral diffusion rates of fluorescent lipid analogs, *Dev. Biol.*, 85, 195, 1981.
97. **Eldridge, C. A., Elson, E. L., and Webb, W. W.,** Fluorescence photobleaching recovery measurements of surface lateral mobilities on normal and SV-40 transformed mouse fibroblasts, *Biochemistry*, 19, 2075, 1980.
98. **Searls, D. and Edidin, M.,** Lipid composition and lateral diffusion in plasma membranes of teratocarcinoma-derived cell lines, *Cell*, 24, 511, 1981.
99. **de Latt, S., Van der Saag, P. T., Elson, E. L., and Schlessinger, J.,** Lateral diffusion of membrane lipids and proteins is increased specificially in neurites of differentiating neuroblastoma cells, *Biochim. Biophys. Acta*, 558, 247, 1979.
100. **de Laat, S., Van der Saag, P. T., Elson, E. L., and Schlessinger, J.,** Lateral diffusion of membrane lipids and proteins, during the cell cycle of neuroblastoma cells, *Proc. Natl. Acad. Sci. U.S.A.*, 77, 1526, 1980.
101. **Boonstra, J., Nelemans, S. A., Feijen, A., Bierman, A., Van Zoelen, E., van der Saag, P., and de Laat, S.** Effect of fatty acids on plasma membrane lipid dynamics and cation permeability in neuroblastoma cells, *Biochim. Biophys. Acta*, 692, 321, 1982.
102. **Jacobson, K., Hou, Y., Derzko, Z., Wojcieszyn, J., and Organiscia, K.,** Lipid lateral diffusion in the surface membrane of cells and in multibilayers formed from plasmalipids, *Biochemistry*, 20, 5268, 1981; *Biophys. J.*, 37, 8, 1982.
103. **Boullier, J. A., Melnykovych, G., and Barisas, G.,** A photobleaching recovery study of glucocorticoid effects on lateral mobilities of a lipid analog in S3G HeLa cell membranes, *Biochim. Biophys. Acta*, 692, 278, 1982.
104. **Schlessinger, J., Axelrod, D., Koppel, D. E., Webb, W. W., and Elson, E. L.,** Lateral transport of a lipid probe and labeled proteins on a cell membrane, *Science*, 195, 307, 1971.
105. **Schlessinger, J., Koppel, D. E., Axelrod, D., Jacobson, K., Webb, W. W., and Elson, E.,** Lateral transport on cell membranes: mobility of concanavalin A receptors on myoblasts, *Proc. Natl. Acad. Sci. U.S.A.*, 78, 2409, 1976.
106. **Galla, H.-J. and Luisetti, J.,** Lateral and transversal diffusion and phase transitions in erythrocyte membranes. An excimer fluorescence study, *Biochim. Biophys. Acta*, 596, 108, 1980.
107. **Gerritsen, W., Henricks, P., de Kruijff, B., and van Deenen, L. L. M.,** The transbilayer movement of phosphatidylcholine in vesicles reconstituted with intrinsic proteins from the human erythrocyte membrane, *Biochim. Biophys. Acta*, 600, 607, 1980.
108. **Dragsten, P., Handler, J., and Blumenthal, R.,** Asymmetry in epithelial cells: is the tight junction a barrier to lateral diffusion in the plasma membrane?, *Prog. Clin. Biol. Res. (Membr. Growth Dev.)*, 91, 525, 1982.

109. **Dragsten, P., Blumenthal, R., and Handler, J.,** Membrane asymmetry in epithelia: is the tight junction a barrier to diffusion in the plasma membrane?, *Nature (London)*, 294, 718, 1981.

Chapter 5

TOPOLOGICAL ASYMMETRY AND FLIP-FLOP OF PHOSPHOLIPIDS IN BIOLOGICAL MEMBRANES

L. D. Bergelson and L. I. Barsukov

TABLE OF CONTENTS

I. INTRODUCTION

To fulfill their vital functions biological membranes must operate differentially on the compartments they separate, hence, the lipids and proteins of the outer and inner surfaces must differ in respect to their transversal distribution and/or orientation. While the vectorial orientation and distribution of proteins is always asymmetric, the membrane asymmetry of lipids is less absolute: mostly the same molecular species can be found at both surfaces, but their quantities in the outer and inner leaflets are different and may change during the life of the cell. Nonetheless, the topological asymmetry of lipids influences a great number of membrane properties and functions, such as charge distribution, ion fluxes, enzyme activities, binding of external molecules, etc. Therefore, lipid asymmetry of biological and artificial membranes is an item of considerable interest. A number of reviews have appeared on this subject,[1-6] however, because of the high activity in the field they already partly are outdated. The present chapter does not intend to be an exhaustive review, but merely to serve as an introduction into the field with emphasis on the factors influencing lipid asymmetry including transmembrane dislocation ("flip-flop") of lipid molecules. We also discuss critically the main techniques which are used in studies of lipid asymmetry and evaluate their possibilities and limitations.

II. FACTORS INFLUENCING LIPID ASYMMETRY AND FLIP-FLOP MOVEMENT

The transmembrane distribution of lipids depends on a large number of diverse factors that can be divided into two main groups: (1) physical factors and (2) factors of biosynthesis and catabolism.

A. Physical Factors

The transversal distribution of lipids depends to some extent on differences of the conditions at both sides of the membrane, e.g., ion concentrations, pH, presence of different solutes, extracellular macromolecules, etc. An important role may be attributed to the membrane proteins, which always are oriented in an asymmetrical fashion. If those proteins are located asymmetrically and possess binding sites specific for some lipid classes, this will clearly influence the transbilayer distribution of the bound lipid species.

Another factor which has to be considered is the curvature of the membrane. In regions with high curvature the packing conditions for lipids differing by charge and relative head group area are different in the outer and inner monolayer. Finally, the transmembrane distribution of lipids will depend on their polymorphism which, in turn, is determined by the relative steric requirements of the polar and apolar moieties of the lipid molecules.

A phenomenon tightly connected with the topological distribution of lipids is the transmembrane translocation of lipid molecules (flip-flop). Experiments performed with model systems have demonstrated that in pure lipid systems the flip-flop rate depends on numerous physico-chemical factors including the structure,[7] phase behavior[8] of the lipids, presence of lipid peroxidation products,[9] and of phospholipids species which tend to form nonbilayer structures.[10,11] The latter factor seems to be of special importance and it has been suggested[10,12] that the transition of lipids from the bilayer configuration into inverted micelles or hexagonal (H_{II}) phase is the general mechanism underlying flip-flop.

Of special interest is the influence of intrinsic proteins on phospholipid flip-flop. It has been reported that flip-flop of phosphatidylcholine (PC) and lyso-PC is markedly enhanced by two major intrinsic membrane proteins of the erythrocyte: glycophorin[13,14] and band 3 protein.[15] On the other hand, another membrane protein, mitochondrial cytochrome oxidase, was shown to have no influence on the PC flip-flop in reconstituted systems.[16] It should be noted that the choice of the above proteins for phospholipid flip-flop studies is not justified logically, because both the erythrocyte membrane and the inner mitochondrial membrane are characterized by extremely low flip-flop rates[17,18] comparable to those in pure lipid bilayers. A more relevant object are microsomes where phospholipid flip-flop occurs very fast. As will be discussed in Section IV.C.4, the rapid phospholipid flip-flop in microsomal membranes appears to be catalyzed by the important membrane-bound enzyme, cytochrome P-450.

B. Factors of Membrane Biosynthesis and Catabolism

Many biological membranes have the capacity to synthesize some (or all) of their phospholipids. On the other hand, a number of enzymatic lipid degradations, modifications, and interconversions are known to occur within intact membranes. Since the corresponding enzymes are localized asymmetrically, they will affect the transmembrane distribution of the phospholipids. For example, in the erythrocyte membrane two methyltransferases are responsible for transformation of phosphatidylethanolamine (PE) into PC: one transferring a methyl group to PE is located at the inner surface, the other, which methylates phosphatidyl-*N*-monomethylethanolamine, faces the exterior.[19] This specific location of the methyltransferases may contribute to the asymmetric distribution of PC and PE in the erythrocyte membrane.

Some membranes, such as the plasma membrane of eukaryotic cells, have rather low biosynthetic activity and acquire their phospholipids as well as other constituents by fusion with small intracellular vesicles or by protein-mediated lipid transfer. In the first case the original membrane asymmetry of the vesicles would be inverted in the target membrane. In the second case lipid asymmetry would be generated by the highly asymmetric mode of interaction of the lipid-transfer proteins with the donor and acceptor membranes (see Section III.E).

Other factors influencing the regulation and maintenance of lipid membrane asymmetry might involve catabolism-dependent selective translocation of phospholipids from the outer to the inner shell and vice versa as well as lipid exchange between the membrane and the surrounding medium. The final topological asymmetry of the membrane lipids will be the outcome of the interplay and delicate balance of all factors mentioned.

III. DETERMINATION OF TRANSVERSAL LIPID DISTRIBUTION AND FLIP-FLOP. A CRITICAL EVALUATION OF THE METHODS

A. General Remarks

Determination of the transversal distribution of lipids in artificial or biological membranes is based on the application of different probes or immunochemical procedures. The latter will not be considered in this chapter. Correct evaluation of the results obtained by membrane probes requires that (1) the probe interacts selectively with lipids in the outer or inner shell of the membrane and (2) the original asymmetry of the membrane under investigation persists during the experiment. The first condition means that the probe should not induce lysis or penetrate the membrane, and the second one implies that flip-flop rates should be slow in comparison with the time required for the experiment. Actually these conditions are rarely fulfilled completely. For this reason reliable conclusions can be drawn only if several independent methods are applied. The following is a short critical survey of the most frequently used probing methods.

B. Chemical Labeling

In order to recognize lipids localized on the outer or inner surface of a vesicular membrane the lipids are labeled with nonpenetrating or penetrating chemical reagents, respectively. The main problems encountered with this approach are

1. "Impermeability" is never absolute and depends not only on the nature of the reagent and the membrane, but also on the concentration of the reagent and other reaction conditions. Thus, the degree of impermeability must be checked in each particular case.
2. Absence of interaction of an "impermeant" reagent with some part of the membrane lipids may be caused by shielding and cannot be taken as an unambiguous proof that the unreactive lipids are located on the opposite side of the membrane. If the reagent is negatively charged its reaction with charged lipid species will always be incomplete.
3. The reagents available interact only with phospholipids bearing reactive polar head groups (OH, NH_2) but not with cholinephosphatides or cholesterol.

The most frequently used impermeant and permeant reagents are 2,4,6-trinitrobenzenesulfonate (TNBS) and 1-fluoro-2,4-dinitrobenzene (FDNB), respectively. These reagents have been used to study the transverse distribution of aminophospholipids in liposomes,[20-22] erythrocytes,[23-28] platelets,[29] fibroblasts,[30-32] synaptosomes,[33,34] microbes,[25,26,35-41] lipid-containing viruses,[42,43] mitochondria,[44,45] retinal rod outer segments,[46-48] sarcoplasmic reticulum,[49] and myoblasts.[50]

The reaction of TNBS with aminophospholipids proceeds comparatively slow and thus cannot be applied to membranes with high phospholipid flip-flop rates. In such cases it is sometimes possible to use the much faster reacting fluorescamine.[51] The reagent has been employed in studies of the transmembrane distribution of phosphatidylserine (PS) and PE in sarcoplasmic reticulum[52-54] and inner mitochondrial membranes[56] as well as in mixed composition liposomes.[22] The impermeability of the reagent can be increased by using its complex with cycloheptaamylose.[54,55]

C. Phospholipase Digestion

If phospholipase hydrolysis is restricted to the outer surface of a membrane, it may reveal the transmembrane distribution of the phospholipids. The main advantage of this approach is that phospholipase hydrolysis occurs under physiological conditions that are much milder than those used in chemical labeling. In contrast to the latter technique which concerns only specific lipid classes, all lipid components can be subjected to enzymatic digestion. However, interpretation of the results is frequently less equivocal because: (1) the hydrolysis products differ fundamentally from the substrates and thus may induce heavy alterations of the membrane structure and asymmetry; (2) phospholipases from different sources frequently reveal different substrate specificity and lytic properties;[56-59] (3) the action of phospholipase depends on physical factors which change during the digestion process. For these reasons application of phospholipases in studies of membrane asymmetry requires careful control of possible artifactual changes of the membrane structure.

In order to make the interpretation of phospholipase A_2 digestion results more reliable, some methodological improvements have been developed in our laboratory. In studies of artificial phospholipid membranes the influence of the lysophospholipids and fatty acids formed is studied routinely by the paramagnetic ion-NMR technique (see Section III.D). With that method we established that incorporation of even 20 mol% of lyso-PC into PC liposomes will not destroy the membrane and will not lead to formation of large pores which would render the membrane permeable for macromolecules.[60] The flip-flop of PC is not enhanced by significant amounts of lysolecithin.[9] At the same time the PC bilayer becomes, however, permeable for inorganic ions because of formation of narrow ion channels. These conclusions later have been confirmed by others.[61-64]

In natural membranes the situation may be more complicated. For example, whereas the membranes of rat liver microsomes and brush border vesicles retain their general packing characteristics and remain impermeable for phospholipase A_2 even after hydrolysis of up to

60% of their phospholipids,[65,66] the cytoplasmic membrane of *Micrococcus lysodeikticus* proved to be much less stable, especially after pronase treatment prior to phospholiase digestion.[67] At a high degree of hydrolysis, penetration of macromolecules including phospholipases becomes possible. In order to exclude rigorously the possibility of digestion of the "inside" phospholipids we use phospholipase A_2 covalently bound to Sepharose 4B.[68] Although such immobilization lowers the enzyme activity, the ability to hydrolyze different phospholipid substrates in micellar or membrane-bound form is retained.[68,69]

An important obstacle in the application of phospholipases is that in venoms from different sources they exist as mixtures of isoenzymes differing by their activity and substrate specificity. Results obtained with unpurified venoms are often irreproducible because of the presence of various nonenzymatic lytic factors or other enzyme activities which may alter the membrane structure. For this reason we used in our experiments highly purified phospholipase A_2 from bee venom (*Apis mellifica*).[68,69] The advantages of this enzyme include its low substrate specificity and its low sensibility towards such factors as surface pressure, lipid packing, and phase state. In contrast, the widely used pancreatic phospholipase A_2 and *Bacillus cereus* phospholipase C are able to attack phospholipid monolayers only at low surface pressures.[70]

The method of phospholipase digestion is widely applied in phospholipid asymmetry studies. Phospholipases A_2, C, D, and sphingomyelinases from various sources have been used successfully in this type of experiment (for a detailed description of the method see Roelofsen and Zwaal[71]). Membranes studied by this technique include erythrocytes,[70,72] liver microsomes,[69,73-76] brush border membranes,[66] and the membranes of various bacteria.[37,40,67,77,78]

D. NMR Spectroscopy

When routine instruments are used the NMR signals from the inward- and outward-facing phospholipid molecules are partly or completely overlapping despite differences in their environment. Much better separation of the "outer" and "inner" signals can be achieved by using paramagnetic ions as shift or broadening reagents. When such ions are applied from the outside, the lipids of the outer shell which are in contact with the paramagnetic ions are broadened or shifted away, thus revealing the unchanged signal from the lipids located in the inner shell.[79] In this way it is possible to differentiate between the phospholipids in the inner and outer monolayer. Since the method can be used not only in ^{1}H but also in ^{13}C and ^{31}P NMR measurements,[79-81] it significantly increases the possibilities of NMR in studies of lipid asymmetry (reviewed by Bergelson[82]). The main advantage of this simple method is that the data easily can be quantitated. For example, quantitation can be carried out by determining the integral intensity ratios of the "outer" and "inner" signals for a given class of phospholipids. However, when ^{1}H and ^{13}C NMR are used, quantitation is possible only for cholinephosphatides whose *N*-methyl protons give rise to sufficiently narrow and intense signals. The number of phospholipids whose transmembrane distribution can be assessed with the paramagnetic ion NMR technique can be increased by using ^{31}P-NMR, since the phosphorus resonances of different phospholipid classes usually differ by their chemical shifts. However, with ^{31}P NMR accurate estimation of the signal intensity is complicated because of the Nuclear Overhauser Effect. With any resonance sufficiently accurate integral intensity determinations can be achieved with the error of 5 to 10% for the major phospholipid components of the membrane. When the content of a given component is below 20% the error increases to 30% or more. In order to increase the accuracy of the paramagnetic ion NMR technique for minor components of the phospholipid mixture, we developed a modification of that method based on measurements of the chemical shifts induced by the paramagnetic ions.[9] In binary mixtures of neutral and negatively charged phospholipids the magnitude of the induced shift strongly depends on the composition of the mixture. In some cases the concentration dependence is linear and can be used for direct determination of the surface concentration of the negatively charged component in the outer shell of the membrane.[9] The chemical shift approach is characterized by high sensitivity and accuracy. For example, using ^{31}P NMR and Yb^{3+} as paramagnetic probe, it is possible to determine accurately 1 to 2% phosphatidylinositol in the outer shell of mixed phosphatidylinositol-PC liposomes. The determination can be carried out with vesicles containing from 1 to 30% of the negative component. At higher concentrations of the latter the paramagnetic ions usually induce flocculation. In such cases the transverse distribution of the two components can more safely be deduced from integral intensity measurements.

Another problem arises from the fact that the magnitude of the induced shift depends not only on the chemical nature and concentration of the phospholipid head group, but also on their conformation. Therefore, it must be ascertained that the conformation of the phospholipid head groups is the same in the samples to be compared. This conveniently can be done by measurement of the ratio of the chemical shifts of the proton signals from the $POCH_2$, N^+CH_2, and $N^+(CH_3)_3$ moieties of the phosphocholine head group.[83] With the very low paramagnetic ion concentrations usually employed in experiments of that type, the above ratio remains unchanged,[83] however, higher ion concentrations may influence the phosphocholine head group conformation.[84]

Summarizing, combinations of integral intensity and induced chemical shift measurements of NMR signals separated by paramagnetic ions permit to determine phospholipid asymmetry with membranes widely differing in their phospholipid composition. With ^1H NMR an important limitation of the method is that it is confined to small unilamellar vesicles, because with larger systems the resonances are broadened as a consequence of slow tumbling rates. This difficulty can be overcome by using ^{13}C or ^{31}P NMR. The paramagnetic ion NMR technique has been used successfully for measuring the transverse distribution of lipids in sonicated liposomes[7,9,85] as well as in the sarcoplasmic reticulum membrane.[86] Other applications of the method involve monitoring lipid flip-flop in PC vesicles after asymmetric perturbation performed on one leaflet by action of phospholipase D or by protein-mediated lipid exchange.[7,87]

E. Intermembrane Phospholipid Exchange

The exchange of lipids between membranes can be catalyzed by phospholipid exchange proteins, i.e., polypeptides, that are capable of transferring lipids from one membrane to another. The phospholipid exchange proteins recognize the polar head group of phospholipids and are relatively insensitive to their fatty acid composition. Two types of such proteins exist, those which exchange specifically only one type of phospholipid and those which are less specific, although mostly they still prefer a given phospholipid class. The potential of protein-mediated lipid exchange in studies of membrane asymmetry was first demonstrated in 1974.[88] We showed that pure or crude lipid exchange proteins (1) do not penetrate or perturb membranes, (2) do not induce lysis, (3) do not enhance flip-flop under conditions of a one-for-one exchange, and (4) affect only the outer leaflet lipids. Thus, lipid exchange proteins are suitable tools to study the transbilayer localization of phospholipids provided that flip-flop movements are slow and that the lipids of the outer leaflet are freely accessible to the protein. These conclusions were confirmed by several other investigators.[18,31,89-92] The major advantages of the method are that the exchange is carried out under mild physiological conditions, that the phospholipid composition remains unaltered when unlabeled lipids are exchanged for labeled ones and vice versa, and that no foreign ions have to be introduced. It is important that not only specific individual lipid exchange proteins but also partly purified cytosolic exchange fractions can be used in this type of experiments. The main limitation of the lipid exchange method is due to the existance in the outer leaflet of different lipid pools that undergo exchange at different rates, whereas tightly bound (boundary) lipids may not exchange at all.[67] Sometimes the amount of exchangeable lipid can be increased by preliminary proteolysis.[67,94]

F. A Kinetic Approach to the Study of Transmembrane Distribution of Phospholipids

The above approaches to the study of phospholipid asymmetry using a nonpenetrating probe give satisfactory results if the flip-flop rate is not high ($\tau_{1/2}$ several hours or days). Under conditions of rapid flip-flop correct data can be obtained only with consideration of the kinetics of the transmembrane migration process.

Quantitative analysis of the probing kinetics can be based on a model of two pools between which rapid exchange of phospholipid molecules occurs.

$$\underset{\text{inner pool}}{\boxed{b}} \underset{k_{ba}}{\overset{k_{ab}}{\rightleftharpoons}} \underset{\text{outer pool}}{\boxed{a}} \xrightarrow{k_0} \text{interaction with a probe}$$

According to this model the outer phospholipid pool (a) is easily accessible for the probe, whereas phospholipid molecules of the inner pool (b) become available for the probe only after their migration across the membrane. The sizes of pools (r_a and r_b) as well as the half-time of flip-flop ($\tau_{1/2}$) can be represented through the rate constants of the elementary steps of the probing process as shown in Equations 1 to 3:

$$r_a = k_{ab}/(k_{ba} + k_{ab}) \tag{1}$$

$$r_b = k_{ba}/(k_{ba} + k_{ab}) \tag{2}$$

$$\tau_{1/2} = 0.693/(k_{ba} + k_{ab}) \tag{3}$$

If flip-flop is a rate-limiting stage the kinetic curves observed should have a biphasic character, evidencing that the probe interacts initially with those phospholipid molecules which are located on the outer membrane surface, and the slower probing phase corresponds to gradual transition of phospholipids from the inner membrane side to the outer one. Biphasic kinetic curves can be approximated by the sum of two exponentials referring to the rapid and slow probing phases. From the parameters of these exponentials the rate constants of the separate steps of the probing process can be calculated and the exact values of the sizes of the outer and inner phospholipid pools and flip-flop rates can be determined using Equations 1 to 3.

The kinetical approach was employed in topological investigations of the rat erythrocyte membrane with the use of phospholipid exchange proteins.[91,98,99] But since the flip-flop rate in this membrane is not high ($\tau_{1/2} \sim 2$ to 7 hr for PC), the results obtained did not differ significantly from the data obtained without kinetic analysis. The advantages of the kinetical approach become visible, at the same time, with membranes with rapid flip-flop, i.e., in cases where the routine methods of elucidation of transmembrane asymmetry turn out to be inefficient or even lead to erroneous conclusions. Thus, for example, only with the use of the kinetical approach we were able to prove the existence of phospholipid asymmetry in rat liver microsomes.[69] Another example is the brush border membrane of the rabbit small intenstine for which the kinetic analysis also allowed to establish asymmetric transmembrane distribution of the phospholipids under conditions of rapid phospholipid flip-flop.[66] These examples will be considered in the next section.

IV. TRANSVERSAL DISTRIBUTION OF PHOSPHOLIPIDS IN VARIOUS MEMBRANES

At present complete or partial data about the topological asymmetry of phospholipids in several dozens of membrane preparations are available. Partially, these data are contradictive, which can be explained by inadequacy of the applied techniques or by the absence of the control experiments. Here we shall deal only with a few types of the membranes which are the best studied with respect of the transbilayer phospholipids distributions.

A. Liposomal Membranes

The topological asymmetry of phospholipids in small unilamellar vesicles (SUV) with diameter less than 500 Å was studied in detail by the paramagnetic ion-NMR technique.[9,85] It was established that in SUV composed of mixtures of PC with acidic phospholipids the latter ones were localized predominantly in the inner monolayer. At the same time the transmembrane distribution of phospholipids in SUV depends on the size of vesicles as well as on the medium. Thus, in SUV consisting of PC and PS the latter phospholipid is located predominantly in the inner monolayer at low pH values but prefers the outer monolayer at high pH.[85] At given pH values and sizes of vesicles the final transmembrane distribution depends both on the relative area occupied by the polar head groups of each phospholipid component and on their surface charges.

B. Membranes of Bacteria and Lipid-Containing Viruses

The existence of phospholipid asymmetry in a bacterial membrane was discovered with the cytoplasmic membrane of *M. lysodeikticus*.[94,95] The membrane of this microorganism contains only three acidic phospholipids: phosphatidylinositol (PI), phosphatidylglycerol (PG), and cardiolipin. It was shown that the former of these lipids is located predominantly on the outer surface of the membrane, the second one on the inner side, and the third phospholipid is almost evenly distributed between both membrane surfaces. Later the topological asymmetry of phospholipids in various bacterial membranes was studied with different techniques, but no general principle of the transmembrane distribution was established.[35-41] At the same time the investigations performed on bacterial membranes allowed to reveal a relationship between the lipid topological asymmetry, flip-flop, and localization of the sites of the phospholipid biosynthesis. For *Bacillus megaterium* membranes, for example, it has been shown that PE is synthesized on the cytoplasmic side of the membrane, and only after a certain period of time this phospholipid appears on the outer side due to its transmembrane migration.[35,36]

Lipid-containing viruses require their membrane by budding out from the host cell and hence should have transverse phospholipid distributions identical with those of the host plasma membrane. Indeed, different viruses grown on similar cells (influenza virus and vesicular stomatitis virus) showed similar phospholipid distributions, PE, being preferentially localized in the inner layer and cholinephosphatides in the outer one.[43,112] However, in a later study using the same influenza virus and the same host cell, completely the opposite distribution was reported.[92] For the bacteriophage PM2 it has been shown that PG and PE are distributed asymmetrically, the former lipid prefering the outer layer, and the latter the inner one.[100] Large differences were found in the phospholipid composition between host cell and phage membranes, suggesting that active and selective processes are involved in the formation of the phage membrane.

C. Membranes of Eukaryotic Cells

1. Plasma Membranes

The transbilayer distribution of phospholipids in erythrocyte membranes was studied most thoroughly. The data obtained with different independent techniques coincide in that the choline-containing phospholipids are located predominantly in the outer monolayer of the erythrocyte membrane, whereas the aminophospholipids are present on the inner surface.[17,23,24,70,72,98] A similar asymmetry was established for the plasma membranes of platelets[101] and fibroblasts,[30-32] but not for that of myoblasts[50] and intestinal brush border[66] (see Section IV.C.3). Therefore, it is not clear, as yet, how widely this principle is valid for membranes of other eukaryotic cells.

2. *The Inner Membrane of Mitochondria*

The main phospholipids of the inner mitochondrial membrane are PC, PE, and cardiolipin. By independent data of two research groups it was established that PC is localized predominantly on the cytoplasmic surface of the membrane, wheras PE and cardiolipin are present predominantly on the membrane surface exposed to the mitochondrial matrix.[45,46] Although the latter two phospholipids in model systems can adopt a nonbilayer (hexagonal II) configuration,[10] they are present practically entirely in the bilayer configuration in the inner mitochondrial membrane at physiological temperature[15] and the rate of flip-flop is low ($\tau_{1/2}$ = 24 hr).[18]

3. *The Brush Border Membrane of Rabbit Small Intestine*

The microvillus membrane of the small intestine is a highly specialized part of plasma membrane responsible for the digestive and absorptive functions of enterocytes. In collaboration with the Laboratorium of Biochemistry of ETH (Zurich, Switzerland) we studied the phospholipid asymmetry of the brush border membrane of the rabbit small intestine.[66,102] Bee venom phospholipase A_2 and PC exchange protein from beef liver were used as membrane probes. The transbilayer migration of phospholipids in the brush border membrane proved to be relatively fast ($\tau_{1/2} < 20$ min). Quantitative analysis of the kinetics of phospholipase A_2 hydrolysis and intermembrane phospholipid exchange demonstrated that the two major phospholipids of the brush border membrane, PC and PE, are asymmetrically distributed. The major part (~75%) of both phospholipids is located on the inner side of the membrane. The established transmembrane distribution of phospholipids is stable and does not depend on temperature. As a result of the asymmetrical phospholipid distribution, the outer surface of the enterocyte plasma membrane exposed to the lumen of the small intestine is poor in lipids and enriched with proteins. One may surmise that the low content of lipids on the outer surface and rapid phospholipid flip-flop in the microvillus membrane play an essential role in fat transport across the brush border membrane of intestinal enterocytes.

4. *Rat Liver Microsomes*

The literature data concerning phospholipid asymmetry in the rat liver microsomal membrane are highly contradictory. Thus, Nilson and Dallner[73] concluded that all of the PE and 55% of the PC are located on the outer surface of the microsomal membrane. A quite different phospholipid distribution was proposed by Higgins and Dawson[75] who suggested that PE was located predominantly on the inner membrane surface and PC on the outer one. Finally, Sundler et al.[74] arrived to the conclusion that there was no asymmetrical distribution of phospholipids in the microsomal membrane. The difficulties of topological investigations of the microsomal membrane and the reasons for so much conflicting results seem to be connected with the extremely rapid phospholipid flip-flop in this membrane.[76,103] We have found that conclusions about the asymmetrical distribution of phospholipids in microsomes can be drawn on the basis of the kinetical analysis of data of phospholipase hydrolysis.[69] Using this approach we established that the two major phospholipids of microsomes are distributed asymmetrically: the main part of PC (60%) is located in the inner side of the membrane, whereas $^2/_3$ of PE is present on its outer surface. At the same time phospholipid molecules exchange very rapidly between the outer and inner surfaces with $\tau_{1/2} \sim 1.5$ to 3.5 min at 23°C. The flip-flop rate is strongly dependent on the temperature, but the transmembrane phospholipid distribution does not change with temperature.

In order to elucidate the factors causing such rapid phospholipid flip-flop in liver microsomes we modified chemically and enzymatically both the lipids and proteins of the microsomal membrane.[69] The transmembrane migration of phospholipids was not affected by lipid peroxidation,[104] by formation of lyso-PC and fatty acids during phospholipase hydrolysis, and by selective removal from the microsomal membrane of more than 90% of PE, which tends to form nonbilayer structures in membranes.[11] The high flip-flop rate of PC was retained also after dinitrophenylation of the membrane with 1-fluoro-2,4-dinitrobenzene and after modification of 40 to 50% of SH-groups in membrane proteins with chemical reagents. At the same time partial digestion of membrane proteins with pronase resulted in a marked decrease of the rate of PC transmembrane migration. Total inhibition of flip-flop was achieved after treatment of microsomes with the cross-linking reagent, 1,3-difluoro-4,6-dinitrobenzene.[69]

Thus, our experiments demonstrated that various modifications of the lipid phase of the microsomal membrane did not influence significantly the transmembrane migration of PC, whereas by modifying protein components of the membrane it is possible to slow down or even to inhibit completely the flip-flop process. On the basis of these findings it seems probable that the extremely rapid PC flip-flop in the microsomal membrane requires participation of some integral membrane protein(s).

In order to identify the proteins responsible for the rapid phospholipid flip-flop, we solubilized the microsomal membrane by detergents and fractionated them on Sepharose 4B. The protein fractions obtained differed markedly in their ability to stimulate the PC flip-flop in liposomes made up from total microsomal lipids. Maximal flip-flop stimulation was observed only with fractions containing cytochrome P-450.

In order to prove that cytochrome P-450, is, indeed, able to enhance phospholipid flip-flop, we incorporated purified cytochrome P-450 into liposomes consisting of total microsomal phospholipids and studied the obtained proteoliposomes with phospholipase A_2 and PC exchange protein.[105,106]

The results demonstrated that in protein-free liposomes as well as in proteoliposomes containing cytochrome b_5, the flip-flop rate was low and those PC molecules which were located on the inner side of the bilayer remained practically nonaccessible for extrinsic probes. On the contrary, in the presence of cytochrome P-450 all the PC was easily accessible for the probes due to a marked increase of the rate of phospholipid translocation across the membrane. The striking acceleration of flip-flop in the presence of cytochrome P-450 in model systems suggests this protein to be responsible for the rapid transmembrane migration of phospholipids in liver microsomes.

5. *Rat Hepatoma Microsomes*

It is well known that malignant transformation of cells results in significant changes of phospholipid composition and enzymatic activity of cellular membranes.[107] Whether these changes are accompanied with alterations of the topology of the membrane phospholipids is not known. Therefore, we undertook a study of the transmembrane distribution and flip-flop of PC and PE in microsomes of rat hepatoma 27 (a transplanted hepatoma initially induced by *N*-nitrosoethylamine).[108]

The results of probing hepatoma microsomes with phospholipase A_2 revealed an asymmetrical distribution of PC and PE across the membrane. However, the pattern of PC distribution was opposite to that in rat liver microsomes: in the hepatoma microsomes this phospholipid was found predominantly on the outer side of the membrane. The most striking difference was observed in comparison of the flip-flop rates. As can be judged from the data of phospholipase treatment, the rate of flip-flop in hepatoma microsomes was more than two orders of magnitude lower ($\tau_{1/2} > 5$ hr) than in microsomes of normal liver.

The reasons why the phospholipid transmembrane distribution and their flip-flop differ so much in microsomes from liver and hepatoma are not clear. Possibly they are connected with changes in the lipid phase of the membrane, since the hepatoma and liver microsomes differ in their lipid composition.[107,108] This proposal is supported by the fact that the microviscosity of the hepatoma microsomal membrane[109] and the packing of phospholipid molecules in that membrane[69] are different from that of rat liver microsomes. On the other hand, the possibility cannot be excluded that the differences observed are connected with changes in character of protein-lipid interactions in hepatoma microsomes. This interpretation can be supported by the changes in activity of several membrane-bound enzymes in hepatoma microsomes.[110] In particular, no cytochrome P-450 activity was detected in the hepatoma 27 microsomes.[111]

V. CONCLUDING REMARKS

A fairly large number of membranes have been studied with respect to the transmembrane distribution of their phospholipids. In most cases the distribution was found to be asymmetric; sometimes conflicting data have been reported. So, far, the most consistent results have been obtained for the erythrocyte membrane where the cholinephosphatides unequivocally have been demonstrated to prefer the outer surface, whereas the aminophospholipids were found preferentially in the inner monolayer. Evidence has been presented that the same type of phospholipid asymmetry occurs also in plasma membranes of some other blood cells. Asymmetrical transmembrane distributions are now established also for the phospholipids of intracellular membranes. In inner mitochondrial membranes PC is found predominantly at the cytoplasmic side and PE at the matrix side. In rat liver endoplasmic reticulum the distribution of the two phospholipids is opposite: a major part of PE is found at the cytoplasmic surface and PC is located predominantly at the luminal side. The controversial results reported earlier were apparently due to nonadequate techniques used in some cases as well as to the rather fast flip-flop movement in the microsomal membranes.

The dependence of phospholipid flip-flop on the presence of various proteins seems to be a general phenomenon, although up to now only a few membrane proteins have been studied in this respect. Rat liver microsomes seem to be a highly interesting object for such studies, due to the exceptionally high rate of flip-flop, which is catalyzed by cytochrome P-450. Recent results obtained with liver microsomes and intestinal brush border vesicles suggest that at least for metabolically active membranes phospholipid asymmetry should be considered as a dynamic event. The possibility that the transmembrane distribution in those membranes may change with development of the cells, changing physiological conditions and enzymatic activities should be taken into consideration.

REFERENCES

1. **Bergelson, L. D. and Barsukov, L. I.,** Topological asymmetry of phospholipids in membranes, *Science*, 197, 224, 1977.
2. **Rothman, J. E. and Lenard, J.,** Membrane asymmetry, *Science*, 195, 743, 1977.
3. **Op den Kamp, J. A. F.,** Lipid asymmetry in membranes, *Ann. Rev. Biochem.*, 48, 47, 1979.
4. **Etemadi, A.-H.,** Membrane asymmetry. A survey and critical appraisal of the methodology. II. Methods for assessing the unequal distribution of lipids, *Biochim. Biophys. Acta*, 604, 423, 1980.
5. **van Deenen, L. L. M.,** Topology and dynamics of phospholipids in membranes, *FEBS Lett.*, 123, 3, 1981.
6. **Krebs, J. J. R.,** The topology of phospholipids in artificial and biological membranes, *J. Bioenerg. Biomembr.*, 14, 141, 1982.

7. **de Kruijff, B. and Wirtz, K. W. A.,** Induction of a relatively fast transbilayer movement of phosphatidylcholine in vesicles. A ^{13}C NMR study, *Biochim. Biophys. Acta,* 468, 318, 1977.
8. **de Kruijff, B. and van Zoelen, E. J. J.,** Effect of the phase transition on the transbilayer movement of dimiristoyl phosphatidylcholine in unilamellar vesicles, *Biochim. Biophys. Acta,* 511, 105, 1978.
9. **Barsukov, L. I., Victorov, A. V., Vasilenko, I. A., Evstigneeva, R. P., and Bergelson, L. D.,** Investigation of the inside-outside distribution, intermembrane exchange and transbilayer movement of phospholipids in sonicated vesicles by shift reagent NMR, *Biochim. Biophys. Acta,* 598, 153, 1980.
10. **Cullis, P. R. and de Kruijff, B.,** Lipid polymorphism and the functional roles of lipids in biological membranes, *Biochim. Biophys. Acta,* 559, 399, 1979.
11. **Noordam, P. C., van Echteld, C. J. A., de Kruijff, B., and de Gier, J.,** Rapid transbilayer movement of phosphatidylcholine in unsaturated phosphatidylethanolamine containing model membranes, *Biochim. Biophys. Acta,* 646, 483, 1981.
12. **Seelig, J.,** ^{31}P-nuclear magnetic resonance and the head group structure of phospholipids in membranes, *Biochim. Biophys. Acta,* 515, 105, 1978.
13. **de Kruijff, B., van Zoelen, E. J. J., and van Deenen, L. L. M.,** Glycophorin facilitates the transbilayer movement of phosphatidylcholine in vesicles, *Biochim. Biophys. Acta,* 509, 537, 1978.
14. **van Zoelen, E. J. J., de Kruijff, B., and van Deenen, L. L. M.,** Protein-mediated transiblayer movement of lysophosphatidyl-choline in glycophorine-containing vesicles, *Biochim. Biophys. Acta,* 508, 97, 1978.
15. **Gerritsen, W. J., Henricks, P. A. J., de Kruijff, B., and van Deenen, L. L. M.,** The transbilayer movement of phosphatidylcholine in vesicles reconstituted with intrinsic proteins from the human erythrocyte membrane, *Biochim. Biophys. Acta,* 600, 607, 1980.
16. **De Corleto, P. E. and Zilversmit, D. B.,** Exchangeability and rate of flip-flop of phosphatidylcholine in large unilamellar vesicles, cholate dialysis vesicles and cytochrome oxidase vesicles, *Biochim. Biophys. Acta,* 552, 114, 1979.
17. **van Meer, G., Poorthuis, B. J. H. M., Wirtz, K. W. A., op den Kamp, J. A. F., and van Deenen, L. L. M.,** Transbilayer distribution and mobility of phosphatidylcholine in intact erythrocyte membranes. A study with phosphatidylcholine exchange protein, *Eur. J. Biochem.,* 103, 283, 1980.
18. **Rousselet, A., Colbeau, A., Vignais, P. M., and Devaux, P. F.,** Study of transverse diffusion of spin-labeled phospholipids in biological membranes. II. Inner mitochondrial membrane of the rat liver: use of phosphatidylcholine exchange protein, *Biochim. Biophys. Acta,* 426, 372, 1976.
19. **Hirata, F. and Axelrod, J.,** Enzymatic synthesis and rapid translocation of phosphatidylcholine by two methyltransferases in erythrocyte membranes, *Proc. Natl. Acad. Sci. U.S.A.,* 75, 2348, 1978.
20. **Litman, B.** Lipid model membranes. Characterization of mixed phospholipid vesicles, *Biochemistry,* 12, 2545, 1973.
21. **Lentz, B. R. and Litman, B. J.** Effect of head group on phospholipid mixing in small, unilamellar vesicles: mixtures of dimiristoylphosphatidylcholine and dimiristoylphosphatidylethanolamine, *Biochemistry,* 17, 5537, 1978.
22. **Litman, B. J.,** Determination of molecular asymmetry of the phosphatidylethanolamine surface distribution in mixed phospholipid vesicles, *Biochemistry,* 13, 2844, 1974.
23. **Bretcher, M. S.,** Phosphatidylethanolamine: differential labeling in intact cells and cell ghosts of human erythrocytes by a membrane impermeable reagent, *J. Mol. Biol.* 71, 523, 1972.
24. **Gordesky, S. E. and Marinetti, G. V.,** The asymmetric arrangement of phospholipids in the human erythrocyte membrane, *Biochem. Biophys. Res. Commun.,* 50, 1027, 1973.
25. **Gordesky, S. E., Marinetti, G. V., and Love, R.,** The reaction of chemical probes with erythrocyte membrane, *J. Membr. Biol.,* 20, 111, 1975.
26. **Marinetti, G. V. and Love, R.,** Differential reaction of cell membrane phospholipids and proteins with chemical probes, *Chem. Phys. Lipids,* 16, 239, 1976.
27. **Haest, C. W. M. and Deuticke, B.,** Experimental alteration of phospholipid-protein interactions within the human erythrocyte membrane. Dependence on glycolytic metabolism, *Biochim. Biophys. Acta,* 401, 468, 1975.
28. **Haest, C. W. M., Kamp, D., and Deuticke, B.,** Penetration of 2,4,6-trinitrobenzenesulfonate into human erythrocytes. Consequences for studies on phospholipid asymmetry, *Biochim. Biophys. Acta,* 640, 535, 1981.
29. **Schick, P. K., Kurica, K. B., and Chacko, G. K.,** Location of phosphatidylethanolamine and phosphatidylserine in the human platelet plasma membrane, *J. Clin. Invest.,* 57, 1221, 1976.
30. **Mark-Malhoff, D., Marinetti, G. V., Hare, J. D., and Meisler, A.,** Elevation of a threonine phospholipid in polyoma virus transformed embryo fibroblasts, *Biochem. Biophys. Res. Commun.,* 75, 589, 1977.
31. **Sandra, A. and Pagano, R. F.,** Phospholipid asymmetry in LM cell plasma membrane derivatives: polar head group and acyl chain distributions, *Biochemistry,* 17, 332, 1978.
32. **Fontaine, R. N. and Schroeder, F.,** Plasma membrane aminophospholipid distribution in transformed murine fibroblasts, *Biochim. Biophys. Acta,* 558, 1, 1979.

33. **Fontaine, R. N., Harris, R. A., and Schroeder, F.,** Neuronal membrane lipid asymmetry, *Life Sci.*, 24, 394, 1979.
34. **Fontaine, R. N., Harris, R. A., and Schroeder, F.,** Aminophospholipid asymmetry in murine synaptosomal plasma membrane, *J. Neurochem.*, 34, 269, 1980.
35. **Rothman, J. E. and Kennedy, E. P.,** Asymmetrical distribution of phospholipids in the membrane of *Bacillus megaterium*, *J. Mol. Biol.*, 110, 603, 1977.
36. **Rothman, J. E. and Kennedy, E. P.,** Rapid transmembrane movement of newly synthesized phospholipids during membrane assembly, *Proc. Natl. Acad. Sci. U.S.A.*, 74, 1821, 1977.
37. **Bishop, D. G., op den Kamp, J. A. F., and van Deenen, L. L. M.,** The distribution of lipids in the protoplast membranes of *Bacillus subtilis*. A study with phospholipase C and trinitrobenzenesulphonic acid, *Eur. J. Biochem.*, 80, 381, 1977.
38. **Shimada, K. and Murata, N.,** Chemical modification by trinitrobenzenesulfonate of a lipid and proteins of intracytoplasmic membranes isolated from *Chromatium vinosum* and *Azotobacter vinelandi*, *Biochim. Biophys. Acta*, 455, 605, 1976.
39. **Paton, J. C., May, B. K., and Elliot, W. H.,** Membrane phospholipid asymmetry in *Bacillus amyloliquefaciens*, *J. Bacteriol.*, 135, 393, 1978.
40. **Demant, E. J. F., op den Kamp, J. A. F., and van Deenen, L. L. M.,** Localization of phospholipids in the membrane of *Bacillus megaterium*, *Eur. J. Biochem.*, 95, 613, 1979.
41. **Kumar, G., Kalva, V. K., and Brodic, A. F.,** Asymmetric distribution of phospholipids in membranes of *Mycobacterium pheli*, *Arch. Biochem. Biophys.*, 198, 22, 1979.
42. **Fong, B. S., Hunt, R. C., and Brown, J. C.,** Asymmetric distribution of phosphatidylethanolamine in the membrane of vesicular stomatitis virus, *J. Virol.*, 20, 658, 1976.
43. **Fong, B. S. and Brown, J. C.,** Asymmetric distribution of phosphatidylethanolamine fatty acyl chains in the membrane of vesicular stomatitis virus, *Biochim. Biphys. Acta*, 510, 230, 1978.
44. **Marinetti, G. V., Senior, A. E., Love, R., and Broadhurst, C. I.,** Reaction of amino-phospholipids of the inner mitochondrial membrane with fluorodinitrobenzene and trinitrobenzenesulfonate, *Chem. Phys. Lipids*, 17, 353, 1976.
45. **Crain, R. C. and Marinetti, G. V.,** Phospholipid topology of the inner mitochondrial membrane of rat liver, *Biochemistry*, 18, 7407, 1979.
46. **Smith, H. G., Fager, R. S., and Litman, B. J.,** Light activated calcium release from sonicated bovine retinal rod outer segment disks, *Biochemistry*, 16, 1399, 1977.
47. **Crain, R. C., Marinetti, G. V., and O'Brien, D. F.,** Topology of amino phospholipids in bovine rod outer segment disk membranes, *Biochemistry*, 17, 4186, 1978.
48. **Drenthe, E. H. S., Klompmarkes, A. A., Bonting, S. L., and Daemen, F. J. M.,** Transbilayer distribution of phospholipids in photoreceptor membranes studied with trinitrobenzene sulfonate alone and in combination with phospholipase D, *Biochim. Biophys. Acta*, 603, 130, 1980.
49. **Vale, M. G. P.,** Localization of the amino phospholipids in sarcoplasmic reticulum membranes revealed by trinitrobenzene sulfonate and fluorodinitrobenzene, *Biochem. Biophys. Acta*, 471, 39, 1977.
50. **Sessions, A. and Horwitz, A. F.,** Differentiation-related differences in the plasma membrane phospholipid asymmetry of myogenic and fibrogenic cells, *Biochim. Biophys. Acta*, 728, 103, 1983.
51. **Udenfriend, S., Stein, S., Böhlen, P., Dairman, W., Leimgruber, W., and Weigele, M.,** Fluorescamine: a reagent for assay of amino acids, peptides, proteins, and primary amines in the picomole range, *Science*, 178, 871, 1972.
52. **Hasselbach, W., Migala, A., and Agostini, B.,** The location of the calcium precipitating protein in the sarcoplasmic membrane, *Z. Naturforsch. Teil C*, 30, 600, 1975.
53. **Hasselbach, W. and Migala, A.** Arrangement of proteins and lipids in the sarcoplasmic membrane, *Z. Naturforsch. Teil C*, 30, 681, 1975.
54. **Hidalgo, C. and Ikemoto, N.,** Disposition of proteins and aminophospholipids in the sarcoplasmic reticulum membrane, *J. Biol. Chem.*, 252, 8446, 1977.
55. **Nakaya, K., Yabuta, M., Inuma, F., Kinoshita, T., and Nakamura, Y.,** Fluorescent labeling of the surface proteins of erythrocyte membrane using cycloheptaamylose-fluorescamine complex, *Biochem. Biophys. Res. Commun.*, 67, 760, 1975.
56. **Krebs, J. J., Hauser, H., and Carafoli, E.,** Asymmetric distribution of phospholipids in the inner membrane of beef heart mitochondria, *J. Biol. Chem.*, 254, 5308, 1979.
57. **Marinetti, G. V.,** Hydrolysis of cardiolipin by snake venom phospholipase A_2, *Biochim. Biophys. Acta*, 84, 55, 1964.
58. **Okyama, H. and Nojima, S.,** Studies on hydrolysis of cardiolipin by snake venom phospholipase A, *J. Biochem. (Tokyo)*, 57, 529, 1965.
59. **Adamich, M. and Dennis, E. A.** Exploring the action and specificity of cobra venom phospholipase A_2 towards hyman erythrocytes, ghosts membranes and lipid mixtures, *J. Biol. Chem.*, 253, 5121, 1978.

60. **Barsukov, L. I., Parfenjeva, A. M., Victorov, A. V., Shapiro, Yu. E., Bystrov, V. F., and Bergelson, L. D.**, Study of the membrane-lytic activity of lysolecithin and its diol analogs, *Biofizika (Russian)*, 19, 456, 1974.
61. **Sundler, R., Alberts, A. W., and Vagelos, P. R.** Phospholipases as probes for membrane sidedness. Selective analysis of the outer monolayer of asymmetric bilayer vesicles, *J. Biol. Chem.*, 253, 5299, 1978.
62. **Wilschut, J. C., Regts, J., and Scherphof, G.** Action of phospholipase A_2 on phospholipid vesicles. Preservation of the membrane permeability barrier during asymmetric bilayer degradation, *FEBS Lett.*, 98, 181, 1979.
63. **Wilbers, K. H., Haest, C. W. M., von Bentheim, M., and Deuticke, B.**, Influence of enzymatic phospholipid cleavage on the permeability of the erythrocyte membrane. I. Transport of non-electrolytes via the lipid domain, *Biochim. Biophys. Acta*, 554, 388, 1979.
64. **van Meer, G., de Kruijff, B., op den Kamp, J. A. F., and van Deenen, L. L. M.**, Preservation of bilayer structure in human erythrocytes and erythrocyte ghosts after phospholipase treatment. A ^{31}P-NMR study, *Biochim. Biophys. Acta*, 596, 1, 1980.
65. **Barsukov, L. I., Kulikov, V. I., Ivanova, V. P., Bergelson, L. D., Victorov, A. V., Vasilenko, I. A., and Evstigneeva, R. P.**, ^{31}P NMR study of the dynamic structure and cation permeability of microsomal membranes, *III Sov. Swed. Symp. Phys.-Chem. Biol. (Tbilisi)*, Abstr. p. 66, 1981.
66. **Barsukov, L. I., Spiess, M., and Bergelson, L. D.**, Transmembrane distribution of phosphatidylcholine and phosphatidylethanolamine in the microvillus membrane of rabbit small intestine (Russian), *Dokl. Akad. Nauk S.S.S.R.*, 266, 1014, 1982.
67. **Barsukov, L. I., Kulikov, V. I., and Bergelson, L. D.**, Study of molecular organization of biological membranes with use of lipid transfer proteins. Phospholipid asymmetry in the cytoplasma membrane of Micrococcus lysodeiktocus (Russian), *Biokhimia*, 42, 1539, 1977.
68. **Barsukov, L. I., Kisel, M. A., Ivanova, V. P., and Bergelson, L. D.** Immobilized phospholipase A_2 as a membrane probe (Russian), *Bioorg. Khim.*, 6, 923, 1980.
69. **Barsukov, L. I., Kulikov, V. I., Ivanova, V. P., and Bergelson, L. D.**, Phospholipid dynamics and transbilayer distribution in rat liver and hepatoma microsomes, *Stud. Biophys.*, 90, 147, 1982.
70. **Zwaal, R. F. A., Roelofsen, B., Comfurius, P., and van Deenen, L. L. M.**, Organization of phospholipids in human red cell membranes as detected by the action of various purified phospholipases, *Biochim. Biophys. Acta*, 406, 83, 1975.
71. **Roelofsen, B. and Zwaal, R. F. A.**, The use of phospholipases in the determination of asymmetric phospholipid distribution in membranes, in *Methods in Membrane Biology*, Vol. 7, Korn, E. D., Ed., Plenum Press, New York, 1976, 147.
72. **Verklej, A. J., Zwaal, R. F. A., Roelofsen, B., Comfurius, B., Kastelijn, D., and van Deene, L. L. M.**, The asymmetric distribution of phospholipids in the human red cell membrane, *Biochem. Biophys. Acta*, 323, 178, 1973.
73. **Nilson, O. S. and Dallner, G.**, Enzyme and phospholipid asymmetry in liver microsomal membranes, *J. Cell. Biol.*, 72, 568, 1977.
74. **Sundler, R., Sarcione, S. L., Alberts, A. W. and Vagelos, R. P.**, Evidence against phospholipid asymmetry in intracellular membranes from liver, *Proc. Natl. Acad. Sci. U.S.A.*, 74, 3550, 1977.
75. **Higgins, J. A. and Dawson, R. M. C.**, Asymmetry of the phospholipid bilayer of rat liver endoplasmic reticulum, *Biochim. Biophys. Acta*, 470, 342, 1977.
76. **van den Besselaar, A. M. H. P., de Kruijff, B., van den Bosch, H., and van Deenen, L. L. M.**, Phosphatidylcholine mobility in liver microsomal membranes, *Biochim. Biophys. Acta*, 510, 242, 1978.
77. **Bevers, E. M., Singal, S. A., op den Kamp, J. A. F., and van Deenen, L. L. M.**, Recognition of different pools of phosphatidylglycerol in intact cells and isolated membranes of *Acholeplasma laidlawaii* by phospholipase A_2, *Biochemistry*, 16, 1290, 1977.
78. **Paton, J. C., May, B. K., and Elliot, W. H.**, Membrane phospholipid asymmetry in *Bacillus amyloliquefaciens*, *J. Bacteriol.*, 135, 393, 1978.
79. **Bergelson, L. D., Barsukov, L. I., Dubrovina, N. I., and Bystrov, V. F.**, Differentiation of inner and outer surfaces of phospholipid membranes of NMR spectroscopy (Russian), *Dokl. Akad. Nauk S.S.S.R.*, 194, 708, 1970.
80. **Bystrov, V. F., Shapiro, Yu. E., Viktorov, A. V., and Barsukov, L. I., and Bergelson, L. D.**, ^{31}P-NMR signals from inner and outer surfaces of phospholipid membranes, *FEBS Lett.*, 25, 337, 1972.
81. **Shapiro, Yu. E., Viktorov, A. V., Volkova, V. I., Barsukov, L. I., Bystrov, V. F., and Bergelson, L. D.**, ^{13}C NMR investigation of phospholipid membranes with the aid of shift reagents, *Chem. Phys. Lipids*, 14, 227, 1975.
82. **Bergelson, L. D.**, Paramagnetic hydrophilic probes in NMR investigations of membrane systems, in *Methods in Membrane Biology*, Vol. 9, Korn, E. D., Ed., Plenum Press, New York, 1978, 275.
83. **Barsukov, L. I., Shapiro, Yu. E., Victorov, A. V., Volkova, V. I., Bystrov, V. G., and Bergelson, L. D.** Conformation of phosphorylcholine groups in phospholipid membranes (Russian), *Bioorg. Khim.*, 2, 1404, 1976.

84. **Seelig, J., Gally, H. U., and Wohlgemuth, R.**, Orientation and flexibility of the choline head group in lecithin bilayers, *Biochim. Biophys. Acta*, 467, 109, 1977.
85. **Berden, J. A., Barker, R. W., and Radda, G. K.**, NMR studies on phosholipid bilayers. Some factors affecting lipid distribution, *Biochim. Biophys. Acta*, 375, 186, 1975.
86. **de Kruijff, B., Rietveld, A., and van Echtfeld, C. J. A.**, ^{13}C-NMR detection of lipid polymorphism in model and biological membranes, *Biochim. Biophys. Acta*, 600, 597, 1980.
87. **de Kruijff, B. and Baken, P.**, Rapid transbilayer movement of phospholipids induced by an asymmetrical perturbation of the bilayer, *Biochim. Biophys. Acta*, 507, 38, 1978.
88. **Barsukov, L. I., Shapiro, Yu. E., Viktorov, A. V, Volkova, V.I., Bystrov, V. F., and Bergelson, L. D.** Study of intervesicular phospholipid exchange by NMR, *Biochem. Biophys. Res. Commun.*, 60, 196, 1974.
89. **Rothman, J. E. and Dawidowicz, E. A.** Asymmetric exchange of vesicle phospholipids catalysed by the phosphatidylcholine exchange protein. Measurement of inside-outside transitions, *Biochemistry*, 14, 2809, 1975.
90. **Johnson, L. W., Hughes, M. E., and Zilversmith, D. B.**, Use of phospholipid exchange protein to measure inside-out-side transposition in phosphatidylcholine liposomes, *Biochim. Biophys. Acta*, 375, 176, 1975.
91. **Bloj, B. and Zilversmit, D. B.**, Asymmetry and transposition rates of phosphatidylcholine in rat erythrocyte ghosts, *Biochemistry*, 15, 1277, 1976.
92. **Rothman, J. E., Tsai, D. K., Dawidowicz, E. A., and Lenard, J.**, Transbilayer phospholipid asymmetry and its maintainance in the membrane of influenza virus, *Biochemistry*, 15, 2361, 1976.
93. **Barsukov, L. I., Shapiro, Yu. E., Viktorov, A. V.- Volkova, V. I., Bystrov, V. G., and Bergelson, L. D.**, Intervesicular phospholipid exchange. An NMR study, *Chem. Phys. Lipids*, 14, 211, 1975.
94. **Barsukov, L. I., Kulikov, V. I., and Bergelson, L. D.**, Lipid transfer proteins as a tool in the study of membrane structure. Inside-outside distribution of the phospholipids in the protoplasmic membrane of *Micrococcus lysodeikticus*, *Biochem. Biophys. Res. Commun.*, 71, 704, 1976.
95. **Barsukov, L. I., Kulikov, V. I., and Bergelson, L. D.**, Study of the structure of biological membranes with phospholipid transfer proteins (Russian), *Dokl. Akad. Nauk S.S.S.R.*, 228, 974, 1976.
96. **Barsukov, L. I., Kulikov, V. I., Simakova, I. M., Tikhonova, G. V., Ostrovski, D. N., and Bergelson, L. D.**, Lipid composition changes of cellular membranes with the use of lipid transfer proteins. Properties of protoplast membranes of I *Micrococcus lysodeikticus* containing phosphatidylcholine (Russian), *Biokhimia*, 42, 2099, 1977.
97. **Barsukov, L. I., Kulikov, V. I., Simakova, I. M., Tikhonova, G. V., Ostrovski, D. N., and Bergelson, L. D.**, Manipulation of phospholipid composition of membranes with the aid of lipid exchange proteins. Incorporation of phosphatidylcholine into protoplasts of *Micrococcus lysodeikticus*, *Eur. J. Biochem.*, 90, 331, 1978.
98. **Crain, R. C. and Zilversmit, D. B.**, Two nonspecific phospholipid exchange proteins from beef liver. II. Use in studying the asymmetry and transbilayer movement of phosphatidylcholine, phosphatidylethanolamine, and sphingomyelin in intact rat erythrocytes, *Biochemistry*, 19, 1440, 1980.
99. **Kramer, R. M. and Branton, D.**, Retention of lipid asymmetry in membranes on polylysine-coated polyacrylamide beads, *Biochim. Biophys. Acta*, 556, 219, 1979.
100. **Schäfer, R., Hinnen, R., and Franklin, R. M.**, Structure and synthesis of a lipid-containing bacteriophage. Properties of the structural proteins and distribution of the phospholipid, *Eur. J. Biochem.*, 50, 15, 1974.
101. **Chap, H. J., Zwaal, R. F. A., and van Deenen, L. L. M.**, Action of highly purified phospholipases on blood platelets. Evidence for an asymmetric distribution of phospholipids in the surface membrane, *Biochem. Biophys. Acta*, 467, 146, 1977.
102. **Barsukov, L. I., Hauser, H., Hasselbach, H.-J., and Semenza, G.**, Phosphatidylcholine exchange between brush border membrane vesicles and sonicated liposomes, *FEBS Lett.*, 115, 189, 1980.
103. **Zilversmit, D. B. and Hughes, M. E.**, Extensive exchange of rat liver microsomal phospholipids, *Biochim. Biophys. Acta*, 469, 99, 1977.
104. **Barsukov, L. I., Kulikov, V. I., Muzia, G. I., and Bergelson, L. D.**, Transmembrane phosphatidylcholine migration and lipid peroxidation in rat liver microsomes (Russian), *Biokhimia*, 47, 1437, 1982.
105. **Barsukov, L. I., Kulikov, V. I., Bachmanova, G. I., Archakov, A. I., and Bergelson, L. D.**, Cytochrome P-450 facilitates phosphatidylcholine flip-flop in proteoliposomes, *FEBS Lett.*, 144, 337, 1982.
106. **Barsukov, L. I., Kulikov, V. I., Bachmanova, G. I., Arachkov, A. I., and Bergelson, L. D.**, Rapid transmembrane migration of phosphatidylcholine under influence of cytochrome P-450 (Russian), *Biokhimia*, 47, 2055, 1982.
107. **Dyatlovitskaya, E. V., Timofeeva, N. G., Gorkova, N. P., and Bergelson, L. D.**, Cholesterol, sphingomyelin and phosphatidylcholine ratios in cellular membranes of hepatomas and liver (Russian), *Biokhimia*, 40, 1315, 1975.
108. **Schwemberger, I. N.**, *Cell Heredity and Malignant Growth*, (Russian), Nauka Moscow, 1966, 154.

109. **Dyatlovitskaya, E. V., Einisman, L. I., Golostchapov, A. N., and Burlakova, E. B.,** Study of microviscosity in intracellular membranes of rat liver and hepatoma (Russian), *Biofizika*, 23, 1104, 1978.
110. **Dyatlovitskaya, E. V., Lemenovskaya, A. F., and Bergelson, L. D.,** Use of protein-mediated lipid exchange in the study of membrane-bound enzymes. The lipid dependence of glucose-6-phosphatase. *Eur. J. Biochem.*, 99, 605, 1979.
111. **Dyatalovitskaya, E. V., Petkova, D. X., and Bergelson, L. D.,** Study of the lipid dependence of the activity of the rat liver cytochrome P-450 with use of beef liver phosphatidylcholine exchange protein (Russian), *Biokhimia*, 47, 1366, 1982.
112. **Tsai, K. H. and Lenard, J.,** Asymmetry of influenza virus bilayer membrane demonstrated with phospholipase C, *Nature*, 253, 554, 1975.

Chapter 6

MEMBRANE FLUIDITY: MOLECULAR BASIS AND PHYSIOLOGICAL SIGNIFICANCE

Giorgio Lenaz and Giovanna Parenti Castelli

TABLE OF CONTENTS

I. INTRODUCTION

Since the formulation of the fluid mosaic model of membrane structure,[1] it is customary and fashionable to talk about membrane "fluidity": membranes are considered bi-dimensional fluids[2] constituted by amphipathic lipids (phospho- and glycolipids) in a bilayer arrangement, where proteins and other components diffuse and rotate unless hindered by specific restrictions.[3] Overwhelming evidence has accumulated stating that membrane fluidity is involved in the control of an increasing number of physiological processes and that derangements of normal fluidity are involved in the development of pathological states.[4]

Unfortunately, the term "fluidity" is elusive in its physical meaning and is often misused as it is implicated from studies using different techniques.[5] On one hand, the term is employed as a bulk property, related to the overall thermodynamic or geometric changes occurring during cooperative phase transitions; on the other hand, fluidity is considered at the level of molecular resolution either as a dynamic property related to the motion of the individual components or as a static feature related to the arrangement or order of the molecules. Each of several techniques may detect only one out of these different aspects (Figure 1).

II. THERMOTROPIC BEHAVIOR OF AMPHIPATHIC LIPIDS

Amphipathic lipids have a bilayer arrangement[6] and nonlamellar structures, if present, represent a quantitative minor phase[7] (even if possibly important on a qualitative basis). Many properties of natural membranes depend on such a bilayer structure, and are mimicked by artificial membranes constituted of purified lipids.[8,9] Several types of model lipid membranes are available to study phospholipids and their interactions with proteins.[10] The most used are liposomes, which are either multilamellar systems or single bilayer vesicles of different sizes,[9] and planar lipid bilayers, or black lipid films,[11] i.e., bilayers formed in a hole separating two aqueous compartments. Liposomes are suited to study physical and biochemical properties of membrane components and for reconstitution of membrane-bound enzymes or transport systems, whereas black films are particularly suited to study their electrical properties.

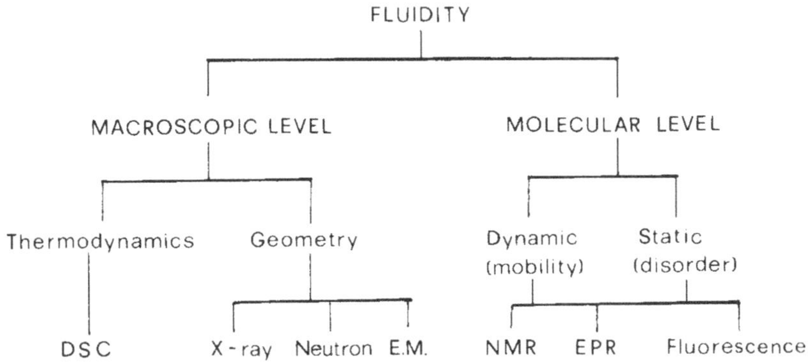

FIGURE 1. Different aspects of lipid fluidity.

A. Phase Transitions and Separations

When a pure crystalline lipid, either anhydrous or hydrated, is heated, an endothermic transition occurs at a well-defined temperature, at which the molecules undergo a change from a relatively immobile gel structure to a disordered liquid-like phase maintaining long-range order but exhibiting considerable disorder at the molecular scale. In particular, NMR (nuclear magnetic resonance) has shown that the molecules pass from a preferential *all-trans* extended disposition to a kinked arrangement of the fatty acyl chains.[12] The transition is directly detected by differential scanning calorimetry (DSC);[13] at constant pressure, the free energy change of melting is zero and the entropy increase (disorder) is accomplished at the expense of an enthalpy change (heat absorption). In DSC, the temperature of the sample and reference material is increased in such a way as to be kept equal; what is measured is the differential power needed to maintain a sample temperature equal to the reference temperature; this power is related to the excess specific heat absorbed during the endothermic transition (Figure 2).

Phase transitions are also detected by X-ray diffraction;[14] studies of pure phospholipid bilayers below and above the transition reveal two well-defined patterns: crystalline packing is characterized by a sharp reflection at 4.15 Å, while in the fluid state a broad maximum occurs at 4.5 Å with considerable spreading. The transition temperature depends on both the polar and apolar phospholipid moieties.[15] The transition is lowered by decreasing the length of the fatty acyl chains or by increasing the number of *cis* double bonds, indicating that Van der Waals attractions among the chains are the main stabilizing factor for cohesion. On the other hand, different lipid classes, even having the same fatty acid composition, melt at different temperatures, indicating that the nature of the headgroup affects the cohesion of the chains[13] (Table 1).

In the case of binary phospholipid mixtures, two transitions may appear, with solid and fluid domains existing in the same bilayer in the region between the two endothermic peaks[13] (Figure 3). This phase separation has been shown directly by freeze-fracture electron microscopy;[16] phospholipids above the phase transition show smooth fracture faces, whereas, below the transition, periodic bands appear; for binary mixtures, bands and smooth regions coexist in the same bilayer.

Phase separations have also been detected with the aid of spin labels and fluorescent probes that partition differently from water to fluid or solid bilayers;[10] the spectroscopic properties of such probes are different in water and in the lipids, thereby offering a simple device for detecting phase changes. A representative example is the spin label TEMPO;[17] it has a higher partition coefficient in fluid than in rigid lipids; its EPR spectrum is different in polar and nonpolar environments, with a splitting of the upper field line when the label is present in both environments. By plotting the ratio of the two peak heights against

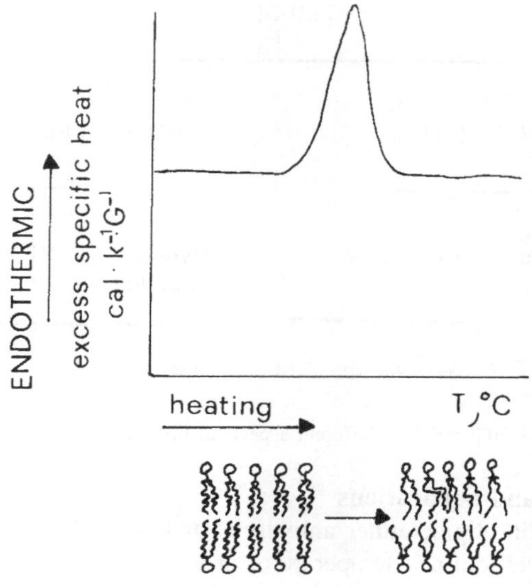

FIGURE 2. Thermal transition of a pure amphipathic lipid, showing a DSC scan and the state of the hydrocarbon chains above and below the transition.

Table 1
TRANSITION TEMPERATURES OF DIFFERENT PHOSPHOLIPIDS

Phospholipid	T_m (°C)	ΔH (kcal/mol)	Phospholipid	T_m(°C)	ΔH (kcal/mol)
PC di 12:0	−1.8	1.7	PC di 18:1 Δ9	−21	7.7
PC di 14:0	23.9	5.4	PC di 18:1 Δ12	−8	7.9
PC di 16:0	41.4	8.7	PC di 18:1 Δ14	7	8.6
PC di 18:0	54.9	10.6	PC di 18:1 Δ16	35	9.6
PC di 18:1 Δ2	41	9.6	PE di 12:0	30.5	3.6
PC di 18:1 Δ4	23	8.2	PE di 14:0	49.5	5.8
PC di 18:1 Δ6	1	7.8	PE di 16:0	63.8	9.6

temperature, two discontinuities are observed in binary mixtures of phospholipids having different transition temperatures. These discontinuities are associated with the onset and the completion of phase separation. The temperature region between the two breaks in a phase diagram corresponds to a temperature range where fluid and solid lipids coexist (Figure 4).

B. Electrostatic Effects

The effect of water on the thermotropic behavior is complex;[15] water tends to increase the cooperative effects and to decrease the transition temperature. A certain amount of water is bound to the polar headgroups and behaves differently from bulk water;[18,19] bound water represents about 20% of the lipid water system. About 10 mol water per mole phospholipid do not freeze at 0°C and are assumed to be bound, albeit with different strength. The extent of hydration depends on the nature of the phospholipids and on the presence of other membrane components: cholesterol, for example, increases the extent of hydration.[20]

The phase transition temperature is also strongly affected by the charge of the headgroup. The transition temperature of synthetic diacylphosphatidic acids falls when the phosphate group attains two negative charges ($pK_2 = 9.5$).[21] Likewise, changes of the electrical charge by salts and binding of polyvalent cations to acidic phospholipids affect their thermotropic

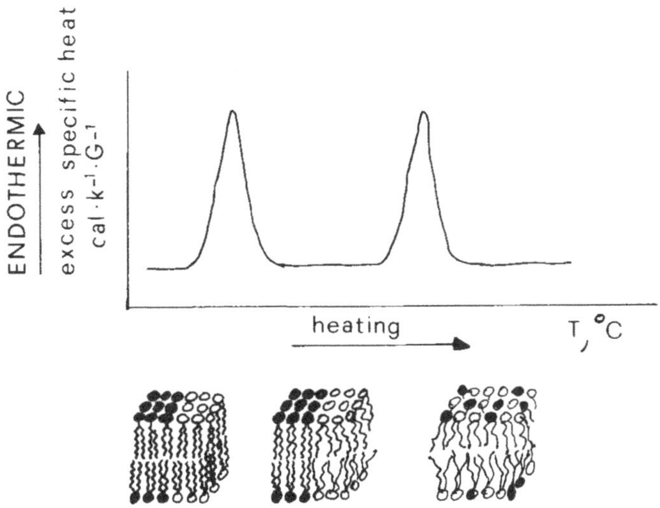

FIGURE 3. Phase separation obtained in a binary mixture of phospholipids, showing a DSC scan and the states of the hydrocarbon chains in the solid, mixed, and fluid region, respectively.

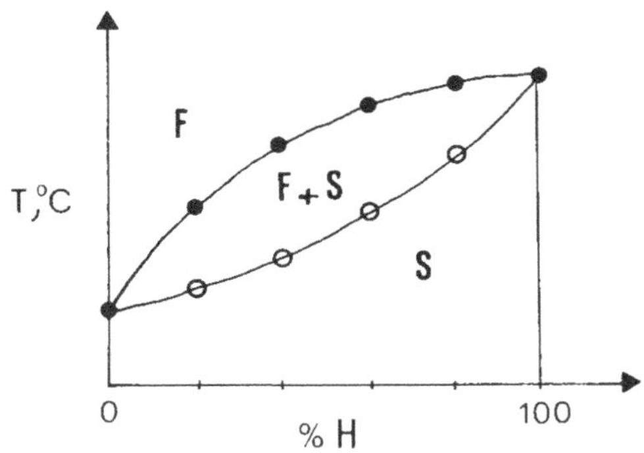

FIGURE 4. Phase diagram constructed from onset and completion temperatures obtained from either DSC or TEMPO partition experiments in binary phospholipid mixtures. In the abscissa, %H indicates the percentage of the higher-melting component; F, fluid; S, solid. In the region between the two curves, fluid and solid bilayers coexist.

properties. Similarly, pH changes or cations induce isothermal phase separations; Ca^{2+} induces phase separation of acidic from neutral phospholipids by binding adjacent anionic groups.[22]

C. Effects of Membrane Constituents on Thermotropic Behavior

Biological membranes have an extremely complex thermotropic behavior, being composed of very heterogeneous lipid mixtures, with further internal heterogeneity of the fatty acyl chains. Moreover, they contain additional components such as cholesterol in many and proteins in all; a variety of other molecules like isoprenoid compounds etc. are also contained in most biomembranes.

Table 2
EFFECTS OF VARIOUS COMPOUNDS ON THERMOTROPIC PROPERTIES OF PHOSPHOLIPID BILAYERS

Compound	ΔH	T_m	Permeability
Cholesterol	−	0	−
Ribonuclease	+	0	+
Polylysine	+	+	+
A_1 myelin protein	−	−	+
Cytochrome *c*	−	−	+
Myelin proteolipid	−	0	+
Gramicidin A	−	0	+
Integral proteins	−	0	+

Note: − = Decrease; + = increase; 0 = no effect.

The effects of all such components have been tested in model systems in the attempt to elucidate the individual reasons for the complex behavior of natural membranes.

1. Cholesterol

Cholesterol is incorporated in the bilayer with its hydrophilic OH group oriented toward water and the hydrophobic sterol ring in contact with the acyl groups. In model lipid systems, cholesterol is incorporated up to a molar ratio of 1:1.[10] Adding cholesterol to dipalmitoyl-lecithin (DPL) progressively decreases the heat absorbed, so that no transition is detected at 1:1 ratio, indicating loss of cooperativity[23] (Table 2). The calorimetric behavior was interpreted as a progressive phase separation of discrete regions of equimolar cholesterol-to-lipid complexes, leaving clusters of free phospholipids which freeze at the normal temperature, until at equimolar ratio no free phospholipids are present. However, computer simulation of sensitive calorimetric studies[24] has shown a nonuniform random array of close-packed cholesterol and phospholipid molecules.[25]

2. Proteins

The interaction of peptides and proteins affects lipids in substantially three ways[26] (Table 2). Basic proteins showing strong electrostatic binding increase the transition enthalpy with little change of the transition temperature; their effect is shared by divalent cations and is due to immobilization of the polar heads. Another group, such as cytochrome *c*, also shows ionic binding, but decreases both the ΔH and temperature of the transition; these proteins may partly penetrate and perturb the bilayer. A third group, including the peptide gramicidin A, and most integral proteins, decreases ΔH without any effect on transition temperature: the effect is mimicked by cholesterol; such proteins undergo complete penetration with hydrophobic interactions to a statistically limited number of lipid molecules, leaving the rest of the bilayer unperturbed. Only the behavior of the unperturbed bilayer is detected by DSC.[10]

3. Others

Among natural membrane molecules, ubiquinone (Q) has been investigated in its thermotropic behavior, for its relevance to its function in energy-conserving membranes.[27] Quinn[28] studied the effects of Qs of different isoprenoid chain length on the thermotropic behavior of pure lipids and found that they affect differently the transition enthalpies. The results suggest that physiological ubiquinones at low percentage are molecularly mixed with phospholipids, but above 20% aggregate in separate phases.

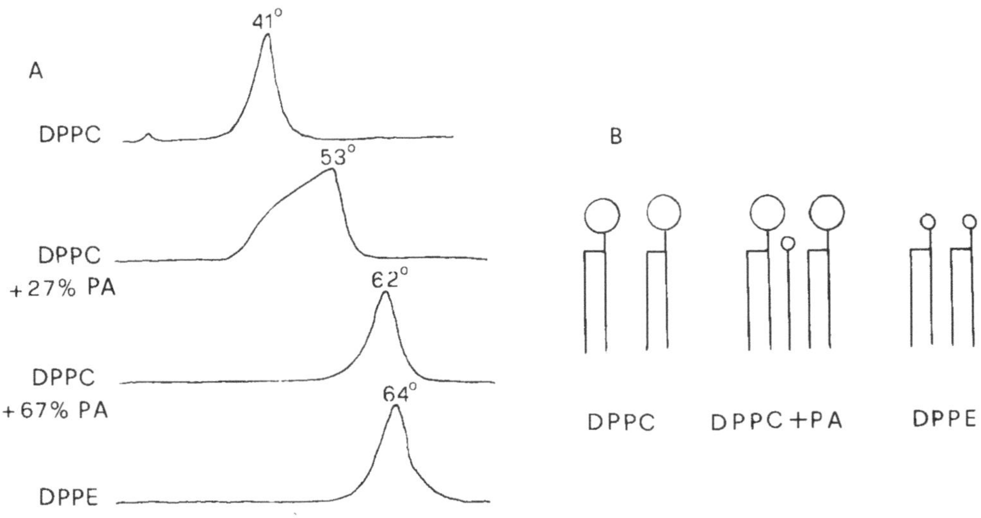

FIGURE 5. (A) Effect of PA on the thermotropic behavior of PC in comparison with PE. DPPC, dipalmitoyl phosphatidylcholine; DPPE, dipalmitoyl phosphatidylethanolamine; PA, palmitic acid. (B) Scheme of the molecular packing of PC, PC + PA, and PE.

D. Effect of Foreign or Nonphysiological Components

Organic solvents and anesthetics dissolve in membranes inducing fluidization; such fluidization is analogous to the depression of freezing point induced by different solutes in water:[29]

$$\Delta T = \frac{RT^2}{Q}(C_1 - C_2)$$

where C_1 and C_2 are the concentrations of the solute in the liquid and the solid, respectively, and Q is the transition enthalpy. This is the basis for the entropic theory of anesthesia: anesthesia is the result of any treatment that increases membrane disorder, and is produced when the free energy of a nonaqueous phase has been changed by a critical amount, irrespective of the method used to induce this change[29] (see Section IV.D).

Contrary to the depression of freezing point induced by small organic solvents, longer hydrocarbon chains increase the transition temperature.[13] For example, while DPL has a transition at 41°C, addition of free palmitic acid (PA) increases the transition up to 62°C at 67% PA in DPL; this temperature agrees well with 64°C for dipalmitoylphosphatidylethanolamine (DPPE); a molecular explanation is given in Figure 5.

III. THE CONCEPT OF MEMBRANE MICROVISCOSITY

A. The Anisotropy of Membrane Systems

Since the lipid bilayer is substantially two-dimensional, its anisotropic nature can be presented, according to Shinitzky,[30] as a combination of two principal viscosity vectors, perpendicular and parallel to the plane of the bilayer. At the level of molecular resolution, the bilayer has further elements of complexity: (1) asymmetric distribution between the two monolayers, (2) lateral phase separation of specific lipid domains, (3) a gradient of segmental motion going from the headgroup to the core of the bilayer. It may be, therefore, somewhat disturbing to consider a membrane isotropic as the term viscosity implies; the need to take account of the local anisotropy in applying macroscopic terms (viscosity) at the molecular level necessarily assumes the bilayer as an isotropic fluid where the microscopic details are averaged out. This is how Shinitzky[30] defines the submacroscopic level, at which a membrane can be treated analogously to an isotropic fluid, despite the complexities mentioned above. The term "microviscosity",[31] therefore, represents the macroscopic simulation of the averaged viscosity in the hydrocarbon core of the bilayer.

The optimal technique for detecting thermal motion and organization of lipids at the molecular level is NMR.[12,32,33] The submacroscopic level, on the other hand, is conveniently investigated by extrinsic probes inserted into membranes and having spectroscopic reporter groups yielding informations on the nature of their environment.[10] To such categories belong the widely used *spin labels* and *fluorescent probes*. To yield significant information, probes must be distributed homogeneously in the system and must not perturb the environment they are designed to explore. This is not always necessarily the case.

B. Viscosity and Order

There is often a fundamental misunderstanding between the concepts of *order* and *viscosity* (and conversely, *disorder* and *fluidity*) of a membrane. Although the two phenomena may well go together, an observed disordering does not necessarily imply an increase in fluidity, and vice versa.[34] Disorder is an increased statistical probability for more distorted chain conformations, while fluidity is proportional to the total number of possible chain conformations. In other words, disorder is a space-averaged property, whereas fluidity is a time-averaged property. Magnetic resonance and fluorescence techniques may alternatively detect one or the other aspect of lipid chain molecular conformation. The rather vague concept of fluidity usually involves both the degree of organization and the rates of movement of the lipids; these must be determined separately if the properties of the lipids are to be described unambiguously.

Motion is usually described by the correlation time τ_c, related to viscosity by:

$$\tau_c = \frac{4}{3} \pi r^3 \frac{\eta}{kT}$$

where r is the rotation radius and η the viscosity.

For the spatial lipid organization, the order parameter S describes the deviation from rigid order within the membrane (fatty acyl chains parallel in *all-trans* configuration). By this definition the order parameter of a crystal is 1 and that corresponding to completely isotropic motion is 0. Since order parameters are geometrical values, not directly related to motion but indicative of space-averaging properties, they cannot yield direct measures of viscosity.

C. Techniques Employed to Investigate Membrane Fluidity

1. NMR

NMR is increasingly used to probe the physical state of membrane lipids.[12,32-34] Any nucleus with a nonzero magnetic moment interacts with electromagnetic radiation in the presence of an external magnetic field and absorbs energy. It is possible to study selected nuclei independent of others (e.g., ^1H, ^2H, ^{13}C, ^{15}N, ^{31}P) and NMR can theoretically give information about the environment and molecular freedom of most of the atoms present in the structure of a membrane. Different nuclei at a given frequency will absorb at different values of magnetic field intensity; in addition, the local magnetic field for a given nucleus is modified by local molecular shielding effects leading to chemical shifts. An effect important in molecules having restricted motion, as in membranes, is dipole-dipole coupling; such coupling is anisotropic and leads to a splitting, depending on the angle with the external magnetic field. Anisotropy also affects the chemical shifts, with different resonance positions at different angles. A completely random solid gives a broad "powder spectrum" due to summation of all possible crystal spectra. Order parameters can easily be determined from

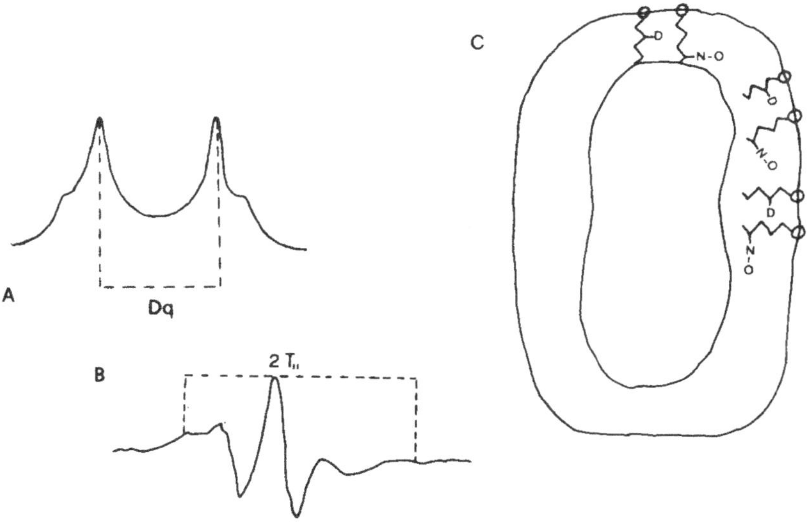

FIGURE 6. Powder spectra in ^{2}H-NMR and in nitroxide EPR. (A) A ^{2}H-NMR spectrum showing the quadrupole splitting Dq when the C–D bond is oriented at all possible angles; (B) EPR spectrum of a nitroxide fatty acid with the N→0 group oriented at all possible angles; (C) a lipid vesicle is represented showing how the C–D bonds and the N→0 bonds assume different orientation.

the separation of the anisotropic spectra. The most interesting property in membrane studies is linewidth: the effect of molecular motion is to time-average the angular dependence of the low-frequency dipole interactions between neighboring nuclei and of chemical shift anisotropy and then to narrow the linewidth (Figure 6).

Various pulse techniques are available to measure relaxation properties of the nuclei following a pulse of external radiation. The lifetime of the spin states is described by the spin-lattice relaxation time T_1, which is the time necessary for the populations of the spin states to come to a new equilibrium in a pulse experiment. The time T_1 is one of the main factors contributing to linewidth in a steady-state NMR absorption; T_1 is grossly inversely proportional to the rotational correlation time (i.e., it is directly proportional to mobility), since the nuclei in a molecule relax more efficiently by decreasing their tumbling rate. Linewidth is, therefore, sensitive to motions defined by the relaxation time; for this reason NMR cannot provide informations on very fast motions.

One advantage of NMR is that it does not need to introduce foreign groups or molecules that can perturb the membrane; ^{13}C and ^{2}H NMR can make use of phospholipids enriched in the two isotopes, whose natural abundance is low. This turns out to be an advantage because selective labeling can provide specific informations on the environment of any one particular atom.

2. EPR and Spin Labels

Unpaired electrons absorb electromagnetic radiation at resonance with an external magnetic field;[32] unfortunately, most molecules in membranes are diamagnetic and cannot give any information by EPR. The drawback has been overcome by use of paramagnetic probes ("spin labels") which are usually nitroxide derivatives. The unpaired electron spin is located in an atomic p orbital on the N atom; the axis of the orbital is normal to the plane defined by the N and O atoms. Since the interaction of the unpaired electron with the nuclear spin of N produces a hyperfine splitting of the spectrum into three lines, their relative positions differ with orientation. As with NMR, if individual labels are randomly oriented in a solid sample, they give rise to a "powder spectrum" with considerable line broadening due to complete summation of all possible orientations. If the spin labels are rotating very rapidly in solution, the anisotropy is completely averaged and a simple three-line spectrum is observed from which a (pseudoisotropic) rotational correlation time can be derived. The motions to which EPR is sensitive are usual in the range of 10^{-9} sec; when tumbling times are slower than 10^{-7} sec, motion is out of range of detection. It is common in spin label studies to have randomly oriented partially immobilized nitroxides yielding spectra intermediate between powder spectra and average isotropic spectra, and order parameters can be obtained (Figure 6).

The spin labels most used for "fluidity" studies are nitroxide derivatives of fatty acids, phospholipids, and sterols, in which the reporter nitroxide group can be situated at different depths in the bilayer.[35-37a]

3. Fluorescence Polarization

Fluorescent probes are molecules having emission properties which respond to the nature of their environment;[38] absorption, relaxation, and emission processes are sensitive to the molecular surroundings of the fluorophore, giving the basis for use of fluorescence to study membrane structure.[39,40] Most investigations concerning membrane fluidity employ fluorescence polarization:[31] if a fluorescent molecule is excited by polarized light, it has a choice to reorient during the time between excitation and emission, and the orientation of the emitted light depends on the degree of molecular reorientation taking place. Polarization is proportional to viscosity according to the Perrin equation:

$$\frac{1}{p} = \frac{1}{3}\left(\frac{1}{p_0} - \frac{1}{3}\right)\left(1 + \frac{RT\tau}{\eta V_0}\right)$$

where p and p_o are the observed polarization and the maximal theoretical value, τ is the excitation lifetime, η is the viscosity, and V_0 is the volume of an equivalent sphere. Fluorescence polarization of diphenylhexatriene (DPH) and other probes is widely employed to detect fluidity in lipid systems and natural membranes.

Another technique used for membrane fluidity is the formation of pyrene excimers (excited dimers);[41] a characteristic emission peak, due to excimer formation, is proportional to the probability of pyrene molecules to undergo collisions, i.e., to fluidity.[42]

4. Other Techniques

Also, infrared[43] and Raman spectroscopy[44,45] have been used to study lipid mobility in bilayers and membranes. The IR band at 720 cm^{-1}, indicating *trans* CH_2 groups, decreases in intensity on going from crystal to liquid crystal. The Raman spectra show abrupt variations in intensity in the 1100-cm^{-1} region at the transition.

D. Intramolecular Motion and Order of Lipids in Bilayers

The intramolecular motion of phospholipids in pure bilayers or biomembranes has been largely investigated by EPR spin labels and either ^{13}C or ^{2}H NMR.[33,37] The presence of a large number of C–C single bonds allows at each position two *gauche* and one *trans* conformer; the relative populations of each conformer may vary greatly with position along the chain. In the crystalline state the chains are in an *all-trans* conformation, in which they are packed as long and thin as possible; above the transition, the assumption of several *gauche* rotamers induces lateral expansion and decrease of thickness (Figure 7). The state of molecular organization of the chain can be quantitated in terms of the bond order parameter S which is the ensemble average of the function $(3\cos^2\vartheta - 1)/2$, where ϑ is the angle between a C–H bond and the long axis of ordering.[12] Of course the order parameter does not provide insight into the *rates* at which the bonds rotate and the entire molecule moves.

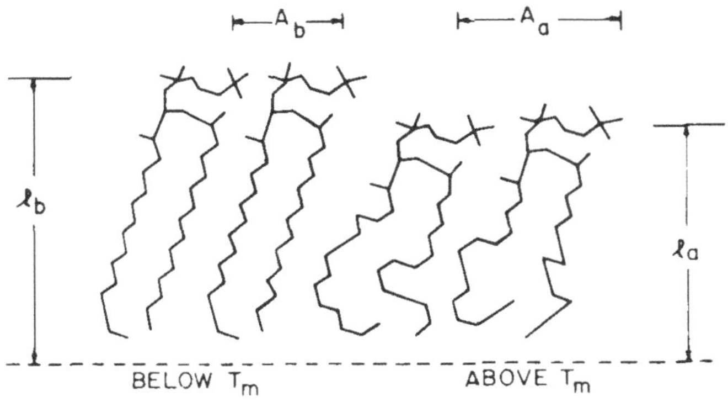

FIGURE 7. Representation of the bilayer below Tm showing the extended *all-trans* chains, and above Tm showing some *gauche* rotamers. Note the lateral expansion and decrease in thickness above Tm. (From Wilkinson, P. A. and Nagle, J. F., *Liposomes: From Physical Structure to Therapeutic Applications*, Knight, C. G., Ed., Elsevier, Amsterdam, 1981, chap. 9. With permission.)

Magnetic resonance methods have revealed the existence of a *fluidity gradient* from the surface to the core of the bilayer;[34] the most reliable data have been obtained with deuterium NMR of selectively deuterated lipids. For both saturated and unsaturated phospholipids, the first eight to ten segments from the carbonyl atom exist in a state of relatively constant order, whereas the order parameter decreases gradually to almost zero at the methyl terminal group of the chain. In unsaturated phospholipids the order profile is slightly shifted to lower order, with a sudden trough at the position of the double bond (Figure 8).[34] This does not mean that the effect of the double bond is to "fluidize" the membrane, because the decrease in order is a consequence of the altered geometry for intrinsic deviation from the *all-trans* configuration around the double bond. Correction for the natural bend induced by the planar *cis* double bond yields an order profile indistinguishable from that of a saturated fatty acid;[46] actual T_1 relaxation time measurements reveal that the *mobility decreases* in proximity of the double bond.

The data obtained with pure deuterated phospholipids and with biological membranes from microorganisms supplemented with specifically deuterated fatty acids are strikingly similar,[47,48] indicating that pure lipid bilayers are remarkably good models for biological membranes.

IV. FACTORS AFFECTING MEMBRANE FLUIDITY

Lipid fluidity is not a constant attribute but can be modulated by a variety of endogenous and exogenous factors[4] (Table 3). Most of such factors can have physiological importance or be involved in the development of pathological states. In general terms, physical factors are homogeneous and instantaneous, whereas chemical factors exert regional effects after minutes or hours.[49]

A. Physical Factors

Viscosity decreases with increasing temperature according to the equation

$$\eta = A\ e^{\Delta E/RT}$$

where ΔE is the flow activation energy, i.e., the energy required to dissociate the flowing unit from the bulk; in a plot of η vs. reciprocal temperature, ΔE can be calculated from the

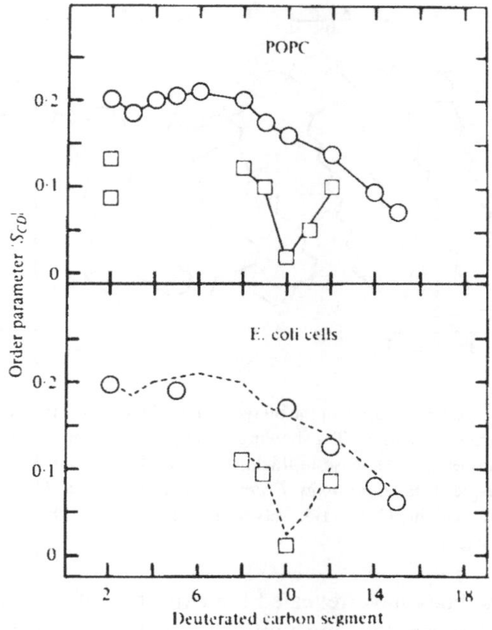

FIGURE 8. Comparison between a biological membrane and a synthetic unsaturated lipid. Variation of the deuterium order parameter $|S_{CD}|$ with segment position. POPC, 1-palmitoyl-2-oleoyl-PC. (○), Deuterium attached at palmitic acyl chains; (□) deuterium attached at oleic acyl chains. (Reprinted with permission from Reference 47. Copyright 1979 American Chemical Society.)

slope[50] (Figure 9). Temperature plots of fluorescent probes or spin label motion parameters are usually linear in pure phospholipids, except for a sudden change at the transition temperature. Discontinuities in the temperature dependence of viscosity or related parameters have also been observed in natural membranes and have often been related to phase transitions or separations[51-53] (Figure 10); however, lack of coincidence with the calorimetric transitions throws doubt on such simple interpretation[54] (see later).

Viscosity is enhanced by pressure in a similar exponential manner; both hydrostatic pressure and intracellular osmotic pressure change η.[49] The pressure effect of counteracting the action of general anesthetics has been taken as an indication that anesthetics act by enhancing membrane fluidity.[55]

Electrostatic effects influence the transition temperature by changing the packing density of bilayer phospholipids caused by electrostatic repulsion or attraction or, in the case of uncharged lipids, by changes of hydration water. The decrease of packing density by hydration is compensated by a tilting of the hydrocarbon chains[56] with decrease of bilayer thickness. Such chains undergo the phase transition at a lower temperature. Likewise, changes in pH in the physiological range induce changes in membrane viscosity.[49]

Lipid viscosity also increases with membrane potential[57] in a nonlinear manner, irrespective of the potential direction; this significant change has been interpreted either as an energy barrier of the potential to the lipid flow or a decrease in lipid-free volume due to the pressure induced by the electrical potential. Changes in lipid viscosity mediated by transient changes in membrane potential may regulate the physiological processes associated with nerve conductance.

Table 3
FACTORS AFFECTING LIPID FLUIDITY IN MEMBRANES

Intrinsic	Indirect
Lipid composition[13,49]	Hormones[49,54,154,187-189,244]
Chain length	Metabolic processes[245]
Unsaturation	Various cellular processes[3,49,156,191-194]
Polar head	Neoplasia[193,195,213-215]
Cholesterol[10,49,50,240]	
Ubiquinone[28,60,61]	Physical
Lipid-soluble vitamins[241]	Temperature[49-54]
Integral proteins[69-83,242,243]	Pressure[49,55]
Peripheral proteins[10,26]	Membrane potential[57]
Cytoskeleton[3]	
Hydration[15,56]	
Ions[22,97]	Extrinsic
pH[21,49]	Detergents[97]
Lysophospholipids[207]	Anesthetics[29,53,97-105]
Fatty acids[49,207]	Synthetic polymers[190]

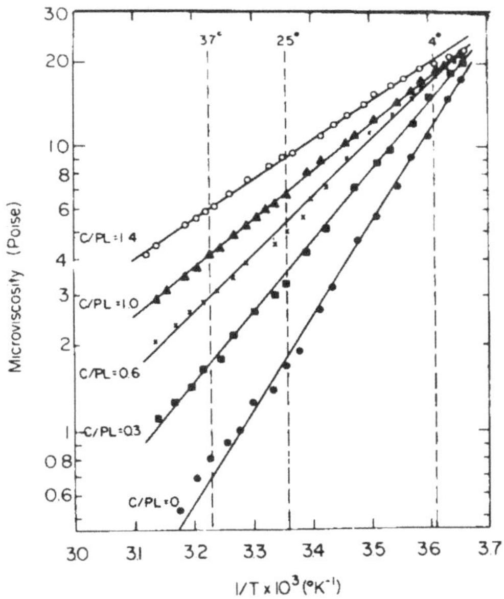

FIGURE 9. Temperature dependence of microviscosity in liposomes made at different cholesterol-to-phospholipid ratios. (Reproduced from Shinitzky, M. and Inbar, M., *Biochim. Biophys. Acta*, 433, 133, 1976. With permission.)

B. Intrinsic Chemical Factors

The main chemical modulators of membrane fluidity are cholesterol, the degree of unsaturation of the acyl chains, the level of sphingomyelin, and the amount of protein.[49] The stationary levels of these modulators in a membrane can change in response to a regulatory signal or stress; the change can be either due to incorporation of a chemical factor from the exterior (e.g., cholesterol from plasma) or by modified membrane biogenesis.[30] Lipid fluidity can also be manipulated in vitro (e.g., by cholesterol or fatty acids[49]). Such methods have shown the possibility to correct a fluidity change by use of another chemical modulator, which would preserve the natural function. Similar physiological mechanisms exist in the

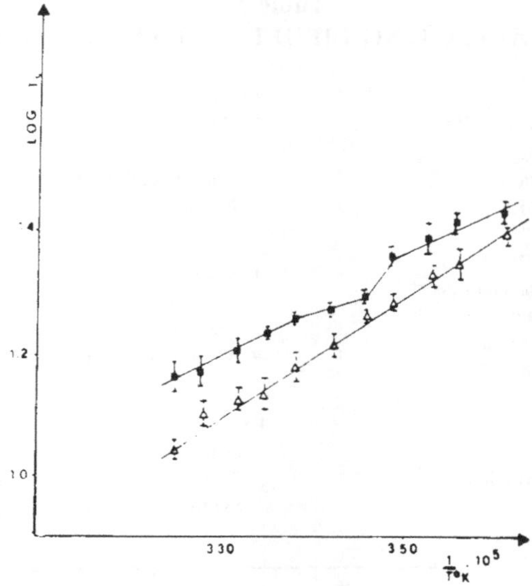

FIGURE 10. Temperature dependence of the correlation time τ_c of 16-doxylstearate in mitochondria; τ_c is in 10^{-10} sec. (■), Control; (Δ), + butanol, 150 m*M*. Protein-dependent discontinuities in Arrhenius plots of the motion of spin labels have also been described by others.[242,243] (Reproduced from Lenaz, G., Curatola, G., Mazzanti, L., Zolese, G., and Ferretti, G., *Arch. Biochem. Biophys.*, 223, 369, 1983. With permission.)

intact cells (Section VII): when the biosynthesis of one of the chemical modulators (e.g., cholesterol) is impaired, it can be compensated by another modulator (e.g., degree of unsaturation).

Cholesterol is well known to increase membrane viscosity;[10] deuterium NMR has shown that cholesterol induces a large increase of molecular ordering of the acyl chains in the first ten to twelve carbons, as expected from its known location with the rigid nature of the A-D ring system and the flexibility of the alkyl tail.[58] The order parameter for cholesterol in egg lecithin bilayers containing 50 mol% cholesterol is close to the value of 1 corresponding to perfect order. On a submacroscopic level, both fluorescence polarization and spin labels show decreases of fluidity induced by cholesterol. In temperature profiles of lipid viscosity, the flow activation energy ΔE is decreased by cholesterol in liposomes, while the actual viscosity is increased, indicating that the sterol decreases the mass of the flowing unit[50] (cf. Figure 9). Unexpectedly, in bilayers of dielaidoyl-PE, cholesterol addition enhances dramatically the T_1 relaxation times indicating faster reorientation rates of the *trans* double bonds.[59]

In mitochondria ubiquinone may be a physiological modulator of lipid fluidity;[60] Q-homologs having either short or long isoprenoid chains in their oxidized forms enhance the fluidity of lipid bilayers, although to different extents, whereas shorter homologs in their reduced forms make the bilayer more rigid.[61] Spin label studies in mitochondrial membranes depleted of their endogenous Q_{10} show a significant decrease of fluidity, particularly in the membrane hydrophobic core; readdition of Q_{10} restores the original fluidity, whereas addition of Q_3 does not[60] (Table 4). The disordering effect of Q_{10} could be the result of the side chain intercalating in the hydrocarbon region of the bilayer, whereas the Q_3 molecules, having the approximate length of a lipid molecule, are inserted in their full length between the fatty acyl chains, inducing a cholesterol-like effect, albeit extended into the bilayer core.

Table 4
EFFECTS OF DIFFERENT UBIQUINONE HOMOLOGS ON FLUIDITY OF MITOCHONDRIAL MEMBRANES, STUDIED WITH SPIN LABELS

Mitochondria	5-Doxylstearate (S_n ± S.D.)	16-doxylstearate (τ_c, nsec, ±S.D.)
Lyophilized	0.66 ± 0.02[a]	1.25 ± 0.06[b]
Extracted	0.72 ± 0.02	1.78 ± 0.21
+Extract	0.67 ± 0.02[b]	1.35 ± 0.12[a]
$+Q_3$	0.71 ± 0.02	1.54 ± 0.14
$+Q_{10}$	0.71 ± 0.02	1.46 ± 0.20[b]

[a] $p < 0.05$.
[b] $p < 0.01$ (Statistical significance with respect to extracted mitochondria).

On the basis of these findings it is tempting to speculate that the high Q_{10} levels found in mitochondria, and in other membranes as well, have the function of modulating the fluidity of the lipids. The decreased levels of Q_{10} found in several diseases and the beneficial effects of clinical use of the quinone have some relevance to this phenomenon.[62]

The effects of physiological or external perturbants on fluidity may be direct on the hydrocarbon chain disposition, but also related to the interfacial water organization as already discussed for the electrostatic effects (Section II.B). For example, it is known that cholesterol tends to increase the extent of hydration of the polar heads,[20] but its effect is considerably reduced in the case of unsaturated lecithins, which are more hydrated per se,[63] and are strongly condensed by cholesterol. In reversed lipid micelles in benzene, two classes of bound water have been identified, and the more tightly bound class is displaced by lipophilic molecules as Q_{10}.[64]

C. Effects of Proteins: Lipid-Protein Interactions

The types of interactions between lipids and proteins depend on the extent of penetration.[10] The determination of the primary structure of integral membrane polypeptides has shown that they possess one or more hydrophobic stretches intercalated with segments of random polarity. The number of hydrophobic residues in one stretch is sufficient to form a transbilayer structure if the segment is coiled in a α-helical conformation (at least 20 amino acids for 30 to 35 Å thickness[5]). It is, therefore, assumed that membrane proteins contain one or more transmembrane helical segments (from one to ten), with interconnecting segments forming globular regions protruding into the aqueous phases. The best characterized transport protein is bacteriorhodopsin of the purple membrane of *Halobacterium halobium;* examination of membrane crystals by electron microscopy and image reconstruction has allowed us to recognize seven helical segments folding across the bilayer;[65] the sequence has been obtained[66,67] and attempts made to predict the folding in the bilayer, based on the principle of keeping charged amino acids outside and considering the data from labeling with non-penetrating reagents to identify segments of the protein outside the bilayer.[68]

1. The Controversy of the Boundary Lipids

Spin label studies of delipidated cytochrome oxidase at increasing lipid supplements[69] showed that highly immobilized lipids persisted up to 0.2 mg lipid per milligram protein; above this value a progressive increase of motional freedom suggested that, on replenishing the lipid-poor complex with new lipids, the amount of immobilized lipids remains constant

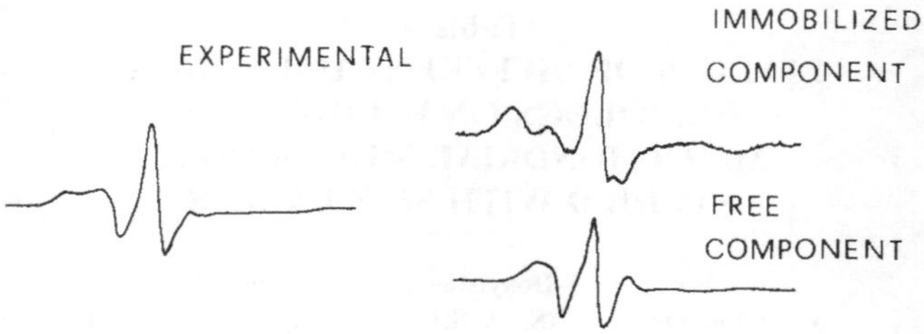

FIGURE 11. (Left) Representative experimental EPR spectrum obtained at 5°C from membranous Na,K-ATPase with a negatively charged methylphosphate lipid spin label; (right) spectral components obtained from the experimental spectrum at the left obtained as described by Brotherus et al.[76] (Modified from Brotherus, J. R., Jost, P. C., Griffith, O. H., Keana, J. F., and Hokin, L. E., *Proc. Natl. Acad. Sci. U.S.A.*, 77, 272, 1980.)

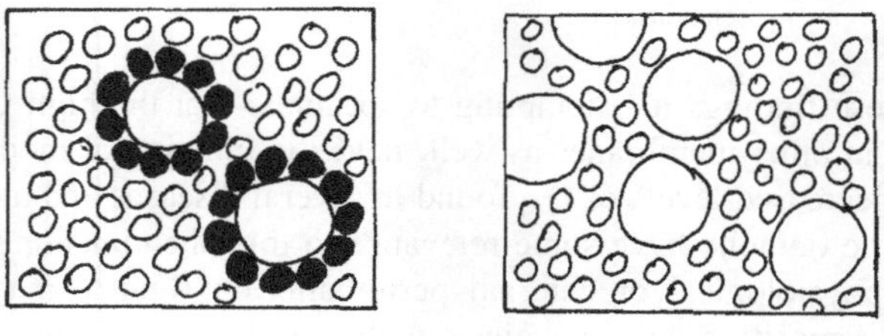

FIGURE 12. Two alternative views of lipid immobilization by proteins. (Left) Boundary lipids; (right) trapping of lipids between proteins.

and a new class of lipids appears having the properties of free bilayer. It was initially calculated that only one layer around the protein is highly immobilized, but theoretical calculations have later shown that the modification extends two to three layers from the protein.[70] These immobilized lipids were called *boundary lipids* and then described by others as the lipid *annulus*. The presence of boundary lipids was claimed in several other membranes.[71-74] The rate of exchange between boundary and bulk lipids was found relatively slow[75] and it was argued that the single lipid shell is long-lived and endowed with remarkable selectivity. Anionic phospholipids are selectively immobilized by Na,K-ATPase while cationic ones are repelled[76] (Figure 11). Such opposite changes are mediated to zero by strongly increasing the ionic strength of the medium. Similarly, cholesterol appears selectively excluded from the lipid annulus of the sarcoplasmic reticulum ATPase.[77]

While recognizing the immobilizing effect of proteins on lipids, Chapman[78] has criticized the concept of the lipid annulus. The extent of immobilization increases with increasing protein concentration in the bilayer; at low lipid/protein ratios each lipid molecule experiences multiple contacts with proteins, whereas at high ratios single lipid-protein contacts are dominant. The immobilizing effect at high protein concentration would be due to the trapping of lipids in highly concentrated protein regions (Figure 12). A maleimide spin label covalently attached to rhodopsin through a fatty acid, and presumably incorporated into the boundary layer, has revealed strong motion restriction if the membrane was partially delipidized, but the probe mobility at low protein concentration was indistinguishable from that of the bulk lipid.[79] On the other hand, in sarcoplasmic reticulum (SR) ATPase a covalently bound spin label is immobilized, when linked by a short-chain arm, and has a fluid environment when linked by a long-chain arm.[80] It is still difficult to come to precise conclusions on these contradictory results, since the binding of the covalent labels to the protein is not characterized and their location with respect to the lipids is not clear. On the other hand, a short-arm spin

Table 5
^{2}H-NMR MEASUREMENTS OF RSR MEMBRANES IN COMPARISON WITH PURE LIPIDS. LIPID USED [9,10 $^{2}H_2$] DOPC[82]

	Quadrupole splitting Dq (KHz)		T_1 (msec)	
Temp (°C)	DOPC	RSR	DOPC	RSR
4	15.8	14.1	7.7	7.4
14	14.0	13.0	10.6	10.2
24	13.2	11.7	13.8	11.0

label could be more sensitive to the rotational motion of the whole protein rather than that of the individual group.[81]

The concept of boundary lipids was mainly supported by spin label experiments. On the other hand, NMR did not show the occurrence of two types of lipids: on the contrary, a continuity between the bulk phase and the "boundary layer" was found, with ready diffusion of the two types of lipids, at least in the slow NMR scale;[34,78,82] however, being the EPR timescale is much shorter than that of NMR, a motion which is fast enough to average out anisotropy in NMR may appear as frozen in the shorter exposure time of EPR.[33] Indeed, pulsed NMR shows that proteins induce immobilization of the lipids; ^{2}H and ^{13}C T_1 relaxation time measurements show a decrease of the rate of segment reorientation in the presence of protein, suggesting an actual increase of viscosity[82,83] (Table 5).

On the other hand, the quadrupole splitting of deuterated lipids is decreased by proteins (Table 5), indicating a disordering effect of the irregular protein interface; the lipids, being flexible molecules, follow the irregular shape of the protein, becoming more distorted.[82] The reduction of the ^{2}H order parameter by the protein is equivalent to a temperature *rise* of the pure bilayer of as much as 20 to 30°C.[34] It has to be stressed that the disordering effect does not imply an increase in mobility; it is likely that the total number of chain configurations is reduced by the protein (= rigidization) with increased statistical probability of the more distorted chain conformations (= disorder). It can be concluded that lipids are disordered by interacting with membrane proteins, but their viscosity is increased because of movement restrictions.

A conclusion is still not reached on this problem, but it is not unreasonable that, although immobilization is partly due to lipid trapping in concentrated protein regions, also a direct effect of the protein surface on the lipids is likely, and selective attractions for specific lipids may be present. Electrostatic attractions of specific amino acid residues on proteins with charged lipids may have significant roles. Such attractions usually involve stoichiometries of few molecules per molecule of integral protein. Such lipids are difficult to extract and belong to the never well-characterized class of the *tightly bound lipids*[10] that remain in the lipid-deficient preparations or in isolated lipid-poor membrane protein complexes. Although tightly bound lipids are presumably immobilized, not all immobilized lipids have to be tightly bound.[2] The concept of the boundary lipids extended to 50 or more lipid molecules around the protein, whereas tightly bound lipids are few molecules only. In cytochrome oxidase there are four to five molecules of tightly bound cardiolipin: three of such molecules are absolutely required for activity.[84] Tightly bound cardiolipin was shown also in complexes I and III of the mitochondrial respiratory chain.[85] A critical evaluation of the controversy of the boundary lipids is presented in Chapter 8.

2. Vertical Displacement of Proteins

Shinitzky and Inbar[50] observed that the flow activation energy ΔE calculated from tem-

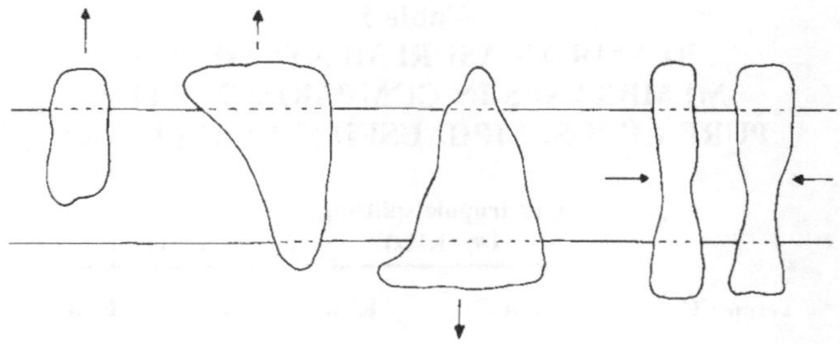

FIGURE 13. Vertical displacement of membrane proteins induced by increased lipid viscosity: arrows indicate possible dislocations.[49]

perature plots in several membranes falls in the narrow range of 7.5 kcal/mol and is independent of lipid viscosity as modulated by the cholesterol levels. This behavior strongly contrasts that of protein-free liposomes, where ΔE ranges from 15 to 5 kcal/mol by increasing cholesterol, and strongly depends on lipid viscosity (Section IV.B). It was assumed that the position of the membrane proteins is displaced towards the aqueous phase when the bilayer becomes more viscous, and this process opposes the ΔE changes; in other words, the proteins "buffer" membrane viscosity by their ability to move vertically with respect to the membrane plane. Spectral and chemical methods showed that alterations in lipid fluidity modulate the extent of exposure of membrane proteins to the aqueous media.[86] According to Shinitzky,[49] a membrane protein exists in an unstable equilibrium of opposing hydrophobic and hydrophilic forces as the results of its amino acid sequence; alteration of lipid fluidity might force this balance to a new equilibrium; when fluidity is decreased, the new equilibrium position will be of an overall weaker protein-lipid interaction, with corresponding greater protein-water interaction (Figure 13). The process can be mimicked by the partition of small amphipaths in the lipid phase, which is viscosity dependent. The changes associated with vertical movements must be extremely complex and the properties of interfacial bound water[15] cannot be overlooked, since they might be involved both as a cause and as a consequence of the vertical changes. On the other hand, interfacial changes can cause cohesion changes within the hydrocarbon chains (Section IV.B) so that the behavior of a membrane as a function, e.g., of temperature, becomes extremely complex. Discontinuities in the temperature dependence of motion of spin labels[51-54] (Figure 10) could have their reasons in the different temperature dependence of many parameters (viscosity, protein displacement, bound water), all influencing each other in a complex pattern.

3. Lateral Protein Aggregation

Chapman[78,87,88] has used two-dimensional arrays of discs appropriate to the relative cross-sectional areas of lipid chains and helical polypeptide structures as models to simulate the events associated with phase separations in membrane systems. At relatively high lipid contents, cooling below the phase transition results in disordering of the gel structure in the protein region with packing faults radiating along hexagonal axes of closely packed hydrocarbon chains. The disruptive effect of imperfect packing arrangements approaches the average spacing of the proteins, at higher protein concentrations, so that at freezing the protein becomes trapped within the dislocations due to thermal fluctuations developing during crystallization. As a result, a squeezing of the proteins out of the crystalline lipid lattice ensues, forming patches of high protein/lipid content (Figure 14); the lipids within the patches are prevented from crystallizing and are rigid because of the high protein content; they will become mobile by heating above the transition, because the remaining crystalline lipids melt, disaggregating the patches. Calorimetric studies have indeed shown that a proportion of the lipids does not undergo the thermal transition;[89] according to this view, the reduced

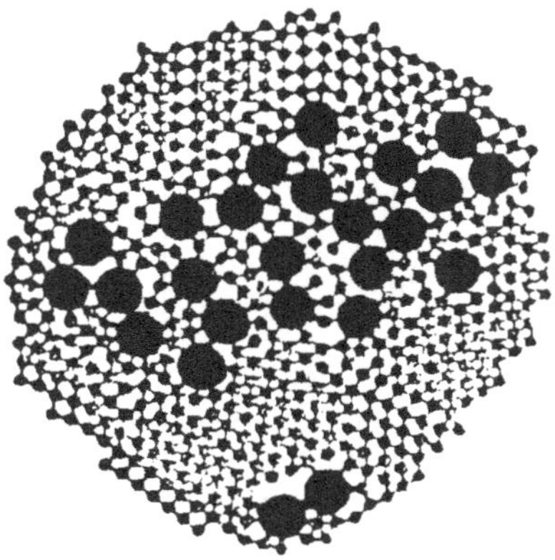

FIGURE 14. Aggregation of membrane proteins during crystallization of lipids.[78] The membrane is depicted in a transversal section with lipids and proteins appearing as small and large circles, respectively.

transition enthalpy (Section II.C) corresponds to the amount of lipids perturbed by trapping in the eutectic lipid/protein mixture. Freeze-fracture[90,91] shows that the intramembrane particles are randomly distributed above the transition; below the transition, however, they are excluded from regions exhibiting a band pattern characteristic of pure frozen lipid. In binary mixtures of lipids having different transitions, proteins are excluded from the gel phase. Not all membranes show protein aggregation below the transition.[92] Cholesterol was shown to have contrasting effects in different studies.[93-96] Nonetheless, it is apparent that protein clustering may take place independently of the phase transition, also as an effect of increased membrane viscosity. As in the case of vertical protein displacement, the properties of interfacial water may have an important part in these changes.

D. Extrinsic Chemical Factors

Membrane fluidity is perturbed by foreign molecules.[97] General anesthetics enhance the fluidity of several membranes at clinically relevant concentrations;[98] at the same concentrations, however, they do not change significantly the fluidity of protein-free phospholipid bilayers. It may be operationally concluded that anesthetics release the immobilization induced by proteins on the lipids; their action appears, therefore, a perturbation of lipid protein interactions rather than mere fluidization.[99]

The "fluidity" theory of anesthesia, once very popular,[100] has been recently criticized[101] on the basis of several lines of evidence, including the observation that a temperature increase diminishes rather than enhances anesthetic action.

Anesthetics tend to abolish the discontinuities in Arrhenius plots of the motion of a spin label in mitochondrial membranes, as well as increasing fluidity in the whole temperature range[53] (Figure 10). Such an effect may well be related to intramembrane movements of proteins (either lateral or vertical displacements) with changes in the environment of the spin label. For example, it can be postulated that an aggregation of proteins in the membrane plane leaves large areas of protein-free bilayer; if such areas contain the majority of the probe molecules, as likely, the probes experience a decreased viscosity because of the smaller number of lipid protein contacts[5] (Figure 15). The discontinuity observed in unperturbed membranes is in line with this interpretation, since a plateau is observed by decreasing the temperature, at which no viscosity increase is apparent.[53] Anesthetics might induce per se protein aggregation, leaving an apparently more fluid membrane at all temperatures. An

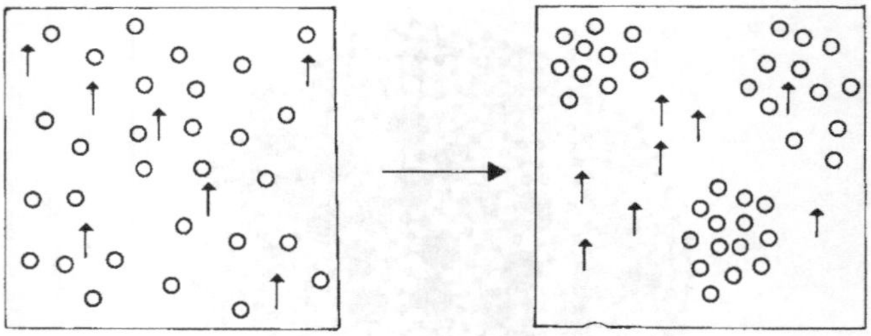

FIGURE 15. Aggregation of intramembrane proteins leaves areas of free lipid bilayers, where probes, designated as ↑, experience increased mobility. The circles indicate the proteins as appearing in a fracture face.

increase in protein aggregation in synaptic membranes has been, indeed, observed by anesthetic treatment in vitro[102] and in vivo.[103] Anesthetics might affect protein distribution as a consequence of their known effects as hydrogen bond breakers[104] or by increasing the membrane thickness.[105] The effects of anesthetics at the lipid-protein-water interface would destabilize proteins, inducing an increased apparent fluidity as a consequence of their redistribution or inducing modifications of lipid packing. The mechanism of anesthesia would be an unidentified change in the activity of a membrane protein(s), following perturbation of its interaction with lipids. The presence of conformational changes in membrane proteins as a consequence of lipid protein interactions and their perturbations will be discussed in Section V.C.

Since lipid fluidity and lipid protein interactions can be artificially changed by several modulators, a pathological deviation induced through a change in fluidity might be corrected by artificially imposing the opposite change by means of an appropriate compound. The therapeutic possibilities inherent in this approach need not to be emphasized.

V. MOLECULAR AND SUPRAMOLECULAR CONSEQUENCES OF MEMBRANE FLUIDITY

A. Mobility of Lipids and Proteins

The concept of membrane fluidity implies that the constituents of all membranes are capable of movements; fluid membranes must, therefore, exist in a dynamic state of relatively free mobility of lipids and proteins. However, membrane fluidity is not distributed homogeneously, but varies within lateral domains in the plane of the bilayer as well as transversally across the membrane. Only few general concepts and applications will be developed here, since the subject is treated elsewhere in this book.

1. Lateral Diffusion

The lateral motion of polar lipids is investigated by EPR spectral broadening due to spin-spin interactions[37] or by NMR[106] and fluorescence[42] techniques, and has allowed us to estimate lateral diffusion coefficients of 10^{-8} cm^2/sec. All evidence points out, as expected, that proteins move at slower rates. The first evidence was obtained by Frye and Edidin,[107] who calculated the rate of intermixing of two fluorescent antigens after fusion of heterokaryons as $D_L = 10^{-10}$ cm^2/sec. The method of fluorescence photobleaching recovery for natural and artificial chromophores[108] has allowed us to measure D_L between 10^{-9} and 10^{-12} cm^2/sec. The variations depend on the type of membrane and of protein; the differences in molecular radii are not sufficient to explain such large differences in D_L, suggesting that mobility in some membranes is hindered by specific restrictions.[3] The elements of the cytoskeleton, the presence of intercellular junctions, and clustered protein regions may all induce restriction of free lateral diffusion.[3]

Protein diffusion is also detected by electron microscopy. Aggregation of antigenic sites on cell surfaces following multivalent antibody binding has been observed;[109] the initially randomly distributed antigens after interaction aggregate into patches; further aggregation to form a "cap" requires metabolic energy. Patch formation is the result of lateral diffusion of aggregated antigen-antibody complexes.[3] Also, freeze-fracture allows detection of clustering of intramembrane particles as a response to different stimuli;[92,110] aggregation is prevented by cytoskeletal constraints.[111] A quantitative method for determining the rate of particle movement has been developed;[112] intramembrane particles in the inner mitochondrial membrane electrophoretically migrate in a single crowded patch facing the anode by inserting a current pulse of few seconds; the rate of return to the homogeneous distribution after the end of the pulse allowed to calculate a D_L of 8.3×10^{-10} cm²/sec.

The cytoskeletal restrictions are not passive constraints to free diffusion of proteins, but represent an active device controlling a specific directed movement of the proteins in the fluid membrane.[3]

2. Rotational Diffusion

Spectroscopic methods to measure rotation[107] depend on photoselection, whereby an oriented population of excited molecules is optically selected from an initially random distribution by excitation with plane-polarized light: only those molecules whose transition dipole moment for absorption is parallel to the electric vector of the incident light are excited. By excitation with a brief pulse of light (flash photolysis) the initial emission or absorption anisotropy decays as the molecules again become randomized. Since protein rotation is relatively slow, it is necessary to use a long-lived spectroscopic state; this is the case of the "triplet probes"[113] which are attached covalently to the protein. Another method for slow rotational movement is saturation transfer EPR with covalently linked nitroxides.[114] By such methods, the rotational mobility of several proteins was found to range between D_R of 20 μsec for rhodopsin,[115] in accordance to the very fluid nature of the retinal rod outer segment membrane[116] to complete immobility of bacteriorhodopsin in the purple membrane.[117] In SR ATPase, D_R was found to undergo a discontinuity at 15 to 20°C,[118] suggesting the presence of a quaternary or conformational change in this temperature region. In the inner mitochondrial membrane, the state of aggregation of cytochrome oxidase depends on the assay conditions;[119] the membrane contains, in addition to freely diffusing redox components, nonspecific protein aggregates. Chemical cross-linking[120] or the binding to cytoskeletal elements[121] also restrict free rotation of membrane proteins.

3. Transbilayer Mobility

The movement of lipids from one leaflet to the other of the bilayer is relatively slow;[15] if molecules migrate to the opposite bilayer leaflet, the polar residue is forced to traverse the hydrocarbon interior, which is energetically unfavorable. The increase in free energy required for transbilayer motion can be reduced by cation binding to the phosphate group, decreasing the overall polarity of the molecule. The ability of phospholipids to act as ionophores[122] may be related to such polarity decrease. The transbilayer diffusion rate of phospholipids in biological membranes (e.g., see van der Bessalaar et al.[123]) appears considerably faster than the rate in lipid vesicles (e.g., see Kornberg and McConnell[124]), indicating that proteins may strongly affect the flip-flop movements. Lipids are asymmetrically disposed in several membranes, and the cytoskeletal elements are involved in the maintenance of the established asymmetry.[15,125] It has been shown that disruption of the cytoskeleton in red blood cells is accompanied by loss of lipid asymmetry due to enhanced flip-flop rates.[126]

Contrary to lipids, proteins are assumed to undergo no transbilayer motion whatsoever for the thermodynamic hindrance imposed by their manifold amphipathic character.[10]

B. Permeability

Passive permeability has been widely investigated in liposomes containing trapped polar solutes. In general, as studied by radioactive tracer techniques, most liposomes are much more permeable to Cl^- than to K^+ and Na^+; these ion fluxes are, however, quite low compared with most biological membranes.[127] Fluxes of 0.4×10^{-15} mol/cm^{-2}sec^{-1} for Na^+ through lecithin liposomes correspond to leakages of 0.05% of the total captured Na^+ per hour; the fluxes are markedly temperature sensitive, with activation energies of up to 30 kcal/mol. Permeability depends on the nature of the phospholipids, for example, negative liposomes have much higher rates of K^+ flux compared to Na^+; liposomes consisting of negative lipids are quite sensitive to the effect of divalent cations on their permeability to other substances.[128]

The permeability of liposomes to nonelectrolytes is higher than toward ions and that to water is extremely high and close to the values in biological membranes.[129] Within a certain size range, permeability is increased by increasing lipid solubility. The mechanism of passive permeability involves fluctuating structural transitions of the phospholipid molecules, which create defects in the bilayer allowing transbilayer transport. No wonder, therefore, that membrane fluidity is a major factor in controlling permeability. Temperature affects the molecular motion, as seen previously, and will, therefore, strongly affect membrane permeability, which is lower in the solid form and higher in the fluid form, but with a maximum in the region of the phase transition.[130] This effect can be explained by the existence of a phase separation when both solid and fluid regions coexist; the boundaries of these regions have maximal instability increasing the permeability to small molecules.

Cholesterol increases the viscosity of fluid lipid bilayers, and the enhanced packing of the molecules is accompanied by decreased permeability to cations, anions, and nonelectrolytes (Table 2). A variety of soluble and membrane proteins can interact with liposomes to cause increased permeability[26] (Table 2). Independently of the existence of specific carriers forming polar channels or other structures suitable for transmembrane transport, proteins appear to increase permeability by inducing faults in the bilayer packing. This observation is pertinent to biological membranes that show high relatively unspecific permeability and a lower electrical resistance in comparison with liposomes.

Also, divalent cations increase the permeability of liposomes containing negatively charged phospholipids,[128,129] possibly by inducing phase separations or other instability factors: addition of Ca^{2+} to both sides of a planar negative bilayer stabilizes the membrane, in agreement with a decrease in bilayer fluidity.

C. Protein Conformation

1. Lipids and Secondary Structure

Proteins linked to biomembranes will assume their optimal conformation when the highest number of hydrophobic residues are situated at the exterior of the protein molecule in the region in contact with the nonpolar hydrocarbon chains of the lipids; it is reasonable that the same proteins assume rather different conformations in an aqueous medium.[54]

The process of transfer of a nonpolar group from nonpolar solvents to water is accompanied by a large positive ΔG, which does not result from unfavorable enthalpy change, since ΔH is negative, but from a large entropy decrease. Such decrease results from an ordering of the water molecules with an increased intermolecular hydrogen bonding. In such conditions, clustering of nonpolar molecules together will reduce the total hydrophobic surface and the amount of ordered water surrounding it, and will be favored thermodynamically. Hydrophobic interaction is the main source stabilizing the native conformation of proteins in aqueous media. In water, hydrogen bonds are important to stabilize protein structure only in regions sequestered from the medium, whereas on the exterior of a protein molecule it makes little difference whether the groups are bonded with each other or with water molecules.[131] In nonpolar solvents, the lowest free energy is achieved when the maximum number interpeptide hydrogen bonds are formed. Since interpeptide hydrogen bonding is maximized in α-helical conformation, this can be the reason why proteins in nonpolar solvents become largely α-helical.[132] In membranes, intrinsic proteins are in close contact with a hydrophobic medium represented by the fatty acyl chains and a conformation similar to that found in nonpolar solvents should be expected.

Intrinsic membrane proteins are arranged in an amphipathic conformation allowing hydrophobic binding of the nonpolar residues with the lipid chains and hydrophilic interactions with the polar heads of the phospholipids and with water.[131] When lipids are removed membrane proteins may still remain in an insoluble form resembling the native membrane;[133] in such case the gross interactions are still expected to be hydrophobic within the protein skeleton.

Protein conformation in solution and in membranes is investigated by optical rotation measurements, in particular, circular dichroism (CD).[134] The CD data of membrane proteins, when corrected for artifacts arising from the particulate nature,[135] showed substantial amounts of α-helix.[10] Very high α-helix contents were shown in intrinsic proteins dissolved in hydrophobic media[136] or incorporated into lipid bilayers.[137-139]

Studies on synthetic polypeptides[140] and soluble proteins[141] have thrown some light in the conformational changes induced by phospholipid binding; the best evidence available — the lipids modify the conformation of proteins — comes, however, from studies of plasma lipoproteins.[142,143] Studies with membrane proteins are far less numerous. In acetone-extracted mitochondria greatly modified CD spectra in comparison with those of original mitochondria suggested a substantial loss of α-helix.[144] Binding back mitochondrial phospholipids or pure lipid species restored the original spectra of intact mitochondria to various extents, cardiolipin being the most effective species. Also, in mitochondrial complex III phospholipid removal induces a reversible loss of α-helix,[145] and the CD spectra in the visible region showed changes in the conformation of cytochrome b. In submitochondrial particles no noticeable change was observed after phospholipase A_2 treatment and removal of the digestion products by albumin. The necessity of complete removal of phospholipids to show conformational changes is in line with the necessity of a minimum amount of phospholipids for enzymic activity.[69]

Addition of small concentrations of organic solvents and general anesthetics also induces changes similar to those induced by delipidation, indicating decrease of α-helical structure.[98] The same solvents induce perturbation of lipid protein interactions (Section IV.D).

2. *Thermal Effects on Conformation*

A temperature decrease below 20°C in the isolated oligomycin-sensitive ATPase induces a decrease of the CD-negative ellipticity in the 225 to 208-nm region, indicative of a decrease of α-helix[147] (Figure 16). The enrichment of erythrocyte membranes with cholesterol also induces a decrease of α-helix.[4] Both processes have in common an increase of membrane viscosity. It is, however, difficult to prove that the effects are a direct consequence of the viscosity increase. The temperature dependence of the motion parameters of spin labels and other probes showing discontinuities in the physiological region[53] may offer a tempting explanation to such conformational changes. It was shown in Section IV.D that such discontinuities may be the expression of relative changes of protein organization in the membrane, with vertical or lateral displacements. In particular, a vertical displacement at a critical viscosity (low temperature or high cholesterol) may be accompanied by an increased water binding to normally hidden peptide structures, with unfolding and loss of secondary structure.[54,131]

A change in protein organization may be related to viscosity directly or indirectly. Temperature-dependent changes in bilayer thickness[148] or in the conformation of the polar heads,[149]

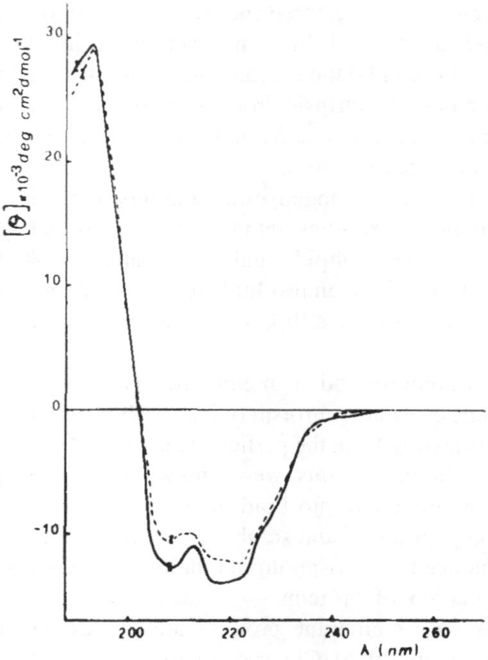

FIGURE 16. Circular dichroism spectra of oligomycin-sensitive ATPase at 25°C (solid line) and 10°C (dashed line). (Reproduced from Curatola, G., Fiorini, R. M., Solaini, G., Baracca, A., Parenti Castelli, G., and Lenaz, G., *FEBS Lett.*, 155, 131, 1983. With permission.)

or even changes in the content and structure of bound water[150-152] or other interfacial phenomena[15] may be concomitant with the viscosity changes and trigger a series of cooperative modifications culminating in the observed secondary structure alterations.

To this purpose, according to Drost-Hansen,[150] "vicinal" water in the form of clathrate-like hydrates, having properties relatively independent of the nature of the surface, exhibits structural transitions at 15, 30, and 45°C. Anomalies in the temperature responses of physiological or biochemical parameters have frequently been reported at the same temperatures at which vicinal water undergoes transitions;[151] such anomalies are proposed to arise from water-induced conformational transitions.[150] When vicinal water undergoes structural transitions, these may well impose conformational constraints upon the proteins.

As briefly outlined in the forthcoming section, conformational changes can be the reason of thermal effects on enzymic and transport activities.

VI. PHYSIOLOGICAL IMPLICATIONS AND PATHOLOGICAL CHANGES OF MEMBRANE FLUIDITY

The physiological implications of membrane fluidity are manifold but often still obscure. There is no cellular process in which membrane fluidity has not been somewhat implicated.[49] Conversely, an abnormal physical state of the membrane has been involved in a number of pathological states and often directly related to the pathogenesis of a disease.

A. Role of Lipid Fluidity in Enzymic and Transport Activities

Lipids are required for activity of membrane-linked enzymes and carriers[10,15,54,153-156] for such functions lipids have two types of roles: one depends on their properties as a bulk

Table 6
EFFECTS OF DIFFERENT TREATMENTS ON MITOCHONDRIAL ATPASE

	Lipid removal	Solvents	Low temp
V_m	Decreased	Decreased	Decreased
K_m (ATP)	Decreased	Decreased	Decreased
Activation energy	Increased	Increased	Increased
ΔG activation			No change
Cooperativity olgomycin inhibition	Decreased	Decreased	
Content α-helix	Decreased	Decreased	Decreased

phase, the other on specific interactions with proteins. In order to perform their bulk function, lipids are required as physical entities and their chemical nature is relatively of minor importance: in this respect, the bilayer represents a dispersing medium where proteins are randomly dissolved. On the other hand, specific interactions of lipid molecules with membrane proteins seem necessary for the catalytic events, possibly for keeping the enzymes in their optimal conformation. It is difficult to discriminate between the two roles, since the more specific lipids are difficult to extract and often remain in the lipid-deficient preparations employed to investigate the role of the lipids.

1. Kinetics of Membrane-Bound Enzymes

Membrane enzymes have kinetic differences in comparison with soluble enzymes or with membrane enzymes after delipidation or solubilization.[54] In most cases, solubilization or delipidation, or perturbation with organic solvents, induces comparable changes, viz. a decrease of both V_m and K_m and changes in allosteric behavior.[98] In mitochondrial ATPase, a decrease of V_m and K_m for ATP is observed after addition of organic solvents and anesthetics or partial delipidation with phospholipase A_2[157] (Table 6), indicating a higher stability of an enzyme-substrate complex or that a decrease of product formation affects both V_m and the apparent affinity of the enzyme for its substrate.

Arrhenius plots of most membrane enzymes show discontinuities or breaks at well-defined temperatures, with increase of activation energy below the break.[158] Detergents and solvents abolish the break in mitochondrial ATPase, increasing E_A over a wide range of temperatures,[157] suggesting that alteration of lipid protein interactions decreases the catalytic power of the enzyme. The temperatures at which the breaks occur are strongly related to the physical state of the lipids, but usually not coincident with the calorimetric transition temperatures, with differences as large as 30°C.[54] In some studies however, the breaks in enzymic activity were correlated to breaks in motion parameters of spin labels in the same membranes.[51-53] In mitochondrial ATPase the plots of V_m and K_m for ATP have distinct breaks,[4] and a similar discontinuity is observed in the intrinsic fluorescence of tryptophans of the isolated enzyme[159] (Figure 17). Comparing the effects of delipidation, organic solvents and low temperatures on ATPase kinetics give surprisingly similar data. In SR ATPase[160] and in mitochondrial ATPase[161] the increase in activation enthalpy below the break is always compensated by increase of activation entropy, resulting in values of free energy of activation which are independent of variations of temperature and lipid composition (Table 6).

2. Role of Lipids in Integrated Enzymic Activities through Protein Mobility or Lipid-Soluble Mediators

The dispersing effect of lipids is well observed in the interaction of membrane protein complexes inserted in multicomponent electron transfer pathways. The studies of Strittmatter[162-164] on the interaction of cytochrome b_5 with NADH-cytochrome b_5 reductase (two proteins involved in the microsomal fatty acid desaturating system) clearly demonstrate

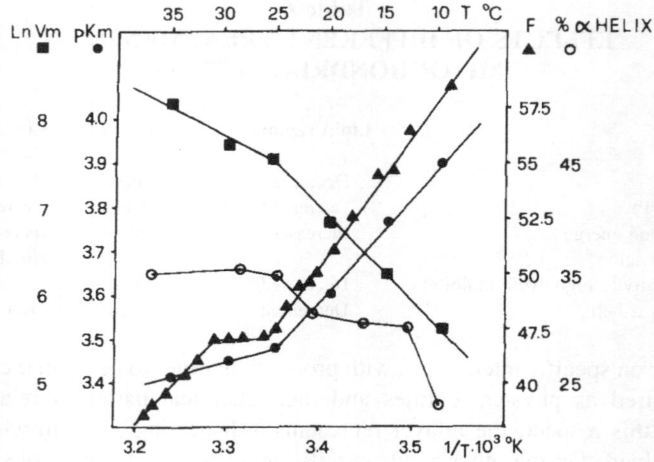

FIGURE 17. Temperature profile of different parameters of mitochondrial oligomycin-sensitive ATPase. The data are presented as ln V_m (■), pK_m for ATP (●), intrinsic tryptophan fluorescence (▲), and % α-helix (○). Redrawn from Lenaz et al.[4] and Curatola et al.[147]

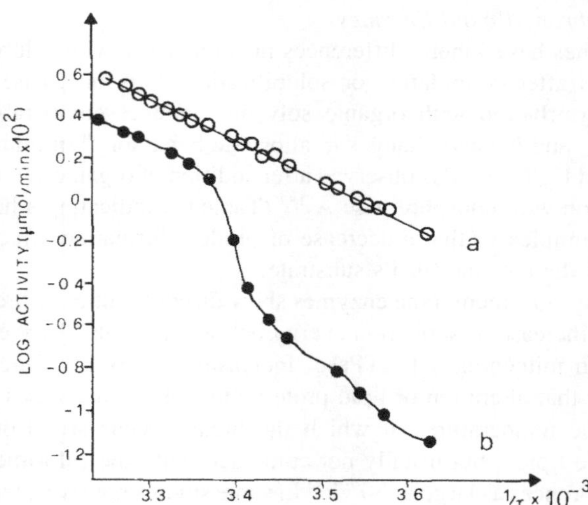

FIGURE 18. Arrhenius plot of NADH-ferricyanide reductase (a) and of NADH-cytochrome b_5 reductase (b). NADH-cytochrome b_5 reductase activity at high protein-to-lipid ratios shows a linear plot with a slope identical to that of NADH-ferricyanide reductase. (Redrawn after Strittmatter, P. and Rogers, M. J., *Proc. Natl. Acad. Sci. U.S.A.*, 72, 2658, 1975.)

such a role. Interaction of the polar heme-bearing moiety of cytochrome b_5 with the catalytic center of the reductase requires translational diffusion of the nonpolar moieties of the proteins in the membrane. The Arrhenius plots of NADH-cytochrome b_5 reductase[164] show a break, whereas the plot of NADH-ferricyanide activity, which is expression of flavoprotein catalysis, is linear. If the protein concentration is increased however, the break in NADH-cytochrome b_5 reductase disappears (Figure 18). A plausible explanation for such a break is a major fall in the process of lateral diffusion of integral proteins when viscosity increases,

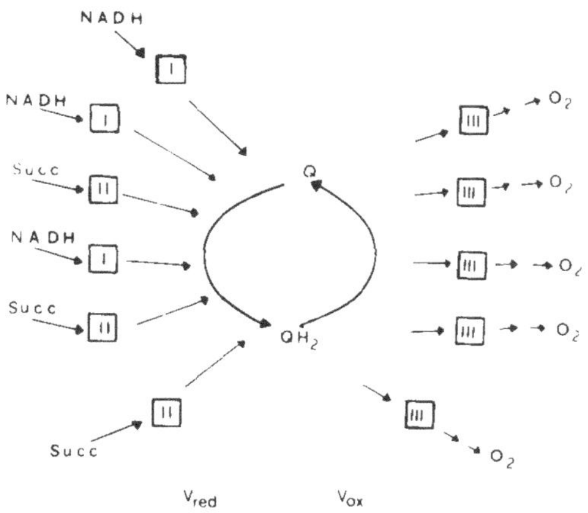

FIGURE 19. The pool function of ubiquinone (Q) in the mitochondrial membrane. I, II, III are the corresponding electron transfer complexes.

rendering diffusion rate-limiting. The increase in collision probability at higher protein concentrations makes the diffusion process never limiting even at low temperatures.

Pathways catalyzed by a series of membrane-bound enzymes involve statistical collisions of the enzymes in the hydrophobic medium. Lateral diffusion and possibly rotation become essential for collisions between the protein molecules. Since lipids are required above their phase transition, it is likely that fluid lipids are required for protein movement in the plane of the membrane.[54] This does not always appear to be the case, and recent experiments on cytochrome oxidase have shown that its aggregation state strongly affects its rotational diffusion but has no effect on enzymic activity.[165]

In mitochondrial and bacterial electron transfer and in chloroplast photosynthetic electron flow, membrane fluidity appears required for the movement of lipophilic redox quinones in the lipid matrix.[166] In mitochondria, ubiquinone was proposed by Green[167] to mediate electron transfer from dehydrogenases to the bc_1 complex by moving in the membrane lipids, transferring electrons between relatively immobile redox complexes. The hypothesis of Q as a mobile pool in the membrane has been tested and confirmed on a kinetic basis[168] (Figure 19). The observation that at low lipid content pool behavior kinetics are lost was explained by Ragan[169] with the formation of complex I/complex III associations as the only competent units in electron transfer, and pool behavior would depend on mobility of complexes and not of ubiquinone. The same results can be explained however, by Q diffusion becoming limiting at low phospholipid concentration.[170] No actual values for D_L of ubiquinone are available in the literature, except some preliminary studies in our laboratory by fluorescence quenching of the probe 12-anthroylstearate in lipid bilayers by Q homologs, showing D_L close to 10^{-6} cm^2/sec[246] (Figure 20). Similar values can be calculated theoretically[171] from the size of the Q molecule and the membrane viscosity. It can be easily calculated that such rates are equivalent to lateral displacements of over 500 nm/msec, compatible with enzymic turnovers as high as 500 sec^{-1} or more[5] for complexes separated by an average distance of 30 nm.[172] Therefore, under normal conditions the diffusion of Q in the fluid membrane is not rate-limiting, in accordance with the kinetic data.

Discontinuities in Arrhenius plots have also been considered to yield information on the mechanisms of membrane transport. The lipophilic ionophores, valinomycin and gramicidin, behave quite differently as a function of temperature;[173] valinomycin acts as a mobile carrier, and accordingly, its activity is drastically reduced below transition; gramicidin, on the contrary, is quite active below the transition, as expected from its channel-forming structure.

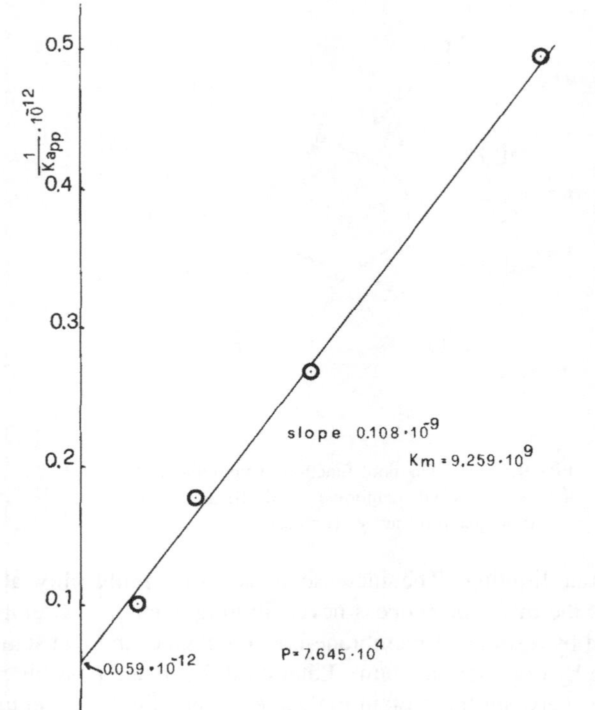

FIGURE 20. A plot of $1/k_{app}$ vs. α_m for ubiquinone-7 added to phospholipid vesicles containing the fluorescent probe 12-anthroyl-stearate. The method of Lakowicz et al.[239] has been used for calculation of D_L of Q; k_{app} is the observed quenching constant from Stern Volmer plots, α_m is the phospholipid concentration. The plot allows calculation of the partition coefficient P of ubiquinone in the membrane, and of k_m, the true quenching constant of ubiquinone in the membrane. From k_m, the diffusion coefficient D_L can be calculated.[239] (From Fato, R. and Lenaz, G., unpublished data.)

Arrhenius plots of physiological transport systems, like bacteriorhodopsin[174] or the proton-translocating membrane portion of mitochondrial ATP-ase,[175] yield contrasting results; however, the presence of a break in a transport activity does not necessarily imply a mobile carrier mechanism, but may be related to many different aspects of the transport mechanism so that extrapolation from simple model systems appears impossible.

3. A Conformational Role of Lipid Fluidity in Membrane-Bound Enzymes

According to Kumamoto et al.,[176] two independent processes having different E_A are required to produce a discontinuity, with the process having the higher E_A operating only at temperatures below the discontinuity. On such a premise a conformational transition in the protein, with the conformation having the higher E_A operating below the critical temperature, will produce a discontinuity. A direct correlation between the inflection point in an Arrhenius plot and a conformational change was observed in the soluble enzyme, D-amino acid oxidase;[177] in such case, the conformational transition is an intrinsic property of the protein molecule, leading to a sharp increase of E_A below the temperature of the conformational change. It was suggested that in membrane enzymes, at least in cases when no diffusion-controlled reaction step should be limited by lipid viscosity, a conformational change may be imposed by constraints from the lipid environment:[54] a phase change in the lipids would, therefore, induce a conformational transition in the protein.

It should be distinguished between studies involving synthetic lipids undergoing phase transitions in the physiological temperature range and those involving natural lipids, whose transitions are usually below 0°C. In the former case, there is no doubt that the lipid phase transition induces a break in enzymic activities.[148,160] The compensation of the increase in activation enthalpy below the break by an increase of activation entropy[160,162] has been explained on the basis of the increased order in the preactivation step in the rigid state, resulting in a greater loss of order during activation in comparison with the fluid state. Phospholipids, therefore, in their frozen condition, stabilize a preactivation state by the highly ordered array of crystalline molecules around the enzyme. It has also been suggested that crystalline lipids inhibit a conformational change during catalysis. A dynamic explanation of the kinetic changes[98] is indeed in alterations of formation of the transition state ES*:

$$E + S \rightleftarrows ES \rightarrow ES^* \rightarrow E + P$$

Lipid removal or presence of cyrstalline lipids hinder ES* formation, in accordance with the increase of E_A; the enzyme-substrate complex ES is more stable, in accordance with the observed lowered K_m.

In the case of breaks in enzymic activity in membranes containing physiological lipids, these proposals appear questionable, since breaks do not coincide with the calorimetric lipid transitions and occur independently of the lipid composition of the membrane or even in detergents.[178,179] The coincidence of the break in mitochondrial ATPase[157] with an α-helix decrease, as observed by the CD studies[147] quoted in Section V.C., is suggestive that the break originates from the conformational change (Figure 18). We have discussed in Section V.C how a conformational change in the protein may be related indirectly to lipid viscosity. The lack of coincidence of the break with the calorimetric transitions, together with the good correlations with breaks in the mobility of spin labels, and the suggestion that such breaks can depend on changes in protein distribution (Section IV.D) favor the working hypothesis depicted in Figure 21. The membrane is represented as a cooperative unit where the most important feature is a suitable overlapping of the polar and nonpolar regions of amphipathic lipids and amphipathic proteins with each other. Any perturbation of such an equilibrium, as, e.g., induced by a decrease in temperature, will induce drastic changes in the organization and conformation of the proteins and, hence, of their activities. Lipid composition affects both the membrane fluidity in the hydrocarbon core and the relative distribution of the polar and nonpolar regions, so that a temperature-dependent enzymic change might be shifted to a different temperature when the lipid composition is altered, the alteration operating not so much to change the fluidity, but the interfacial properties of the system. The better agreement existing between breaks in Arrhenius plots of sarcoplasmic reticulum ATPase with an increased freedom of the phospholipid polar heads[149] than of the methylene groups is in line with this hypothesis. The hydration state of the polar groups is of particular importance.[152] In the case of lipid-dependent C_{55}-isoprenoid alcohol kinase, fluid lipids restore activity only when they have sufficiently hydrated polar groups.[180,181]

B. Effects of Lipids on Hormonal Responses

Polypeptide and other hormones have receptors on the plasma membrane of target cells which transfer the signals elicited by hormone interaction to different effector systems in the cell. Many such effects are mediated through cAMP, synthetized by action of adenylate cyclase located at the inner side of the membrane.

Lipids are not required in hormone binding but are required to transfer the signal from the receptor to the active site of adenylate cyclase on the opposite side of the membrane.[182]

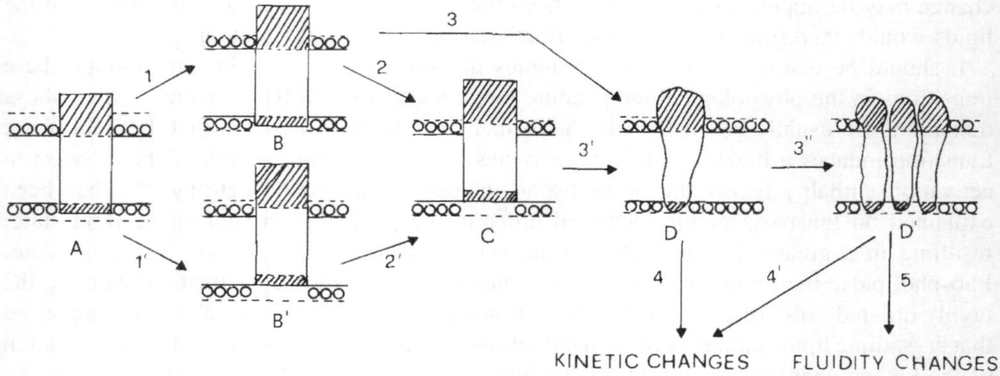

FIGURE 21. A hypothetical scheme showing the perturbation of the interfacial regions of membranes as the cause for activity changes in membrane-bound enzymes. Removal or changes of the lipids, perturbation with solvents, or lowering the temperature may result in a series of structural changes depicted as follows: (1) changes in interfacial properties (depicted here, in general, as a decrease of the bound water layer); (1′) vertical displacement via a primary viscosity change induced by the same treatments; (2, 2′) a state is achieved where protein and lipid polar moieties do not overlap anymore in a thermodynamically favorable state. This state may be reached through either states B or B′ depending on the type of treatment (e.g., solvents may primarily affect the interface whereas lowering temperature may induce both vertical displacements[49] or changes in bound water[150]). Similar results could be obtained by treatments modifying the bilayer thickness; (3) a conformational change now ensues to reduce the free energy of the system; however, (3′) a conformational change may result directly from interfacial changes without the necessity of primary protein movement; (3″) Protein aggregation may also result as the most favorable thermodynamic response in the new equilibrium position; (4,4′) as the result of the structural changes the activity and kinetics of membrane-bound enzymes or transport systems are profoundly modified; finally (5) the aggregation of membrane proteins may be accompanied by an increase of protein-free bilayer areas with an apparent fluidity transition, which is detected by spin labels or other probes in concomitance with the enzymatic changes (cf. Figures 10 and 17).

This observation suggested a model for the receptor-adenylate cyclase system as an allosteric enzyme formed by three subunits:[182] a catalytic site at the interior, a regulatory site being the receptor, and a coupling site needed to transfer the signal and which is lipid-dependent.[183] In the action of the β-adrenergic receptor, the G-protein functions as a coupling site, allowing adenylate cyclase activity when it binds GTP.[184] The nature of the coupling factor is more complex and has even been attributed to the lipid bilayer itself; according to the mobile receptor hypothesis,[185] the main prerequisite for coupling is membrane fluidity, necessary to allow diffusion of independent receptor and effector molecules in the plane of the membrane. The linkage of the hormone to its receptor forms a complex having greater affinity for the effector than the receptor alone. The change in affinity may result from a conformational change in the receptor, in line with the observation of hormone-dependent conformational changes in membranes.[186]

Membrane fluidity also appears to be modified by hormone-receptor interactions;[154] to this purpose Shinitzky[49] proposes the concept of *passive modulation* of membrane receptors; fluidity changes, possibly evoked by the hormone-receptor binding itself, modify the exposure of the receptors by vertical displacement, thereby modulating the binding capacity.

The findings of Axelrod[187,188] have given exciting implications on mechanisms whereby hormone binding is coupled to the physiological effects. Two methyltransferases in series form lecithin in the external monolayer utilizing as substrate PE in the internal monolayer; as methylated PE is translocated to the external surface, a significant increase in membrane fluidity is also observed.[4,187,189] Axelrod[188] has related the fluidity increase with the hormone effects, since β-agonists enhance transmethylation and transmethylation enhances both fluidity and adenylate cyclase activity. Accordingly,[188] several hormone signals interacting with cellular-specific receptors initiate a cascade of biochemical and biophysical changes in local

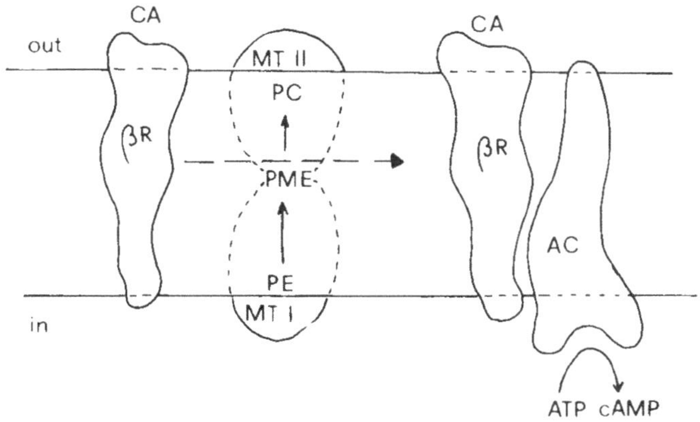

FIGURE 22. Simplified scheme of the possible role of PE methylation by methyltransferases (MTI and MTII) in coupling of β-receptor (βR) bound to cathecolamine (CA) with adenylate cyclase (AC).

domains of the membrane, leading to increased mobility of receptors, increased PE methylation, and generation of cAMP (Figure 22).

The mechanism whereby transmethylation enhances fluidity is not clear; the extent of newly synthetized lecithin appears insufficient to explain the large fluidity change. The fluidity increase is associated with impressive protein aggregation, as seen by freeze-fracture electron microscopy.[189] As explained in Section IV.D, a fluidity increase may be the result of protein aggregation leaving extended areas of free bilayer where the probes experience increased freedom of motion. Similar conclusions were reached in a study of increased fluidity in erythrocyte ghosts induced by synthetic water-soluble polymers.[190] It remains to be explained by which mechanism transmethylation induces protein aggregation.

C. Lipid Fluidity and Other Cellular Processes

Membrane fluidity appears of general importance for all cellular activities, and fluidity changes are associated with cell cycle[191] and growth,[156] cell maturation and differentiation,[3,192] recognition and intercellular contacts,[193] fusion,[194] and with malignant transformation.[195] The time-scale of the changes may vary from days (the changes will involve several cell divisions and distinct changes in membrane composition) to hours (it is likely that the fluidity changes involve chemical or physical modulators).[49]

Escherichia coli fatty acid auxotrophs are unable to grow when the amount of gel state lipid exceeds 50%,[196] and similar observations have been made for the simple organism *Acholeplasma laidlawii B* which is incapable of adaptation to ambient temperature[197] (see Section VII).

It has long been realized that the receptors on the cell surface change their relative responses along with various cellular processes;[198] in some cases this modulation could be accounted for by the turnover and membrane biogenesis, whereas in others active modulation was postulated, namely, by structural changes in the cytoskeleton network to which the receptors are considered to be anchored.[199] This mechanism requires metabolic energy and is blocked by metabolic inhibitors.[3] Shinitzky[49] also postulates a mechanism of passive modulations in which rapid changes in membrane viscosity could be involved in vertical movements of membrane receptors, thus promoting receptor aggregation under certain circumstances.[200]

Also, membrane antigens may be subjected to passive modulation: antigenic expression is enhanced by increase of membrane viscosity and is suppressed by a viscosity decrease, obtained through variations of the cholesterol level.[201]

The phenomena of antibody-induced patch and cap formation have provided evidence for

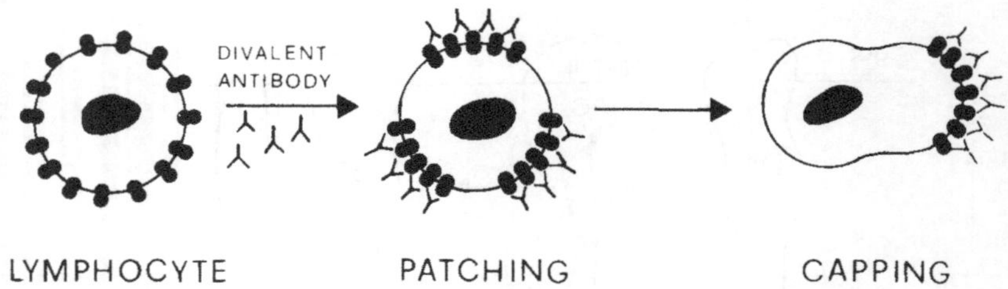

FIGURE 23. Schematic drawing of patch and cap formation.

rapid lateral diffusion of membrane proteins.[63,202] In initial experiments, surface immunoglobulins of mouse spleen cells were labeled with fluorescent antibodies; if the labeling was carried out at 0°C, the fluorescence was observed in patches over the cell surface, whereas at 37°C it was found concentrated in a cap at one pole of the cell. The process can be elicited also with lectin receptors and was investigated by immunoferritin electron microscopy to detect aggregates too small to be seen by light microscopy. Surface antigens are usually random, but interaction with multivalent antibodies or lectins induces clustering to form patches; patch formation depends on lateral diffusion, but the subsequent aggregation to form a polar cap is energy-dependent (Figure 23). Disruption of microtubules but not of microfilaments induces capping by contraction and redistribution by the microfilament system, whereas disruption of both microtubules and microfilaments induces only patches.[3] Such phenomena may be involved in a variety of cell surface phenomena requiring specific controlled membrane movements, such as locomotion, receptor-mediated endocytosis, etc. The mitogenic effect of antigen binding to superficial immunoglobulin receptors[203] or of lectin binding to their receptors,[204] the stimulatory action of proteolytic cleavage,[205] and the intracellular changes preceding cell division, such as those in cyclic nucleotide concentration,[206] raise important questions concerning the role of receptor mobility as a stimulus for triggering intracellular changes.[199]

D. Pathological Alterations of Lipid Fluidity

Abnormalities of membrane fluidity have been observed in a wide range of diseases, ranging from myocardial ischemic damage and infarction,[207] hematological disorders,[208] liver disease,[209] muscular dystrophies,[210-212] respiratory distress syndrome in newborns (quoted by Shinitzky and Barenholz[31]), and neoplasia.[193,195,213—215]

In some cases the fluidity changes were detected in the diseased tissue, whereas in others the changes were found in other membranes, in particular, in the blood cells.[216] Studies of erythrocyte membrane fluidity in several types of nonhematological disorders offer the possibility not only of an investigation of the pathogenesis of systemic diseases, but also to device diagnostic tests through simple assays of little invasiveness. An increase in fluidity of the erythrocyte membrane, observed on the basis of EPR spin label studies in myotonic[210] and Duchenne muscular dystrophy,[211,212] may be a secondary consequence of a systemic altered protein lipid organization in plasma membranes,[210] possibly through cytoskeletal organization changes, and is amenable to investigation in the well-characterized erythrocyte membrane.

In liver disease[209,216] a decreased fluidity of the erythrocyte membrane finds its explanation in the abnormalities in the metabolism and composition of plasma lipoproteins,[217] with consequent incorporation of abnormal lipid ratios in the erythrocyte membrane (enrichment of cholesterol and lecithin, with increased cholesterol/phospholipid and PC/sphingomyelin molar ratios); these changes in membrane lipid composition affect the structure and properties of erythrocytes.[218]

In myocardial damage,[207] abnormalities in lipid metabolism are not only involved at a systemic level in the pathogenesis of atherosclerosis, but also locally in the cardiac cell membranes of ischemic heart. Such changes may be initiated by activation of membrane-bound phospholipases[219] and hydrolysis of membrane phospholipids, with both depletion of the normal phospholipids and accumulation of hydrolysis products, which form mixed micelles and disrupt membranes by a detergent-like action. More subtle changes by smaller amounts of the amphiphilic hydrolysis products could involve fluidity changes and all the consequent modifications that were reported in Section VI.A.

One particular case to which substantial research is addressed is neoplastic disease. Changes in membrane dynamics have been reported upon malignant transformation, and related to membrane fluidity,[193] although the types of changes observed and their relevance to the malignant process are still controversial. While an increased membrane fluidity was observed in leukemic cells, the presence of more rigid membranes, accompanied by increases of cholesterol, is observed in solid tumors as hepatomas.[50] The differences in the levels of cholesterol and consequent fluidity in the two types of neoplasia may depend on the different origin of membrane cholesterol, from serum in the circulating cell types and from endogenous biosynthesis in the solid tumor cells. The malignant cells are also accompanied by a profound derangement of normal lipid metabolism and distribution within the cell; in particular, in tumors the synthesis of cholesterol is deregulated and loses control early after exposure to carcinogenic events and much before the appearance of malignant growth.[220] A large increase of cholesterol in mitochondria,[221] where the sterol is normally absent, has been involved in a series of metabolic changes related to the development of the disease[220] (Figure 24). Furthermore cholesterol enrichment of tumor cell membranes is proposed to potentiate the exfoliation of the tumor cell plasma membrane in the form of antigenically competent, cholesterol-rich microvesicules.[195] Such vesicle shedding is conceived to enhance tumor proliferation by allowing the tumor cell to escape the host immunologic processes which are diverted by the large amount of immunocompetent material present in the circulation.[220]

To this purpose the concept of vertical membrane displacement and passive modulation of membrane receptors[49] can be extended to membrane antigens, opening new avenues to study and correct tumor antigenicity. It is conceivable that an increased exposure of antigenic determinants could facilitate the immunological defenses. On the other hand, autoimmune diseases may be associated with overexposure of membrane antigenic determinants, so that their availability should be suppressed in order to correct the pathological state.[49]

VII. BIOCHEMICAL REGULATION OF MEMBRANE FLUIDITY

In addition to the chemical and physical factors discussed in Section IV which can contribute to a relatively fast passive modulation of membrane responses, membrane fluidity can be regulated by long-range control mechanisms at the genetic and metabolic level.[15]

The biosynthesis of cell membranes involves the process of differentiation in which a preexisting membrane is modified by the addition, removal, degradation, or chemical modification of the respective constituents, lipids, and proteins. Any process that can interfere with membrane differentiation can alter the membrane fluidity. For a detailed analysis of the biochemical control of fluidity see Quinn.[15]

One of the primary sources of unsaturated fatty acyl residues of membrane lipids in animals is diet. On the other hand, nutritional supplementation of the growth medium in microorganisms and cells in tissue culture has been used to manipulate membrane fluidity as an experimental tool to investigate not only the consequences of changed fluidity on membrane processes, but also the biochemical mechanisms responsible for regulating fluidity.[156] Supplementation of polar constituents, fatty acids, and cholesterol or other sterols have all been found to result to various extents into incorporation in the membranes and to modify fluidity and physiological characteristics of the cells.

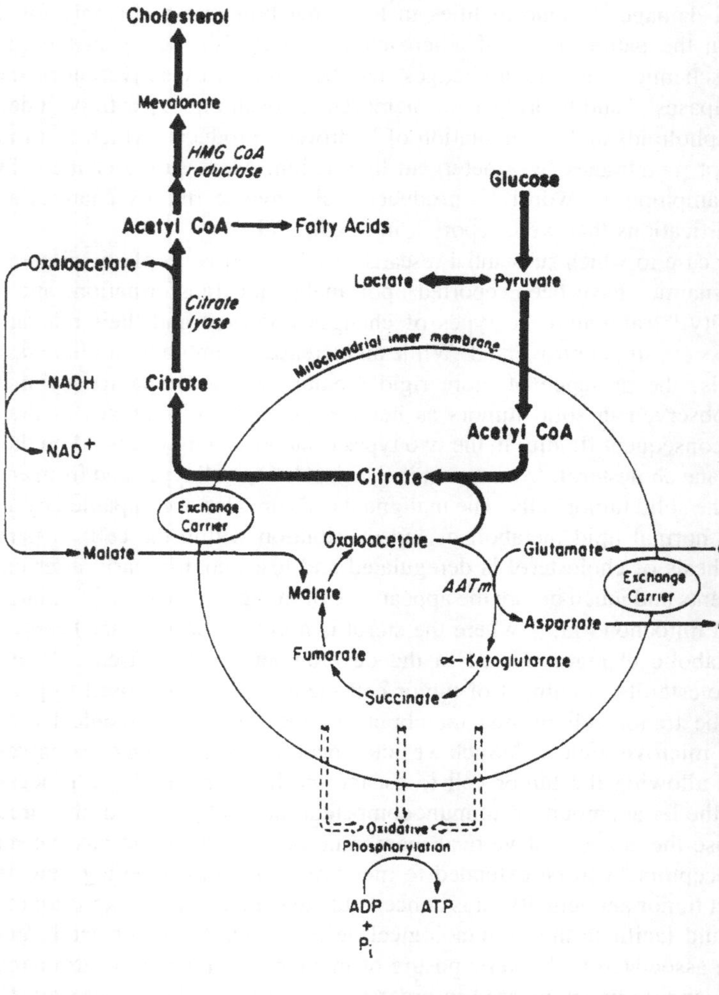

FIGURE 24. Mitochondrial derangements in cancer. Rerouting of glucose carbons and concomitant changes in respiratory metabolic patterns subsequent to the early deregulation of cholesterol synthesis in tumors. Enhanced cholesterol synthesis alters the balance of substrates supporting respiration and phosphorylation. (Reproduced from Coleman, P. S. and Lavietes, B. B., *Crit. Rev. Biochem.*, 11, 341, 1981.)

Studies in higher animals include dietary regimens and the investigation of lipid metabolic pathways under different conditions including the use of drugs affecting lipid metabolism. Laboratory animals respond to a deficiency of the polyunsaturated essential fatty acids of the 3 and 6 series by a marked increase of stearoyl CoA desaturase in an attempt to compensate for lack of polyenoic acids by increased formation of oleic acid.[222] The induction of this enzyme involves obligatory insulin and there is an interdependence of insulin and dietary intake of fats in controlling hepatic desaturase activity.[223] In animals, the incorporation of fatty acids into phospholipids competes with the desaturase reactions, so that no further desaturation can take place once the fatty acid is incorporated into the membrane lipids.[224] The type of fatty acids that are associated with membrane lipids will reflect the amount and type of fatty acid derived from the diet and the extent to which these are subject to metabolic interconversions in the animal before they become incorporated into the lipids. A self-regulatory mechanism for the control of membrane fluidity through the activity of acyl CoA desaturases has been considered.[225] Since the electron transport system required for acyl

CoA desaturase is dependent on the translational motion of the competent proteins in the endoplasmic reticulum membranes[164] (Section VI.A), a decreased fluidity would be expected to reduce the overall rate; nevertheless, if the components were confined to restricted domains of remaining fluid lipids, the probability of collisions would be enhanced. An increase of stearoyl CoA desaturase activity of animals fed an EFA-deficient diet[225] is in line with this hypothesis. In spite of the adaptation mechanisms to maintain unsaturation, EFA-deficient diets are associated with an overall decrease of membrane fluidity, as found in a series of studies by Farias and co-workers.[154] The double bond index/saturation ratio, indicative of fluidity, is decreased in membranes after fat-free diets, and such decrease is accompanied by allosteric changes in several membrane-bound enzymes.

The incorporation of acyl CoAs into phospholipids obeys strict specificities;[226] thus, acylation of C-2 of glycerol occurs mainly with unsaturated fatty acyl CoAs whereas acylation at C-1 occurs mainly with saturated or trans-unsaturated acyl CoAs. The dominant factor controlling the type of fatty acids that are associated with membrane lipids is the availability of the acyl CoAs, which is ultimately controlled by the activity level of the desaturases and the supply of dietary fatty acids. The activity of acyltransferases might be regulated in part by membrane fluidity itself.[15]

Also the rate of biosynthesis and turnover of individual phospholipid classes will affect fluidity; in particular, the lecithin/sphingomyelin ratio is directly proportional to fluidity[227] and can be altered in disease states, e.g., in the erythrocytes in abetalipoproteinemia. The exchange of phospholipids of different composition between subcellular membranes via the phospholipid exchange proteins or between plasma lipoproteins and membranes may be one way how membranes control their fluidity characteristics.[228] The transfer of cholesterol appears of utmost importance since this sterol is easily transferred between membranes or from plasma lipoproteins to membranes. The cholesterol content is usually constant for a specific membrane but can be easily manipulated in any membrane,[49,229] indicating that its content is not due to peculiar features of the recipient membranes, but to the cellular control mechanisms that transfer cholesterol from the site of synthesis in the endoplasmic reticulum.

The genetic approach involves altered biosynthesis of enzymes of lipid metabolism, and has been necessarily limited to microorganisms, both prokaryotes and eukaryotes.[15] Fatty acid auxotrophs of *E. coli* have been particularly useful in studies of the mechanisms regulating fluidity; mutants incapable to synthetize unsaturated fatty acids cannot grow if unsaturated fatty acids are not supplemented in the medium.[156] Such mutants have different lipid compositions when grown at different temperatures, presumably reflecting the adjustments necessary to maintain the optimal membrane fluidity. In yeasts, desaturation requires oxygen so that anaerobic growth in a fermentable substrate results in enhanced unsaturated fatty acid uptake.[230] Also, sterol mutants have shown substantial changes in membrane fluidity.[231]

Extreme examples of adaptation are those concerned with organisms responding to changes in temperature or osmotic strength of the growth environment. In general, prokaryotes respond to changes in growth temperature by synthetizing such lipids to preserve the viscosity of their membranes within physiological limits, so that increasing the growth temperature is associated with synthesis of more saturated or longer fatty acids and decreasing temperature in production of unsaturated or branched fatty acids.[232] The concept of *homeoviscous adaptation* has been advanced from this type of observation.[233]

One curious adaptive phenomenon is represented by thermophilic bacteria[234] possessing an unusual lipid composition; they have, in fact, a covalently linked double monolayer structure in the form of an isoprenoid macrocyclic tetraether virtually resulting by dimerization of two diether molecules; such structure may be particularly suited to resist high temperatures.

Also, poikilothermic animals exhibit adaptive changes in membrane lipids to environmental temperatures;[235] for example, membranes and extracted membrane lipids from cold-acclimated fish are more fluid than those of warm-acclimated fish when measured at an intermediate temperature.[236] Similar changes are observed in hibernating animals; during hibernation an increase of unsaturated fatty acids is observed in depot fats and in membranes.[237]

ABBREVIATIONS

CD, circular dichroism; DOPC, dioleylphosphatidylcholine; DPH, diphenylhexatriene; DPL, dipalmitoyllecithin; DSC, differential scanning calorimetry; E_A, activation energy; EFA, essential fatty acids; EPR, electron paramagnetic resonance; NMR, nuclear magnetic resonance; PA, palmitic acid; PC, phosphatidylcholine; PE, phosphatidylethanolamine; Q, ubiquinone; RSR, reconstituted sarcoplasmic reticulum; SR, sarcoplasmic reticulum; TEMPO, 2,2,6,6-tetramethyl-piperidine-*N*-oxyl; T_m, transition temperature (temperature of melting).

ACKNOWLEDGMENTS

Original studies from our laboratories have been performed with grants from the CNR and the Ministero della Pubblica Instruzione, Rome, Italy. The original figures have been drawn by Mr. M. Crimi and Miss E. Mandrioli.

REFERENCES

1. **Singer, S. J. and Nicolson, G. L.,** The fluid mosaic model of the structure of cell membranes. *Science,* 175, 720, 1972.
2. **Vanderkooi, G.,** Organization of proteins in membranes with special reference to the cytochrome oxidase system, *Biochim. Biophys. Acta,* 344, 307, 1974.
3. **Nicolson, G. L.,** Transmembrane control of the receptors of normal and tumor cells. I. Cytoplasmic influence over cell surface components, *Biochim. Biophys. Acta,* 457, 57, 1976.
4. **Lenaz, G., Curatola, G., Fiorini, R. M., and Parenti Castelli, G.,** Membrane fluidity and its role in the regulation of cellular processes, in *Biology of Cancer (2), Proc. 13th Int. Cancer Congr. Part C,* Mirand, E. A., Hutchinson, W. B., and Mihich, E., Eds., Alan R. Liss, New York, 1983, 25.
5. **Lenaz, G.,** Membrane fluidity, in *Biomembranes: Dynamics and Biology,* Guerra, F. C. and Burton, R. M., Eds., Plenum Press, New York, in press.
6. **Danielli, J. F. and Davson, H. A.,** A contribution to the theory of permeability of thin films, *J. Cell. Comp. Physiol.,* 5, 495, 1935.
7. **Cullis, P. R. and De Kruyff, B.** Lipid polymorphism and the functional roles of lipids in biomembranes, *Biochim. Biophys. Acta,* 559, 399, 1979.
8. **Bangham, A. D.,** Membrane models with phospholipids, *Prog. Biophys. Mol. Biol.,* 18, 29, 1968.
9. **Knight, C. G., Ed.,** *Liposomes: From Physical Structure to Therapeutic Applications,* Elsevier, Amsterdam, 1981.
10. **Lenaz, G.** Lipid properties and lipid protein interactions, in *Membrane Proteins and Their Interactions with Lipids,* Capaldi, R. A., Ed., Marcel Dekker, New York, 1977, chap. 3.
11. **Szabo, G.,** Lipid bilayer membranes, in *Membrane Molecular Biology,* Fox, C. F. and Keith, A., Eds., Sinauer, Stanford, Conn., 1972, 46.
12. **Seelig, J.,** Deuterium magnetic resonance: theory and application to lipid membranes, *Q. Rev. Biophys.,* 10, 353, 1977.
13. **Mabrey-Gaud, S.,** Differential scanning calorimetry of liposomes, in *Liposomes: From Physical Structure to Therapeutic Applications,* Knight, C. G., Ed., Elsevier, Amsterdam, 1981, chap. 5.
14. **Blaurock, A. E.,** Evidence of bilayer structure and of membrane interactions from X-ray diffraction analysis, *Biochim. Biophys. Acta,* 650, 127, 1981.
15. **Quinn, P. J.,** The fluidity of cell membranes and its regulation, *Prog. Biophys. Mol. Biol.,* 38, 1, 1981.
16. **Verkleij, A. J. and Ververgaert, P. H.,** Freeze-fracture morphology of biological membranes, *Biochim. Biophys. Acta,* 515, 303, 1978.

17. **Shimshick, E. J., Kleemann, W., Hubbell, W. L., and McConnell, H.,** Lateral phase separations in membranes. *J. Supramol. Struct.*, 1, 285, 1973.
18. **Small, D. M.,** Phase equilibrium and structure of dry and hydrated egg lecithin. *J. Lipid Res.*, 8, 551, 1967.
19. **Chapman, D., Williams, R. M., and Ladbrooke, B. D.,** Physical studies with phospholipids. VI. Thermotropic and lyotropic mesomorphism of some 1,2-diacyl-phosphatidylcholines (lecithins). *Chem. Phys. Lipids*, 1, 445, 1967.
20. **Lundberg, B., Svens, E., and Ekman, S.,** The hydration of phospholipids and phospholipid-cholesterol complexes. *Chem. Phys. Lipids*, 22, 285, 1978.
21. **Eibl, H. and Blume, A.,** The influence of charge on phosphatidic acid bilayer membranes, *Biochim. Biophys. Acta*, 553, 476, 1979.
22. **Ito, T. and Ohnishi, S.,** Ca^{2+}-induced lateral phase separations in phosphatidic acid-phosphatidylcholine membranes, *Biochim. Biophys. Acta*, 352, 29, 1974.
23. **Ladbrooke, B. D., Williams, R. M., and Chapman, D.,** Studies on lecithin-cholesterol-water interactions by differential scanning calorimetry and X-ray diffraction, *Biochim. Biophys. Acta*, 150, 333, 1968.
24. **Mabrey, S., Mateo, P. L., and Sturtevant, J. M.,** High-sensitivity scanning calorimetric study of mixtures of cholesterol with dimyristoyl and dipalmitoyl phosphatidylcholines, *Biochemistry*, 17, 2464, 1978.
25. **Cornell, B. A., Chapman, D., and Peel, W. E.,** Random close-packed arrays of membrane components, *Chem. Phys. Lipids*, 23, 223, 1979.
26. **Papahadjopoulos, D., Moscarello, M., Eylar, E. H., and Isac, T.,** Effects of proteins on thermotropic phase transitions of phospholipids membranes, *Biochim. Biophys. Acta*, 401, 317, 1975.
27. **Trumpower, B. L., Ed.,** *Function of Quinones in Energy-Conserving Systems*, Academic Press, New York, 1982.
28. **Katsikas, H. and Quinn, P. J.,** The thermotropic properties of coenzyme Q_{10} and its lower homologues, *J. Bioenerg. Biomembr.*, 15, 67, 1983.
29. **Hill, M. W.,** The effect of anesthetic-like molecules on the phase transition in smectic mesophases of dipalmitoyl lecithin, *Biochim. Biophys. Acta*, 356, 117, 1974.
30. **Shinitzky, M. and Yuli, I.,** Lipid fluidity at the submacroscopic level: determination by fluorescence polarization, *Chem. Phys. Lipids*, 30, 261, 1982.
31. **Shinitzky, M. and Barenholz, Y.,** Fluidity parameters of lipid regions determined by fluorescence polarization, *Biochim. Biophys. Acta*, 515, 367, 1978.
32. **Knowles, P. F., Marsh, D., and Rattle, H. W. E.,** *Magnetic Resonance of Biomolecules*, John Wiley & Sons, London, 1976.
33. **Browning, J. L.,** NMR studies of the structural and motional properties of phospholipids in membranes, in *Liposomes: From Physical Structure to Therapeutic Applications*, Knight, C. G., Ed., Elsevier, Amsterdam, 1981, chap. 7.
34. **Seelig, J. and Seelig, A.,** Lipid conformation in model membranes and biological membranes, *Q. Rev. Biophys.*, 13, 19, 1980.
35. **Gaffney, B. J.,** Spin label measurements in membranes, *Methods Enzymol.* 32B, 161, 1974.
36. **Berliner, L., Ed.,** *Spin Labelling: Theory and Application*, Academic Press, New York, 1976.
37. **Marsh, D. and Watts, A.,** ESR spin label studies of liposomes, in *Liposomes: From Physical Structure to Therapeutic Applications*, Knight, C. G., Ed., Elsevier, Amsterdam, 1981, chap. 6.

37a. **Benga, Gh.,** Spin labelling, in *Biochemical Research Techniques*, Wrigglesworth, J., Ed., John Wiley & Sons, London, 1983, 79.

38. **Penzer, G. R.,** Molecular emission spectroscopy (fluorescence and phosphorescence), in *An Introduction to Spectroscopy for Biochemists*, Brown, S. B., Ed., Academic Press, New York, 1980, chap. 3.
39. **Radda, G. K. and Vanderkooi, J.,** Can fluorescent probes tell us anything about membranes?, *Biochim. Biophys. Acta*, 265, 509, 1972.
40. **Beddard, G. S. and West, M. A., Eds.,** *Fluorescent Probes*, Academic Press, London, 1981.
41. **Pownall, H. J. and Smith, L. C.,** Viscosity of the hydrocarbon region of micelles: measurement of excimer formation, *J. Am. Chem. Soc.*, 95, 3136, 1973.
42. **Galla, H. J. and Sackmann, E.,** Lateral diffusion in the hydrocarbon region of membranes: use of pyrene excimers as spectral probes, *Biochim. Biophys. Acta*, 339, 103, 1974.
43. **Chapman, D.,** Some recent studies of lipids, lipid-cholesterol and membrane systems, in *Biological Membranes*, Vol. 2, Chapman, D. and Wallach, D. F. H., Eds., Academic Press, London, 1973, 91.
44. **Gaber, B. P., Yagu, P., and Peticolas, W. L.,** Interpretation of biomembrane structure by Raman spectroscopy, *Biophys. J.*, 21, 161, 1978.
45. **Curatolo, W., Verma, S. P., Sakura, J. D., Small, D. M., Shipley, G. G., and Wallach, D. F. H.,** Structural effects of myelin proteolipid apoprotein on phospholipids: a Raman spectroscopic study, *Biochemistry*, 17, 1802, 1978.
46. **Seelig, J. and Sarcevic, N.,** Molecular order in cis and trans unsaturated phospholipid bilayers, *Biochemistry*, 17, 3310, 1978.

47. **Gally, H. U., Pluschke, G., Overath, P., and Seelig, J.,** Structure of *E. coli* membranes. Phospholipid conformation in model membranes and cells as studied by deuterium magnetic resonance, *Biochemistry,* 18, 5605, 1979.
48. **Davis, J. H., Nichol, C. P., Werks, G., and Bloom, M.,** Study of the cytoplasmic and outer membrane of *E. coli* by deuterium magnetic resonance, *Biochemistry,* 18, 2103, 1979.
49. **Shinitzky, M.,** The concept of passive modulation of membrane responses, in *Physical Chemical Aspects of Cell Surface Events in Cellular Regulation,* DeLisi, C. and Blumenthal, R., Eds., Elsevier, Amsterdam, 1979, 173.
50. **Shinitzky, M. and Inbar, M.,** Microviscosity parameters and protein mobility in biological membranes, *Biochim. Biophys. Acta,* 433, 133, 1976.
51. **Raison, J. K. and McMurchie, E. J.,** Two temperature-induced changes in mitochondrial membranes detected by spin labelling and enzyme kinetics, *Biochim. Biophys. Acta,* 363, 135, 1974.
52. **Morrisett, J. D., Pownall, H. J., Plumlee, R. T., Smith, L. C., Zehner, Z. E., Esfahani, M., and Wakil, S. J.,** Multiple thermotropic phase transitions in *E. coli* membranes and membrane lipids, *J. Biol. Chem.,* 250, 6969, 1975.
53. **Lenaz, G., Curatola, G., Mazzanti, L., Zolese, G., and Ferretti, G.,** Electron spin resonance studies of the effects of lipids on the environment of proteins in mitochondrial membranes, *Arch. Biochem. Biophys.,* 223, 369, 1983.
54. **Lenaz, G.,** The role of lipids in the structure and function of membranes, *Subcell. Biochem.,* 6, 233, 1979.
55. **Trudell, J. R., Hubbell, W. L., and Cohen, E. N.,** Pressure reversal of inhalation anesthetic-induced disorder of spin-labelled phospholipid vesicles, *Biochim. Biophys. Acta,* 291, 328, 1973.
56. **Jahnig, F., Harlos, K., Vogel, H., and Heibl, H.,** Electrostatic interactions of charged lipid membranes. Electrostatically induced tilt, *Biochemistry,* 18, 1459, 1979.
57. **Corda, D., Pasternak, C., and Shinitzky, M.,** Increase in lipid microviscosity of unilamellar vesicles upon creation of transmembrane potential, *J. Membr. Biol.,* 65, 235, 1982.
58. **Davis, J. H., Bloom, M., Butler, K. W., and Smith, I. C. P.,** The temperature dependence of molecular order and the influence of cholesterol in *A. laidlawii* membranes, *Biochim. Biophys. Acta,* 597, 497, 1980.
59. **Ghosh, R. and Seelig, J.,** The interaction of cholesterol with bilayers of phosphatidylethanolamine, *Biochim. Biophys. Acta,* 691, 151, 1982.
60. **Lenaz, G., Curatola, G., and Parenti Castelli, G.,** The role of lipid fluidity in the structure and function of heart mitochondrial membranes, in *Advances in Studies on Heart Metabolism,* Caldarera, C. M. and Harris, P., Eds., CLUEB, Bologna, 1982, 15.
61. **Spisni, A., Masotti, L., Lenaz, G., Bertoli, E., Pedulli, F., and Zannoni, C.,** Interactions between ubiquinones and phospholipid bilayers, *Arch. Biochem. Biophys.,* 190, 454, 1978.
62. **Folkers, K. and Yamamura, Y., Eds.,** *Biomedical and Clinical Aspects of Coenzyme Q,* Vol. 3, Elsevier, Amsterdam, 1981.
63. **Jendrasiak, G. L. and Hasty, J. H.,** Hydration of phospholipids, *Biochim. Biophys. Acta,* 337, 79, 1974.
64. **Boicelli, C. A., Conti, F., Giomini, M., and Giuliani, A. M.,** Water organization in reversed micelles, in *Physical Methods in Biological Membranes,* Conti, F., Ed., Plenum Press, New York, in press.
65. **Unwin, P. N. and Henderson, R.,** Molecular structure determination by electron microscopy of unstained crystalline specimens, *J. Mol. Biol.,* 94, 425, 1975.
66. **Ovchinnikov, Y., Abdulaev, N., Feigira, M., Kiselev, M., and Lobanov, N.,** The structural basis of the functioning of bacteriorhodopsin: an overview, *FEBS Lett.,* 100, 219, 1979.
67. **Khorana, H. G., Gerberg G. E., Herlihy, W. C., Gray, C. P., Anderegg, R. S., and Nineik-Bremann, K.,** Aminoacid sequence of bacteriorhodopsin, *Proc. Natl. Acad. Sci. U.S.A.,* 76, 5046, 1979.
68. **Engelmann, D. M., Henderson, R., McLachlan, A., and Wallace, B. A.,** Path of the polypeptide in bacteriorhodopsin, *Proc. Natl. Acad. Sci. U.S.A* , 77, 2023, 1980.
69. **Jost, P. C., Griffith, O. H., Capaldi, R. A., and Vanderkooi, G.,** Identification and extent of fluid bilayer regions in membranous cytochrome oxidase, *Biochim. Biophys. Acta,* 311, 141, 1973.
70. **Marcelja, S.,** Lipid mediated protein interaction in membranes, *Biochim. Biophys. Acta,* 455, 1, 1976.
71. **Dehlinger, P. J., Jost, P. C., and Griffith, O. H.,** Lipid binding to the amphipathic membrane protein cytochrome b_5, *Proc. Natl. Acad. Sci. U.S.A.,* 71, 2280, 1974.
72. **Boggs, J. M., Vail, W. J., and Moscarello, M. A.,** Preparation and properties of vesicles of a purified myelin hydrophobic protein and phospholipid, *Biochim. Biophys. Acta,* 448, 517, 1974.
73. **Laggner, P. and Barratt, M. D.,** The interaction of a proteolipid from sarcoplasmic reticulum membranes with phospholipids, *Arch. Biochem. Biophys.,* 170, 92, 1975.
74. **Warren, G. B., Bennett, J. P., Hesketh, T. R., Houslay, M. D., Smith, G. A., and Metcalfe, J. C.,** The lipids surrounding a calcium transport protein: their role in calcium transport and accumulation, *FEBS Proc. Meet.,* 3, 1975.
75. **Jost, P. C., Nadakavukaren, K. K., and Griffith, O. H.,** Phosphatidylcholine exchange between the boundary lipid and bilayer domains in cytochrome oxidase containing membranes, *Biochemistry,* 16, 3110, 1977.

76. **Brotherus, J. R., Jost, P. C., Griffith, O. H., Keana, J. F., and Hokin, L. E.,** Charge selectivity of the lipid protein interaction in Na,K-ATPase, *Proc. Natl. Acad. Sci. U.S.A.*, 77, 272, 1980.
77. **Warren, G. B., Houslay, M. D., Metcalfe, J. C., and Birdsall, N. J. M.,** Cholesterol is excluded from the phospholipid annulus surrounding an active calcium transport protein, *Nature (London)*, 255, 684, 1975.
78. **Chapman, D., Gomez-Fernandez, J. C., and Goñi, F. M.,** Intrinsic protein lipid interactions, *FEBS Lett.*, 98, 211, 1979.
79. **Favre, E., Baroin, A., Bienvenue, A., and Devaux, P.,** Spin label studies of lipid protein interactions in retinal rod outer segment membranes. Fluidity of the boundary layer, *Biochemistry*, 18, 1156, 1979.
80. **Nakamura, H. and Martonosi, A. N.,** Effect of phospholipid substitution on the mobility of protein-bound spin labels in sarcoplasmic reticulum, *J. Biochem.*, 87, 525, 1980.
81. **Ariano, B. and Azzi, A.,** Rotational motion of cytochrome oxidase in phospholipid vesicles, *Biochem. Biophys. Res. Commun.*, 93, 478, 1980.
82. **Seelig, J., Tamm, L., Hymel, L., and Fleischer, S.** Deuterium and phosphorus NMR and fluorescence depolarization studies of functional reconstituted sarcoplasmic reticulum membrane vesicles, *Biochemistry*, 20, 3922, 1981.
83. **Stoffel, W., Zierenberg, D., and Scheefers, H.,** Reconstitution of Ca^{2+}-ATPase of sarcoplasmic reticulum with ^{13}C-labelled lipids, *Hoppe-Seyler's Z. Physiol. Chem.*, 358, 865, 1977.
84. **Vik, S. B., Georgevich, G., and Capaldi, R. A.,** Diphosphatidyl glycerol is required for optimal activity of beef heart cytochrome oxidase, *Proc. Natl. Acad. Sci. U.S.A.*, 78, 1456, 1981.
85. **Fry, M. and Green, D. E.,** Cardiolipin requirement for electron transfer in complex I and III of the mitochondrial respiratory chain, *J. Biol. Chem.*, 256, 1874, 1981.
86. **Borochov, H., Abbot, R. E., Schachter, D., and Shinitzky, M.,** Modulation of erythrocyte membrane proteins by membrane cholesterol and lipid fluidity, *Biochemistry*, 18, 251, 1979.
87. **Chapman, D., Cornell, B. A., and Quinn, P. J.,** Phase transitions, protein aggregation and a new method for modulating membrane fluidity, in *Biochemistry of Membrane Transport*, Semenza, G. and Carafoli, E., Eds., Springer, Berlin, 1977, 72.
88. **Quinn, P. J. and Chapman, D.** The dynamics of membrane structure, *Crit. Rev. Biochem.*, 8, 1, 1980.
89. **Blazyk, J. F., and Steim, J. M.,** Phase transitions in mammalian membranes, *Biochim. Biophys. Acta*, 266, 737, 1972.
90. **Shechter, E., Letellier, L., and Gulik, T.,** Relation between structure and function in cytoplasmic membrane vesicles isolated from *E. coli* fatty acid auxotroph, *Eur. J. Biochem.*, 49, 61, 1974.
91. **Kleemann, W. and McConnell, H. M.,** Interactions of proteins and cholesterol with lipids in bilayer membranes, *Biochim. Biophys. Acta*, 419, 206, 1976.
92. **Haest, C. W. M., Verkleij, A. J., De Gier, J., Sheek, R., Ververgaert, P. H. J., and Van Deenen, L. L. M.,** The effect of lipid phase transitions on the architecture of bacterial membranes, *Biochim. Biophys. Acta*, 356, 17, 1974.
93. **Rottem, S., Cirillo, V. P., De Kruyff, B., Shinitzky, M., and Razin, S.,** Cholesterol in mycoplasma membranes: correlation of enzymic and transport activities with physical state of lipids in membranes of *M. mycoides* var. *capri* adapted to grow with low cholesterol concentration, *Biochim. Biophys. Acta*, 323, 509, 1973.
94. **Hui, S. W., Stewart, C. M., Carpenter, M. P., and Stewart, T. P.,** Effects of cholesterol on lipid organization in human erythrocyte membrane, *J. Cell Biol.* , 85, 283, 1980.
95. **Cherry, R. J., Müller, U., Holenstein, C., and Heyn, M. P.,** Lateral aggregation of proteins induced by cholesterol in bacteriorhodopsin-phospholipid vesicles, *Biochim. Biophys. Acta*, 596, 145, 1980.
96. **Mühlebach, C. and Cherry, R. J.,** Influence of cholesterol on the rotation and self-association of band 3 in the human erythrocyte membrane, *Biochemistry*, 21, 4225, 1982.
97. **Lenaz, G., Curatola, G., and Masotti, L.,** Perturbation of membrane fluidity, *J. Bioenerg.*, 7, 223, 1975.
98. **Lenaz, G., Curatola, G., Mazzanti, L., and Parenti Castelli, G.,** Biophysical studies on agents affecting the state of membrane lipids, *Mol. Cell. Biochem.*, 22, 3, 1978.
99. **Lenaz, G., Curatola, G., Mazzanti, L., Parenti Castelli, G., and Bertoli, E.,** Effect of general anesthetics of lipid protein interactions and ATPase activity in mitochondria, *Biochem. Pharmacol.*, 27, 2835, 1978.
100. **Seeman, P.,** The membrane action of anesthetics and tranquillizers, *Pharmacol. Rev.*, 24, 583, 1972.
101. **Franks, N. P. and Lieb, W. R.,** Molecular mechanisms of general anesthesia, *Nature (London)*, 300, 487, 1982.
102. **Pasquali-Ronchetti, I., Curatola, G., Mazzanti, L., Lenaz, G., and Bertoli, E.,** Effect of some general anesthetics on isolated synaptic membranes, *Ultramicroscopy* 5, 394, 1980.
103. **Mazzanti, L. and Lenaz, G.,** Effetti della somministrazione in vivo di ketamina in membrane sinaptiche di cervello di ratto, *Abstr. 28th Congr. It. Soc. Biochem.*, 465, 1982.
104. **Trudeau, G., Cole, K. C., Masruda, R., and Sandorfy, C.,** Anesthesia and hydrogen bonding. A semiquantitative infrared study at room temperature, *Can. J. Chem.*, 56, 168, 1978.

105. **Haydon, D. A., Hendry, B. M., Levinson, S. R., and Raquena, J.,** Anesthesia by the n-alkanes, *Biochim. Biophys. Acta*, 470, 17, 1977.
106. **Cornell, B. and Pope, J. M.,** Low frequency and diffuse motion in aligned phospholipid multilayers studied by pulsed NMR, *Chem. Phys. Lipids*, 27, 151, 1980.
107. **Frye, L. D. and Edidin, M.,** The rapid intermixing of cell surface antigens after formation of mouse-human heterokaryons, *J. Cell Sci.*, 7, 319, 1970.
108. **Cherry, R. J.,** Rotational and lateral diffusion of membrane proteins, *Biochim. Biophys. Acta*, 559, 289, 1979.
109. **Nicolson, G. L.,** Temperature-dependent mobility of concanavalin A sites on tumor cell surface, *Nature (London), New Biol.*, 243, 218, 1973.
110. **Pinto da Silva, P. and Branton, D.,** Membrane intercalated particles: the plasma membrane as a planar fluid domain, *Chem. Phys. Lipids*, 8, 265, 1972.
111. **Elgsaeter, A., Shotton, D. M., and Branton, D.,** Intramembrane particle aggregation in erythrocyte ghosts: the influence of spectrin aggregation, *Biochim. Biophys. Acta*, 426, 101, 1976.
112. **Sowers, A. E. and Hackenbrock, C. R.,** Rate of lateral diffusion of intramembrane particles: measurement by electrophoretic displacement and rerandomization, *Proc. Natl. Acad. Sci. U.S.A.*, 78, 6246, 1981.
113. **Cherry, R. J.,** Measurement of protein rotational diffusion in membranes by flash photolysis, *Methods Enzymol.*, 54, 47, 1978.
114. **Hyde, J. S. and Dalton, L. R.,** Saturation transfer of spectroscopy, in *Spin Labelling, II, Theory and Application*, Berliner, L. J, Ed., Academic Press, New York, 1979, chap. 1.
115. **Cone, R. A.,** Rotational diffusion of rhodopsin in the visual receptor membrane, *Nature (London), New Biol.*, 236, 39, 1972.
116. **Daemen, F. J. M.,** Vertebrate rod outer segment membranes, *Biochim. Biophys. Acta*, 300, 255, 1973.
117. **Naqvi, R. K., Gonzalez-Rodriguez, J., Cherry, R. J., and Chapman, D.,** Spectroscopic technique for studying protein rotation in membranes, *Nature (London), New Biol.*, 245, 249, 1973.
118. **Hoffman, W., Sarzala, M. G., and Chapman, D.,** Rotational motion and evidence for oligomeric structures of sarcoplasmic reticulum Ca-activated ATPase, *Proc. Natl. Acad. Sci. U.S.A.*, 76, 3860, 1979.
119. **Kawato, S., Lehner, C., Müller, M., and Cherry, R. J.,** Protein-protein interactions in cytochrome oxidase in inner mitochondrial membranes, *J. Biol. Chem.*, 257, 6470, 1982.
120. **Baroin, A., Bienvenue, D. D., and Devaux, P. F.,** Spin label studies of protein-protein interaction in retinal rod outer segment membranes. Saturation transfer EPR spectroscopy, *Biochemistry*, 18, 1151, 1979.
121. **Nigg, E. A. and Cherry, R. J.,** The influence of temperature and cholesterol on the rotational diffusion of band 3 in the human erythrocyte membrane, *Biochemistry*, 16, 3457, 1979.
122. **Tyson, C. A., Vande Zande, H., and Green, D. E.,** Phospholipids as ionophores, *J. Biol. Chem.*, 251, 1326, 1976.
123. **Van der Bessalaar, A. M., De Kruyff, B., Van der Bosch, H., and Van Deenen, L. L. M.,** Phosphatidylcholine mobility in liver microsomal membranes, *Biochim. Biophys. Acta*, 510, 242, 1978.
124. **Kornberg, R. D. and McConnell, H. M.,** Inside-outside transitions of phospholipids in vesicles and membranes, *Biochemistry*, 10, 1111, 1971.
125. **Haest, C. W. M.,** Interactions between membrane skeleton proteins and the intrinsic domain of the erythrocyte membrane, *Biochim. Biophys. Acta*, 694, 331, 1982.
126. **Chiu, D., Lubin, B., Roelofsen, B., and Van Deenen, L. L. M.,** Sickle erythrocytes accelerate clotting in vitro. An effect of abnormal membrane lipid asymmetry, *Blood*, 58, 398, 1981.
127. **Papahadjopoulos, D. and Kimelberg, H. K.,** Phospholipid vesicles (liposomes) as models for biological membranes: their properties and interactions with cholesterol and proteins, in *Progress in Surface Science*, Davison, S. G., Ed., Pergamon Press, Oxford, 1974, 41.
128. **Papahadjopoulos, D. and Ohki, S.,** Stability of asymmetric phospholipid membranes, *Science*, 164, 1075, 1969.
129. **Papahadjopoulos, D.** Phospholipid membranes as experimental models for biological membranes, in *Biological Horizons in Surface Science*, Prince, L. and Sears, D. F., Eds., Academic Press, New York, 1973, 60.
130. **Papahadjopoulos, D., Jacobson, K., Nir, S., and Isac, T.,** Phase transitions in phospholipid vesicles. Fluorescence polarization and permeability measurements concerning the effect of temperature and cholesterol, *Biochim. Biophys. Acta*, 311, 330, 1973.
131. **Singer, S. J.,** The molecular organization of biological membranes, in *Structure and Function of Biological Membranes*, Rothfield, L., Ed., Academic Press, New York, 1971, 45.
132. **Fasman, G. D.,** Factors responsible for conformational stability, in *Poly-α-Aminoacids*, Fasman, G. D., Ed., Marcel Dekker, New York, 1967, 499.
133. **Fleischer, S., Fleischer, B., and Stoeckenius, W.,** Fine structure of lipid-depleted mitochondria, *J. Cell Biol.*, 32, 193, 1967.
134. **Holzwarth, G.,** Ultraviolet spectroscopy of biological membranes, in *Membrane Molecular Biology*, Fox, C. F. and Keith, A., Eds., Sinauer, Stamford, 1972, 228.

135. **Urry, D. W. and Long, M. M.,** Ultraviolet absorption, circular dichroism and optical rotatory dispersion in biomembrane studies, in *Physiology of Membrane Disorders*, Andreoli, T., Hoffman, J. F., and Fanestil, D. D., Eds., Plenum Press, New York, 1978, 107.
136. **Sherman, G. and Folch-Pi, J.,** Rotatory dispersion and circular dichroism of brain "proteolipid" protein, *J. Neurochem.*, 17, 597, 1970.
137. **London, Y., Demel, R. A., Van Kassel, G. W., Zahler, P., and Van Deenen, L. L. M.,** The interaction of the "Folch-Lees" protein with lipids at the air-water interface, *Biochim. Biophys. Acta*, 332, 69, 1972.
138. **Laggner, P.,** A highly α-helical structure protein in sarcoplasmic reticulum membranes, *Nature (London)*, 255, 427, 1975.
139. **Segrest, J. P.,** The erythrocyte: topomolecular anatomy of MN-glycoprotein, in *Mammalian Cell Membranes*, Vol. 3, Jamieson, G. A., and Robson, D. M., Eds., Butterworths, London, 1977, 1.
140. **Bach, D., Rosenheck, K., and Miller, I. R.,** Interaction of basic polypeptides with phospholipid vesicles; conformational studies, *Eur. J. Biochem.*, 53, 265, 1975.
141. **Solomon, B. and Miller, I. R.,** Interaction of glucose oxidase with phospholipid vesicles, *Biochim. Biophys. Acta*, 455, 332, 1976.
142. **Jackson, R. L., Morrisett, J. D., Pownall, H. J., and Gotto, A. M.,** Human HDL, apolipoprotein Glutamine II: the immunochemical and lipid binding properties of apolipoprotein Glutamine II derivatives, *J. Biol. Chem.*, 248, 5218, 1973.
143. **Segrest, J. P., Jackson, R. I., Morrisett, J. D., and Gotto, A. M.,** A molecular theory of lipid protein interactions in the plasma lipoproteins, *FEBS Lett.*, 38, 247, 1974.
144. **Masotti, L., Lenaz, G., Spisni, A., and Urry, D. W.,** Effect of phospholipids on the protein conformation of the inner mitochondrial membrane, *Biochem. Biophys. Res. Commun.*, 56, 892, 1974.
145. **Yu, C. A. and Yu, L.,** Structural role of phospholipids in ubiquinol cytochrome c reductase, *Biochemistry*, 19, 5715, 1980.
146. **Zahler, W. L., Puett, D., and Fleischer, S.,** Circular dichroism of mitochondrial membranes before and after extraction of lipids and surface proteins, *Biochim. Biophys. Acta*, 255, 365, 1972.
147. **Curatola, G., Fiorini, R. M., Solaini, G., Baracca, A., Parenti Castelli, G., and Lenaz, G.,** Temperature dependent conformational changes in isolated oligomycin sensitive ATPase, *FEBS Lett.*, 155, 131, 1983.
148. **Johansson, A., Keightley, C. A., Smith, G. A., Richards, C. D., Hesketh, T. R., and Metcalfe, J. C.,** The effect of bilayer thickness and n-alkanes on the activity of the Ca,Mg-dependent ATPase of sarcoplasmic reticulum, *J. Biol. Chem.*, 256, 1643, 1981.
149. **Davis, D. G., Inesi, G., and Gulik, T.,** Lipid molecular motion and enzyme activity in sarcoplasmic reticulum membranes, *Biochemistry*, 15, 1271, 1976.
150. **Drost-Hansen, W.,** Structure of water near solid interfaces, *Ind. Eng. Chem.*, 61, 10, 1969.
151. **Drost-Hansen, W.,** Thermal anomalies in aqueous systems. Manifestations of interfacial phenomena?, *Chem. Phys. Lett.*, 2, 647, 1969.
152. **Sandermann, H.,** Regulation of membrane enzymes by lipids, *Biochim. Biophys. Acta*, 515, 209, 1978.
153. **Fourcans, B. and Jain, K. M.,** Role of phospholipids in enzymatic and transport reactions, *Adv. Lipid Res.*, 12, 147, 1974.
154. **Farias, R. N., Bloj, B., Morero, R. D., Siñeriz, F., and Trucco, R. E.,** Regulation of allosteric membrane-bound enzymes through changes in membrane lipid composition, *Biochim. Biophys. Acta*, 415, 231, 1975.
155. **Bennett, J. P., McGill, K. A., and Warren, G. B.,** The role of lipids in the functions of a membrane protein: the sarcoplasmic reticulum Ca-pump, *Curr. Top. Membr. Transp.*, 14, 127, 1980.
156. **McElhaney, R. N.,** Effects of membrane lipids on transport and enzymic activities, *Curr. Top. Membr. Transp.*, 17, 317, 1982.
157. **Parenti Castelli, G., Sechi, A. M., Landi, L., Cabrini, L., Mascarello, S., and Lenaz, G.,** Lipid protein interactions in mitochondria. VII. A comparison of the effects of lipid removal and lipid perturbation on the kinetic properties of mitochondrial ATPase, *Biochim. Biophys. Acta*, 547, 161, 1979.
158. **Raison, J. K.,** The influence of temperature-induced changes on the kinetics of respiratory and other membrane-associated enzyme systems, in *Membrane Structure and Mechanisms of Biological Energy Transduction*, Avery, J., Ed., Plenum Press, London, 1973, 559.
159. **Parenti Castelli, G., Baracca, A., Fato, R., and Rabbi, A.,** A temperature-dependent structural change in mitochondrial ATPase, *Biochem. Biophys. Res. Commun.*, 111, 366, 1983.
160. **Hidalgo, C., Ikemoto, N., and Gergely, J.,** Role of phospholipids in the calcium-dependent ATPase of sarcoplasmic reticulum, *J. Biol. Chem.*, 250, 4224, 1976.
161. **Parenti Castelli, G., Solaini, G., Baracca, A., Curatola, G., Fiorini, R. M., and Lenaz, G.,** A conformational role of lipids in beef heart mitochondrial ATPase, *J. Mol. Cell Cardiol.*, 15, (Suppl. 1, Abstr. 268), 1983.
162. **Spatz, L. and Strittmatter, P.,** A form of NADH-cytochrome b_5 reductase containing both the catalytic site and additional hydrophobic membrane-binding segment, *J. Biol. Chem.*, 248, 793, 1973.

163. **Rogers, J. M. and Strittmatter, P.,** Evidence for random distribution and translational movement of cytochrome b_5 in endoplasmic reticulum, *J. Biol. Chem.*, 249, 895, 1974.
164. **Strittmatter, P. and Rogers, M. J.** Apparent dependence of interaction between cytochrome b_5 and cytochrome b_5 reductase upon translational diffusion in dimyristoyl lecithin liposomes, *Proc. Natl. Acad. Sci. U.S.A.*, 72, 2658, 1975.
165. **Swanson, M. S., Quintanilha, A. T., and Thomas, D. D.,** Protein rotational mobility and lipid fluidity of purified and reconstituted cytochrome oxidase, *J. Biol. Chem.*, 255, 7494, 1980.
166. **Hauska, G. and Hurt, E.,** Pool function behavior and mobility of isoprenoid quinones, in *Function of Quinones in Energy Conserving Systems*, Trumpower, B. L., Ed., Academic Press, New York, 1982, 87.
167. **Green, D. E.,** The mitochondrial electron transfer system, in *Comprehensive Biochemistry*, Florkin, M., and Stotz, E. H., Eds., Elsevier, Amsterdam, 1966, 309.
168. **Kröger, A. and Klingenberg, M.,** The kinetics of the redox reactions of ubiquinone related to the electron transport activity in the respiratory chain, *Eur. J. Biochem.*, 34, 358, 1973.
169. **Heron, C., Ragan, C. I., and Trumpower, B. L.,** The interaction between mitochondrial NADH-ubiquinone oxidoreductase and ubiquinol-cytochrome c oxidoreductase, *Biochem. J.*, 174, 791, 1978.
170. **Gutman, M.,** Kinetic flux through mitochondrial ubiquinone, in *Coenzyme Q*, Lenaz, G., Ed., John Wiley & Sons, London, in press.
171. **Saffmann, P. G. and Delbrück, M.,** Brownian motion in biological membranes, *Proc. Natl. Acad. Sci. U.S.A.*, 72, 3111, 1975.
172. **Capaldi, R. A.,** Arrangement of proteins in the mitochondrial inner membrane, *Biochim. Biophys. Acta*, 694, 291, 1982.
173. **Krasne, S., Eisenmann, G., and Szabo, G.,** Freezing and melting of lipid bilayers and the mode of action of nonactin, valinomycin and gramicidin, *Science*, 174, 412, 1971.
174. **Racker, E. and Hinkle, P. C.,** Effect of temperature on the function of a proton pump, *J. Membr. Biol.* 17, 181, 1974.
175. **Okamoto, H., Sone, N., Hirata, H., Yoshida, M., and Kagawa, Y.** Purified proton conductor in proton translocating ATPase of a thermophilic bacterium, *J. Biol. Chem.*, 252, 6125, 1977.
176. **Kumamoto, J., Raison, J. K., and Lyons, J. M.,** Temperature breaks in Arrhenius plots: a thermodynamic consequence of a phase change, *J. Theor. Biol.*, 31, 47, 1971.
177. **Massey, V., Curti, B., and Ganther, H.,** A temperature-dependent conformational change in D-amino acid oxidase and its effect on catalysis, *J. Biol. Chem.*, 241, 2347, 1966.
178. **Dean, W. L. and Tanford, C.,** Properties of a delipidated, detergent-activated calcium ATP-ase, *Biochemistry*, 17, 1683, 1978.
179. **Kaizu, T., Kirino, Y., and Shimizu, H.,** A saturation transfer ESR study on the break in the Arrhenius plot for the rotational motion of Ca-dependent ATPase molecules in purified and lipid-replaced preparations of rabbit skeletal muscle sarcoplasmic reticulum, *J. Biochem.*, 88, 1837, 1980.
180. **Sandermann, H.,** The reactivation of C_{55}-isoprenoid alcohol phosphokinase apoprotein by lipids: evidence for lipid hydration in lipoprotein function, *Eur. J. Biochem.*, 43, 415, 1974.
181. **Sandermann, H.,** A possible correlation between lipid hydration and lipid activation of the C_{55}-isoprenoid alcohol phosphokinase apoprotein, *Eur. J. Biochem.*, 62, 479, 1976.
182. **Levey, G. S.,** The role of phospholipids in hormonal activation of adenylate cyclase, *Recent Prog. Horm. Res.*, 29, 361, 1973.
183. **Loh, H. H. and Law, Y. Y.,** The role of membrane lipids in receptor mechanisms, *Annu. Rev. Pharmacol. Toxicol.*, 20, 201, 1980.
184. **Rodbell, M.,** The role of hormone receptors and GTP-regulatory proteins in membrane transduction, *Nature (London)*, 284, 17, 1980.
185. **Cuatrecasas, P.,** Membrane receptors, *Annu. Rev. Biochem.*, 43, 169, 1974.
186. **Rubin, M. S., Swislocki, N. I., and Sonenberg, M.,** Alteration of liver plasma membrane protein conformation by bovine growth hormone in vitro, *Arch. Biochem. Biophys.*, 157, 252, 1973.
187. **Hirata, F. and Axelrod, J.,** Enzymatic methylation of phosphatidylethanolamine increases erythrocyte membrane fluidity, *Nature (London)*, 275, 219, 1978.
188. **Hirata, F. and Axelrod, J.,** Phospholipid methylation and biological signal transmission, *Science*, 209, 1082, 1980.
189. **Fiorini, R. M., Cinti, S., and Curatola, G.,** Modification of erythrocyte membrane structure associated with phospholipid transmethylation, *Int. Wksh. Membr. Transp. Biosystems*, Bari, June 28 to July 2, 1982.
190. **Ohno, H., Shimidzu, N., Tsuchida, N., Sasakawa, S., and Honda, K.,** Fluorescence polarization study on the increase of membrane fluidity of human erythrocyte ghosts induced by synthetic water-soluble polymers, *Biochim. Biophys. Acta*, 649, 221, 1981.
191. **De Laat, S. W., Van der Saag, P. T., and Shinitzky, M.,** Microviscosity modulation during the cell cycle of neuroblastoma cells, *Proc. Natl. Acad. Sci. U.S.A.*, 74, 4458, 1977.

192. **Bruscalupi, G., Curatola, G., Lenaz, G., Leoni, S., Mangiantini, M. T., Mazzanti, L., Spagnuolo, S., and Trentalance, A.**, Plasma membrane changes associated with rat liver regeneration, *Biochim. Biophys. Acta*, 597, 263, 1980.
193. **Inbar, M., Yuli, I., and Raz, M.**, Contact-mediated changes in the fluidity of membrane lipids in normal and malignant transformed mammalian fibroblasts, *Exp. Cell Res.*, 105, 325, 1977.
194. **Papahadjopoulos, D., Poste, G., and Schaeffer, B.**, Fusion of mammalian cells by unilamellar lipid vesicles: influence of lipid surface charge, fluidity and cholesterol, *Biochim. Biophys. Acta*, 323, 23, 1973.
195. **Nicolson, G. L.**, Transmembrane control of the receptors of normal and tumor cells. II. Surface changes associated with transformation and malignancy, *Biochim. Biophys. Acta*, 458, 1, 1976.
196. **Jackson, M. B. and Cronan, J. E.**, An estimation of the minimum amount of fluid lipid required for the growth of *E. coli*, *Biochim. Biophys. Acta*, 512, 472, 1978.
197. **McElhaney, R. N.**, The effect of alterations in the physical state of membrane lipids in the ability of *A. laidlawii B* to grow at various temperatures, *J. Mol. Biol.* 84, 145, 1974.
198. **Raff, M.**, Self regulation of membrane receptors, *Nature, (London)*, 259, 265, 1976.
199. **Edelman, G. M.**, Surface modulation in cell recognition and cell growth, *Science*, 192, 218, 1976.
200. **Shinitzky, M.**, Regulation of membrane functions by lipids: implication for tumor development, in Proc. 13th Int. Cancer Congr., Abstr. 2085, Seattle, September 8 to 15, 1982.
201. **Luly, P. and Shinitzky, M.**, Gross structural changes in isolated liver cell plasma membranes upon binding of insulin, *Biochemistry*, 18, 445, 1979.
202. **Yahara, I. and Edelman, G. M.**, Restriction of the mobility of lymphocyte immunoglobulin receptors by Concanavalin A, *Proc. Natl. Acad. Sci. U.S.A.*, 69, 608, 1972.
203. **Shearer, W. T., Philpott, G. W., and Parker, C. W.** Stimulation of cells by antibody, *Science*, 182, 1357, 1973.
204. **Anderson, J., Edelman, G. M., Moller, G., and Sjoberg, G.**, Activation of B lymphocytes by locally concentrated concanavalin A, *Eur. J. Immunol.*, 2, 233, 1972.
205. **Noonan, K. D.**, Proteolytic modification of cell surface macromolecules: mode of action in stimulating cell growth, *Curr. Top. Membr. Transp.*, 11, 397, 1978.
206. **Olten, J., Johnson, G. I., and Pastan, I.**, Regulation of cell growth by cyclic AMP, *J. Biol. Chem.*, 247, 7082, 1972.
207. **Katz, A. M. and Messineo, F. C.**, Lipid membrane interactions and the pathogenesis of ischemic damage in the myocardium, *Circ. Res.*, 48, 1, 1981.
208. **Rice Evans, C., Rush, J., Omorphos, S. C., and Flynn, D. M.**, Erythrocyte membrane abnormalities in glucose-6-phosphate dehydrogenase deficiency of the Mediterranean and A types, *FEBS Lett.*, 136, 148, 1981.
209. **Owen, J. S., Bruckdorfer, K. R., Day, R. C., and McIntyre, N.**, Decreased erythrocyte membrane fluidity and altered lipid composition in human liver disease, *J. Lipid Res.*, 23, 124, 1982.
210. **Butterfield, D. A., Roses, A. D., Appel, S. H., and Chesnut, D. B.**, ESR studies of membrane proteins in erythrocytes in myotonic muscular dystrophy, *Arch. Biochem. Biophys.*, 177, 226, 1976.
211. **Dellantonio, R., Angeleri, F., Capriotti, M., Lenaz, G., Curatola, G., Mazzanti, L., and Bertoli, E.**, Progressive muscular dystrophy type Duchenne. I. Spin label studies on the physicochemical state of erythrocyte membranes from patients with Duchenne muscular dystrophy, *Ital. J. Biochem.*, 29, 121, 1980.
212. **Sato, B., Nishikida, K., Samuels, L. T., and Tyler, F. H.**, ESR studies of erythrocytes from patients with Duchenne muscular dystrophy, *J. Clin. Invest.*, 61, 251, 1978.
213. **Bartoli, G., Bartoli, S., Galeotti, T., and Bertoli, E.**, Superoxide dismutase content and microsomal lipid composition of tumors with different growth rates, *Biochim. Biophys. Acta*, 620, 205, 1980.
214. **Johnson, S. M. and Robinson, R.**, The composition and fluidity of normal and leukemic or lymphomatous lymphocyte plasma membranes in mouse and man, *Biochim. Biophys. Acta*, 558, 282, 1979.
215. **Chen, H. W., Kandutsch, A. A., and Heiniger, H. J.**, The role of cholesterol in malignancy, in *Prog. Exp. Tumor Res., Vol. 22*, Wallach, D. F. H., Ed., S. Karger, Basel, 1978, 275.
216. **Cooper, R. A.**, Abnormalities of cell membrane fluidity in the pathogenesis of disease, *N. Engl. J. Med.*, 297, 371, 1977.
217. **Day, R. C., Harry, D. S., and McIntyre, N.**, Plasma lipoproteins and the liver, in *Liver and Biliary Disease*, Wright, K., Alberts, K. G., Karran, S., and Millward-Sadler, G. H., Eds., Saunders, London, 1979, 63.
218. **Cooper, R. A., Diloy-Puray, M., Lando, P., and Greenberg, M. S.**, An analysis of lipoproteins, bile acids and red cell membranes associated with target cells and spur cells in patients with liver disease, *J. Clin. Invest.*, 51, 3182, 1970.
219. **Kennett, F. F. and Weglicki, W. B.**, Effects of well defined ischemia on myocardial lysosomal and microsomal enzymes in a canine model, *Circ. Res.*, 43, 750, 1978.
220. **Coleman, P. S. and Lavietes, B. B.**, Membrane cholesterol, tumorigenesis and the biochemical phenotype of neoplasia, *Crit. Rev. Biochem.*, 11, 341, 1981.

221. **Pedersen, P. L.,** Tumor mitochondria and the bioenergetics of cancer cells, in *Prog. Exp. Tumor Res., Vol. 22,* Wallach, D. F. H., Ed., S. Karger, Basel, 1978, 90.
222. **Oshino, N., and Sato, R.,** The dietary control of the microsomal stearyl CoA desaturation enzyme system in rat liver, *Arch. Biochem. Biophys.,* 149, 369, 1972.
223. **Enser, M.,** The role of insulin in the regulation of stearic acid desaturase activity in the liver and adipose tissue from obese hyperglycemic and lean mice, *Biochem. J.,* 180, 551, 1979.
224. **Nervi, A. M., Brenner, R. R. and Peluffo, R. O.,** Effect of arachidonic acid on the microsomal desaturation of linoleic into γ-linolenic acid and their simultaneous incorporation into the phospholipids, *Biochim. Biophys. Acta,* 152, 539, 1968.
225. **Peluffo, R. O., Nervi, A. M., and Brenner, R. R.,** Linoleic acid desaturation activity in liver microsomes of EFA deficient and sufficient rats, *Biochim. Biophys. Acta,* 441, 25, 1976.
226. **Lands, W. E. M. and Hart, P.,** Metabolism of glycerolipids. VI. Specificities of acyl CoA phospholipid acyltransferases, *J. Biol. Chem.,* 240, 1905, 1965.
227. **Borochov, H., Zahler, P., Wilbrandt, W., and Shinitzky, M.,** The effect of lecithin and sphingomyelin mole ratio on the dynamic properties of sheep erythrocyte membranes, *Biochim. Biophys. Acta,* 470, 382, 1977.
228. **King, M. D. and Quinn, P. J.,** The use of phospholipid exchange processes to modulate the fluidity of biological membranes, *Biochem. Soc. Trans.,* 8, 322, 1980.
229. **Madden, T. D., Vigo, C., Bruckdorfer, K. K., and Chapman, D.,** The incorporation of cholesterol into inner mitochondrial membranes and its effect on lipid phase transition, *Biochim. Biophys. Acta,* 559, 528, 1980.
230. **Jollow, D., Kellermann, G. W., and Linnane, A. W.,** The biogenesis of mitochondria. III. The lipid composition of aerobically and anaerobically grown *S. cerevisiae* as related to the membrane systems of the cells, *J. Cell Biol.,* 37, 221, 1968.
231. **Lees, N. D., Bard, M., Kemple, M. D., Haak, R. A., and Kleinhans, F. W.,** ESR determination of membrane order parameter in yeast sterol mutants, *Biochim. Biophys. Acta,* 553, 469, 1979.
232. **McElhaney, R. N. and Souza, K. A.,** The relationship between environmental temperature, cell growth and physical state of the membrane lipids of *B. stearothermophilus, Biochim. Biophys. Acta,* 443, 348, 1976.
233. **Sinensky, M.,** Homeoviscous adaptation, a homeostatic process that regulates the viscosity of membrane lipids of *E. coli, Proc. Natl. Acad. Sci. U.S.A.,* 71, 522, 1974.
234. **Gliozzi, A., Rolandi, R., De Rosa, M., Gambacorta, A., and Nicolaus, B.,** Membrane models of Archaebacteria, in *Transport in Biomembranes: Model Systems and Reconstitution,* Antolini, R., Gliozzi, A., and Gorio, A., Eds., Raven Press, New York, 1982, 39.
235. **Johnston, P. V. and Roots, B. I.,** Brain lipid fatty acids and temperature acclimation, *Comp. Biochem. Physiol.,* 11, 303, 1974.
236. **Cossins, A. R.,** Adaptation of biological membranes to temperature. The effect of temperature acclimation of goldfish upon the viscosity of synaptosomal membranes, *Biochim. Biophys. Acta,* 470, 395, 1977.
237. **Goldman, S.,** Cold resistance of the brain during hibernation. III. Evidence of lipid adaptation, *Am. J. Physiol.,* 228, 834, 1973.
238. **Wilkinson, P. A. and Nagle, J. F.,** Thermodynamics of lipid bilayers, in *Liposomes: From Physical Structure to Therapeutic Applications,* Knight, C. G., Ed., Elsevier, Amsterdam, 1981, chap. 9.
239. **Lakowicz, J. R., Hogen, D., and Omann, G.,** Diffusion and partitioning of a pesticide, lindane, into phosphatidylcholine bilayers, *Biochim. Biophys. Acta,* 471, 401, 1977.
240. **Demel, R. A. and De Kruyff, B.,** The function of sterols in membranes, *Biochim. Biophys. Acta,* 457, 109, 1976.
241. **Lenaz, G. and Degli Esposti, M.,** Physical properties of ubiquinone in model systems and membranes, in *Coenzyme Q,* Lenaz, G., Ed., John Wiley & Sons, London, in press.
242. **Benga, Gh., Hodârnău, A., Böhm, B., Borza, V., Tilinca, R., Dancea, S., Petrescu, I., and Ferdinand, W.,** Human liver mitochondria: relation of a particular lipid composition to the mobility of spin-labelled lipids, *Eur. J. Biochem.,* 84, 625, 1978.
243. **Minetti, M. and Ceccarini, M.,** Protein-dependent lipid phase separation as a mechanism of human erythrocyte ghost resealing, *J. Cell. Biochem.,* 19, 59, 1982.
244. **Dave, J. R., Brown, N., and Knazek, R. A.,** Prolactin modifies the prostaglandin synthesis, prolactin binding and fluidity of mouse liver membranes, *Biochem. Biophys. Res. Commun.,* 108, 193, 1982.
245. **Haest, C. W. M. and Deuticke, B.,** Experimental alteration of phospholipid-protein interactions within the human erythrocyte membrane: dependence on glycolytic metabolism, *Biochim. Biophys. Acta,* 401, 468, 1975.
246. **Fato, R. and Lenaz, G.,** unpublished data.

Chapter 7

LIPID DEPENDENCE OF MEMBRANE ENZYMES

John M. Wrigglesworth

TABLE OF CONTENTS

I. INTRODUCTION

There have been several general reviews on the lipid dependence of membrane enzymes,[1-8] and specific aspects of the subject have also been covered including prokaryote membrane-bound enzymes,[8,9] protein-lipid interactions,[10-13] topology and reconstitution,[14-21] and lipid effects on various transport[22-24] and receptor systems.[25,26] However, it is only over the past few years that the concept of lipid dependence has been clarified sufficiently to rationalize a large number of experimental observations.

The problem has often been one of technique. The traditional approach of isolation of the enzyme followed by reconstitution with specific lipids still provides one of the most useful methods to study the lipid dependence of enzyme activity, but far too often the isolated protein has not been sufficiently purified and the manner of reconstitution has not been sufficiently controlled. An example of the confusion that can result from the use of different enzyme preparations and reconstitution techniques is shown in studies on the lipid specificity of various (Na^+ + K^+)-ATPases. Table 1 summarizes some of these studies. Initial experiments indicated some specificity for phosphatidylserine (PS)[27] but present consensus now suggests a requirement only for a fluid apolar environment with some surface negative charge.[28-30] This can be provided by a number of lipids or even, in some cases[31-33] by a detergent.

The selective replacement and/or modification of endogenous lipid components has also proved useful in assessing the lipid dependence of several membrane enzymes. The methods include the use of specific enzymes such as the various phospholipases and phospholipid-exchange proteins as well as modification of membrane-lipid composition by diet. A brief review of these different strategies will be presented in the first section of the present chapter. The second section is concerned with the specificity of lipid dependence which, I suggest, can now be rationalized into three general types of interaction (see also Figure 1):

1. There can be a *specific binding* of lipid to the protein, effectively modifying enzyme activity by allosteric interactions.
2. The so-called *lipid solvation* effect where the lipid provides a suitable apolar environment for optimum conformation and orientation of the enzyme.
3. *Interfacial regulation* can occur, where polar lipid head groups provide a suitable microenvironment between the polar and nonpolar phases for optimum conformation and charge interaction of the enzyme with its substrate.

Many effects of membrane fluidity and surface potential on the activity of membrane-associated enzymes can now be explained by reference to one or more of these three types of interaction. The general principles behind many of the physical properties of membranes and the effects of changes in membrane structure on enzyme activity are covered in other chapters of this volume. I will be concerned here only with the specific effect of lipids on membrane-associated enzymes.

II. METHODOLOGY

A. Delipidation of Membrane Proteins

The "fluid mosaic" model of Singer and Nicolson[61] provides a useful framework for investigating the functional properties of membrane enzymes. The model envisages close interactions between membrane-proteins and lipids with the so-called "intrinsic" membrane-proteins associating with, and often spanning, hydrocarbon regions of the membrane lipid phase. Because of these close interactions, membrane enzymes usually have to be solubilized by detergents for isolation in aqueous solution.[62-64] Removal of lipid components by organic solvents, unless carried out on lyophylized preparations below 0°C, leads to denaturation of

Table 1
REPORTED LIPID SPECIFICITIES OF (Na^+ + K^+)-ATPases

Source of enzyme	Method of study	Reported lipid specificity	Ref.
Human erythrocyte	Phospholipase treatment of intact membrane	PS, PA	34, 35
	Incorporation of lipid into intact membrane	Modulation by cholesterol	36, 37
	Enzyme modification of intact membrane	Sulfatide lipid	38
Ox brain cortex	Reactivation of purified enzyme	PC	39
		PS	40
		PA, PS, Lys	41
		Cholesterol	42
	Phospholipase treatment of intact membrane	*Not* PS	43
	Reconstitution of purified enzyme into liposomes	Modulation by ''membrane fluidity''	44, 165
Rat brain cortex	Reactivation of purified enzyme	PS	45
	Phospholipase treatment of intact membrane	No specificity	46
Rabbit kidney	Reactivation of purified enzyme	PS, PG, ''membrane fluidity''	47
		PI	24, 167
		Acidic phospholipids	48
	Enzymic and chemical modification of partially purified enzyme	*Not* cholesterol	49
		Acidic phospholipids	50
Rat kidney	Reactivation of purified enzyme	PS, PA, PG, ''membrane fluidity''	51
		PS, PG	52
		PS	45
Pig kidney	Reactivation of purified enzyme	Acidic phospholipids	53
Sheep kidney	Reactivation of purified enzyme	Detergent activation	54
Ox heart	Reactivation of purified enzyme	PS, PG	55
Electric organ of *Electrophorus electricus*	Reactivation of purified enzyme	Cholesterol	56
	Phospholipase treatment of partially purified enzyme	PS, PE	57
	Spin-label binding to partially purified enzyme	Acid phospholipids	58
Electric organ of *Torpedo marmorata*	Delipidation of membrane by organic solvents	Sulfatide lipid	59
Dogfish rectal salt gland	Reactivation of purified enzyme	PC, PE	32
	Phospholipid substitution on purified enzyme preparation	PC	33
T lymphocytes	Membrane lipid alteration in cell culture	Inhibited by saturated fatty acyl side chains	60

Note: Abbreviations are as follows: PA, phosphatidic acid; PC, phosphatidylcholine; PE, phosphatidylethanol; PG, phosphatidylglycerol; PS, phosphatidylserine; Lys, lysolecithin.

the enzyme.[65] Even for ''extrinsic'' membrane enzymes, the detachment into aqueous medium by salt or pH treatment can cause loss of function.[1]

It is not surprising that the activity of membrane-bound enzymes is often diminished or even abolished on isolation. However carefully the method of purification is chosen, the replacement of membrane lipid by detergent will invariably lead to conformational changes in the membrane protein[62,63] with subsequent effects on enzyme activity. It must not be assumed, however, that these changes in catalytic activity are always due to lipid dependence in the strict sense of the term. Solubilization and purification can lead to alterations in normal

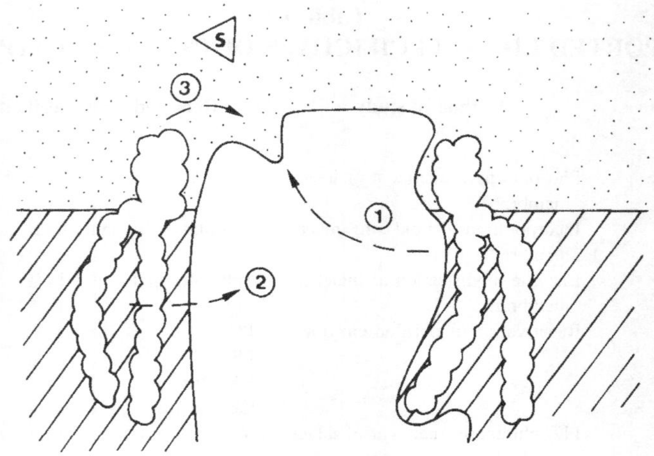

FIGURE 1. Model of the three general types of interaction of lipid with membrane enzymes. In type 1, specific binding of lipid to the protein effectively modifies enzyme activity by allosteric interaction. In type 2 interaction, the lipid provides a suitable apolar environment for optimum conformation of the enzyme. In type 3, interfacial regulation occurs where the polar lipid head groups provide a suitable microenvironment for interaction of the enzyme with its substrate (S).

protein-protein interactions as well as the possibility of subunit disaggregation. For example, mitochondrial cytochrome *c* oxidase can be isolated and purified by ammonium sulfate fractionation of detergent-solubilized mitochondrial proteins.[66,67] Although the purified preparations are free of other cytochromes, most preparations are hydrodynamically polydisperse and frequently vary in the number of different polypeptides as resolved by gel electrophoresis.[68-70] The fit between the measured cytochrome-hem/protein ratio, the molecular weight as determined by sedimentation analysis, and the sum of the subunit molecular weights from gel electrophoresis has not always been consistent.[71] These variations are probably due to dimer formation and/or partial splitting of the multisubunit enzyme by the detergents used in its isolation.[72,73]

Finally, inhibition or activation of the purified enzyme by detergents can often mask the specific effects of lipid dependence. Unfortunately, it is not always possible to solubilize lipid-free preparations of membrane enzymes in the absence of lipids and detergents. Where this can be achieved, for example, in the case of β-hydroxybutyrate dehydrogenase,[74] more reliable reactivation specificity by lipid can be measured.[75]

1. Delipidation by Detergents

The general properties of detergents and their use in the isolation and purification of membrane proteins have been discussed in several papers[76-80] and recent reviews.[62-64,81,82]

Three stages can be discerned in the action of a detergent in solubilizing a biological membrane.[78,83] In the initial stage, the detergent intercalates between lipid molecules in the membrane bilayer. Even at this nonsolubilizing stage, the detergent can directly perturb enzyme activity. For example, Andersen and colleagues[84-86] report that nonsolubilizing concentrations of the detergent $C_{12}E_8$ cause a decrease in the immobilized components of spin-labeled fatty acid derivatives covalently attached to the Ca^{2+}-ATPase of intact sarcoplasmic reticulum (SR). They conclude[85] that direct interaction of the detergent with the enzyme is occurring rather than indirect effects exerted via changes in the lipid phase or protein aggregation.

The second stage of solubilization occurs as the concentration of detergent increases sufficiently for detergent-lipid micelles to break off from the membrane into the surrounding medium. The relative concentrations of detergent and lipid for this to occur have been considered in detail by Tanford[11,63] and others,[87] and depend on the nature of the detergent and the conditions used.

Finally, the detergent solubilizes integral membrane enzymes by exchanging with the lipid molecules associated with the hydrophobic regions of the membrane protein. In order to prevent large-scale conformation changes at this stage, the detergent chosen must be able to simulate the native environment of the protein as much as possible. Above solubilizing concentrations, powerful detergents such as sodium dodecyl sulfate (SDS) will unfold proteins, whereas milder ionic detergents such as cholate and nonionic detergents such as Tween® can be increased in concentration to cause almost complete delipidation without large-scale denaturation of the enzyme and loss of activity.

Preparations of intrinsic membrane-enzymes solubilized by detergent and purified by ammonium sulfate fractionation or column chromatography in the presence of detergent usually contain protein-associated lipid. The amount of lipid remaining after purification depends on the extent of detergent treatment, but also appears to be particular to individual membrane enzymes. This "boundary layer",[88-90] or "annular lipid",[91-93] does exchange with the bulk lipid pool when the enzyme is incorporated in a bilayer, but at a slower rate than is found for the mobility of lipid in the bulk phase. Exchange rates between 10^4 to 10^7 sec^{-1} have been found for lipid associated with most of the integral proteins studied so far[83] and this compares to an exchange rate of greater than 10^7 sec^{-1} for the phospholipids of the bulk lipid pool. The apparent discrepancy between ESR and NMR results which earlier caused some confusion can now be ascribed to the difference in the frequency range of the two techniques.[84] ESR and fluorescence studies, with time windows of around 1 to 10 nsec, indicate an apparent immobilized lipid fraction, whereas in the microsecond range, annular lipid molecules freely exchange with the bulk phase as indicated by NMR studies.[94] Measurements of the wobbling motions of fatty acyl chains on the phospholipids do indicate, however, that chain motion is severely restricted in amplitude at the surface of protein, and some of the chain may even be trapped in grooves on the surface.[95]

Although rapid exchange occurs between lipids in the boundary layer and the bulk phase, the steady-state composition of lipids in the boundary layer does not necessarily reflect that of the membrane bulk lipid. Membrane enzymes are found to interact preferentially with particular lipid species, presumably under the influence of electrostatic and Van der Waals forces. Preferential segregation of lipid molecules has been noted for several membrane-associated enzymes.[83,98] This segregation is often reflected in the lipid composition of the partially purified protein. The boundary-lipid composition of preparations of Ca^{2+}-ATPase,[23,96] $(Na^+ + K^+)$-ATPase,[29] and cytochrome oxidase[88,99-101] has been reported. Often the amount of associated detergent is not included in the analysis, which to some extent makes speculation about the number phospholipid molecules normally present in the boundary layer a fruitless exercise.

It is important for delipidation of partially purified enzyme to use mild conditions to avoid irreversible loss of activity. This loss of activity has to be distinguished from the reversible inhibition that can occur when bound phospholipid is replaced by detergent. Three general approaches have been used:

1. Extraction of bound phospholipid using mild organic solvents such as aqueous acetone[49,102-104]
2. Further treatment with mild detergent followed by isolation using ammonium sulfate fractionation,[105] gel filtration,[72,106] or density gradient centrifugation[107-109]
3. Phospholipase digestion of bound phospholipids followed by removal of digestion products with fat-free albumin or extraction with dry ether[50,57]

A typical example of an intrinsic membrane-protein complex that can be detergent solubilized and then further delipidated by the above treatments is cytochrome oxidase. Most solubilized preparations of this enzyme contain around 50 to 70 associated phospholipid molecules per molecule of enzyme protein (approximately 0.2 mg phospholipid per milligram protein), of which 40 to 50 can be replaced by a suitable detergent such as Tween®-80 without any apparent selectivity in composition[104] and without affecting the activity of the enzyme.[72,104,109,110] The remaining boundary-layer phospholipids have been divided into two categories by Robinson.[111] The first, of six to eight molecules, requires higher detergent concentrations for extraction, but also appears to be nonessential for full activity of the complex. These may represent phospholipid molecules whose fatty acyl chains have a particularly "good fit" on the hydrophobic surface of the protein. The remaining two to three molecules of very tightly bound diphosphatidylglycerol (DPG) do appear to be required for efficient functioning of the bovine enzyme.[111,112] Removal of all phosholipids from cytochrome *c* oxidase results in complete loss of activity. The specificity requirement for reactivation does vary between cytochrome *c* oxidases isolated from different species,[100] as will be discussed in a later section.

2. Delipidation by Organic Solvents

Organic solvents are routinely used for the extraction and purification of lipid from biological membranes. Unfortunately, most procedures involving organic solvents result in the denaturation of membrane proteins to such an extent that recombination experiments are not possible. The low solvent polarity causes denaturation and micelle formation of the protein components. For example, in the Folch procedure[113] and its subsequent modifications,[114,115] denatured membrane protein collects at the interface with further small amounts in the organic phase.[116] Isolation of membrane-associated glycoprotein and its subsequent purification can be achieved from the upper aqueous layer,[117] but, in general, organic solvent procedures are not suited to reversible delipidation of amphipatic proteins.

In some cases, partial delipidation can be done under controlled conditions. For example, acetone with its relatively high dipole moment is a poor lipid solvent but, nevertheless, can be used for specific extraction of some membrane lipid without causing damage to membrane protein.[118] Diethyl ether has also been used in reconstitution studies, for example, with a $(Na^+ + K^+)$-stimulated ATPase,[34] but for successful reconstitution it is essential that the preparation be lyophilized before ether treatment and that dry ether be used.[119] Similarly, no major loss of $(Na^+ + K^+)$-ATPase activity occurs when the lyophilized enzyme is extracted with anhydrous hexane,[49,56] pentane, or butanol.[120] A useful description of the physical properties of the various organic solvents used in membrane research has been given by Zahler and Niggli.[65]

At temperatures appreciably colder than 0°C, organic solvents could potentially be used to remove lipid components while leaving enzymatic activity intact,[122,123] and Noguchi et al.[42] describe a procedure for the lipid extraction of brain tissue by various organic solvents, including chloroform-methanol, at −75°. The activity of Mg^{2+}-dependent ATPase remains essentially intact after this treatment.

3. Covalently Bound Lipid

Finally, it should be remembered that not all lipid-protein interactions are noncovalent. In many membranes, lipid is covalently bound to membrane protein[124,125] and full delipidation by detergent or solvent treatment is not possible. One example is the membrane penicillinase of *Bacillus licheniformis*[126] where the presence of a phospholipopeptide segment attached to the NH_2-terminal residue of the usual exopenicillinase allows for secure anchorage of the enzyme in the membrane.[127]

B. Selective Modification of Lipid

1. Lipid Substitution

The lipid substitution technique developed by Warren et al.[91,128] has been particularly successful in studies on the lipid dependence of Ca^{2+}-ATPase.[91,92,128-130] The technique involves the incubation of partially purified membrane protein with a pool of exogenous lipid of defined composition in the presence of detergent. Substitution by exchange occurs between the exogeneous and endogenous lipid pools. The protein can then be isolated by centrifugation in a detergent-free sucrose density gradient. Using this technique, Warren et al.[128] found that more than 98% of the lipid associated with partially purified Ca^{2+}-ATPase from SR can be replaced by dioleoyllecithin (DOL) without any severe effect on enzyme activity, whereas substitution of cholesterol for the endogenous lipid caused a complete but reversible inactivation of activity.[92] Further work using the same technique has shown the enzyme to be particularly sensitive to the chain length of the substituting lipid.[129] The Warren technique has also been used to good effect on preparations of (Na^+, K^+)-ATPase[33,131,132] and has provided an improved method for introducing photoactive probes into the hydrophobic sector of integral membrane proteins.[133]

Slight modifications of the method, including the use of different detergents, have been introduced.[132] For example, Moore et al.[134] first heavily delipidated a Ca^{2+}-ATPase by the method of Dean and Tanford[108] and then incubated the preparation with a deoxycholate-lipid mixture. Subsequent recentrifugation and dialysis produced liposomal preparations of defined phospholipid composition with enzyme inserted into the bilayer.

The substitution technique does have some disadvantages. Yields of the substituted material are generally low and residual detergent has to be removed, usually by recentrifugation and repeated dialysis. An alternative substitution technique which can be applied to native membranes or to the enzyme incorporated into reconstituted proteoliposomes is by exchange of the endogenous lipid with exogenous lipid added in excess in the form of liposomes of defined lipid composition. Lipid substitution can be promoted by specific phospholipid exchange enzymes, including ones to mediate the translocation of phosphatidylcholine (PC),[135,137] phosphatidylethanolamine (PE),[138] phosphatidylinositol (PI),[138,139] sphingomyelin,[140] and various galactolipids.[141]

In the case of steroid molecules, exchange can occur in the absence of exchange enzymes.[142,143] Madden et al.[144] and Johannsson et al.[130] have managed to alter the cholesterol content of SR by incubating membrane vesicles with cholesterol-rich liposomes and have examined the effect of this substitution on the activity of Ca^{2+}-dependent ATPase. Conflicting results were obtained mainly due to the necessity of preparing the SR in the presence of a thiol-reducing agent to prevent spontaneous uncoupling and activation of the ATPase. When this is done, vesicles of SR can be loaded with cholesterol to at least 20 mol% with respect to endogenous phospholipid without any effect on the ATPase activity.[130] A different effect is reported for adenylate cyclase.[143] Depletion of rat liver membrane cholesterol by incubation with liposomes inhibits adenylate cyclase activity and the membrane becomes more rigid as determined by a fatty acid spin probe. These findings illustrate a main problem of the liposome exchange technique which is that the exchanged lipid may often be excluded from direct interaction with the enzyme under study because of selective interaction of the protein with the shell of annular phospholipid. This has to be remembered when conclusions are made about the specificity of lipid dependence of membrane enzymes using the technique.

2. Enzymic Modification of Endogenous Lipid

Phospholipases and other lipid-degrading enzymes have been used to remove or modify endogenous phospholipids in biomembranes.[4,145] They have the added advantage for some studies of being membrane impermeable, and their action can, therefore, provide information on the transverse distribution of phospholipids over the two faces of the membrane[146] and the "sidedness" of lipid dependence of transmembrane proteins. Phospholipases are available

Table 2
LIPID-DEGRADING ENZYMES USED TO STUDY THE LIPID DEPENDENCE OF MEMBRANE-ASSOCIATED ENZYMES

Enzyme	Substrate preference	Example of use
Phospholipase A_2	PS > PC > PE	Ca^{2+}-ATPase[35,147,148]
		$(Na^+ + K^+)$-ATPase[34,57]
Phospholipase C	PC > PE > PS	Ca^{2+}-ATPase[35,147,149]
		$(Na^+ + K^+)$-ATPase[34,57]
Phospholipase D	PC > PE >> PS	$(Na^+ + K^+)$-ATPase[34]
Spingomyelinase C	SM >> PS, PC, PE	Ca^{2+}-ATPase[35]
		$(Na^+ + K^+)$-ATPase[34]
PS decarboxylase	PS (requires surfactant)[150,151]	$(Na^+ + K^+)$-ATPase[43]

Note: Abbreviations are as in Table 1.

from different origins and each has its own substrate preference. Analysis of the products of enzyme action enables the degree of hydrolysis to be determined. It should also be made sure that commercially available phospholipases are free from proteolytic activity to avoid erroneous results in reactivation studies. Table 2 lists some of the enzymes used to study the lipid dependence of membrane enzymes.

One problem with phospholipid digestion using phospholipases is the necessity for removal of the digestion products. The delipidized preparations can be lyophilized and then extracted with anhydrous diethyl ether.[34,147] This treatment also extracts any endogenous neutral lipid. An alternative, more gentle procedure, is to apply the digesting enzymes together with albumin.[35,148] Swoboda et al.[148] report that the albumin method of delipidation makes it possible to dismantle the annular lipid gradually in a selective way. Lyso-PE and saturated fatty acids are removed relatively slowly, while lyso-PC and unsaturated fatty acids are extracted more rapidly. The lyso-PC molecules appear to be extracted irrespective of their fatty chain structure. It is of interest to compare the effect of this procedure on the activity of partially purified Ca^{2+}-ATPase with the lipid depletion methods involving detergents. Using the latter method, Warren et al.[91,92] found that the preparation remained fully active until a critical amount of residual phospholipids per ATPase was reached. On the other hand, Swoboda et al.[148] report an immediate decline of the residual ATPase activity even at low degrees of delipidation. As these workers point out, with enzyme digestion and treatment with albumin, the unsaturated fatty acids are preferentially removed and the composition of the residual lipid becomes increasingly saturated. The concomitant decrease in fluidity presumably causes the sharp decline of the ATPase activity at room temperature.

3. Modification of Endogenous Lipid by Diet

Changes in diet or growth conditions can cause changes in the fatty acid composition of membrane lipids,[121,152-154] and these often affect the activity of membrane-associated enzymes. Effects have been reported for H^+-ATPase,[80] Ca^{2+}-ATPase,[156,157] (Na^+, K^+)-ATPase,[157] adenylate cyclase,[158,159] stearyl coenzyme A desaturase,[160] and various phospholipid transferases.[161,162]

Because of the general nature of the alterations, it is difficult to come to conclusions about specific lipid dependence. In most cases the effects can be attributed to changes in membrane fluidity.[164]

III. SPECIFICITY OF LIPID DEPENDENCE

A. Type 1 Interactions — Specific Binding

Type 1 interactions, involving high specificity in the binding of lipids to membrane-associated enzymes, are rare. In most cases, a particular lipid associated with an enzyme in the native membrane can be substituted by several other lipids or even suitable detergents without significant loss of activity. This is particularly so when the substitution involves differences in the fatty acyl chain region of phospholipids, as discussed in the next section. When specificity does occur, then interaction mainly involves the polar head group of the lipid. This is especially the case when charged groups are involved. Membrane-associated enzymes which exhibit a high degree of specificity for a particular charged phospholipid include ox heart cytochrome *c* oxidase[121] (with DPG), (Na^+ + K^+)-ATPase[50] (with PS and/or PI), erythrocyte acetylcholinesterase[168] (with DPG), and mitochondrial NADH-ubiquinone reductase[169] (also with DPG). Specific interactions involving neutral lipid molecules include plasma-membrane 5′-nucleotidase[170] (with sphingomyelin and PC) and mitochondrial β-hydroxybutyrate dehydrogenase[171] (with PC).

Immobilization of the head group of a particular phospholipid need not necessarily require the immobilization of the hydrocarbon chains on the same time scale. On the basis of ^{31}P-NMR studies, Yeagle[172] argues for special, long-lived, phospholipid head group-protein interactions in some lipid-requiring enzymes while retaining the possibility that the hydrocarbon chains of the same phospholipid are moving on and off the protein, or from one orientation to another on the protein, in a series of short-term interactions. Thus, specificity for type 1 interactions may reside in one region of the lipid only.

Most information concerning type 1 interactions has been gained from studies on the two mitochondrial enzymes, cytochrome *c* oxidase and β-hydroxybutyrate dehydrogenase. As mentioned in a previous section, isolated ox heart cytochrome *c* oxidase containes two to three molecules of tightly bound DPG per monomer (2 hem complex) which appear to be essential for activity.[111,121] Although the fatty acid composition of DPG does not appear to be the controlling structural feature necessary for regeneration of activity, the overall structure of the phospholipid is important for specific binding since neither phosphatidylglycerol (PG), PS, nor phosphatidic acid (PA), all of which are negatively charged, are able to stimulate activity in lipid-depleted preparations.[111]

The negatively charged phospholipid is thought to be involved in the binding of the substrate, ferrocytochrome *c*, to the enzyme, in particular, the binding of this substrate at the low-affinity cytochrome *c* binding site. Kinetic and binding measurements of cytochrome *c* oxidase activity[173-177] indicate that there are both high- and low-affinity sites for cytochrome *c* on the oxidase complex. Cytochrome *c* is strongly positively charged at neutral pH and electrostatic interactions between positively charged groups around the haem edge of the molecule and negatively charged groups on the oxidase would allow for correct orientation of the molecule for optimum electron transfer. Bisson et al.,[178] using a series of arylazidocytochrome *c* derivatives, found that removal of DPG resulted in loss of the low-affinity binding of cytochrome *c* but did not affect binding to the high-affinity site, although the rate of electron transfer through this site was lowered. It would appear, therefore, that DPG may contribute some of the negative charges in the low-affinity binding site of the enzyme.

The ox heart enzyme contains at least seven major subunits[71] and the complexity of its structure may, in part, determine the high degree of structural specificity for DPG. This may not be the case for cytochrome *c* oxidases from other species. There are reports of oxidases from yeast[179] and *Nitrobacter agilis*[180] where no essential DPG is required for activity. It has also been suggested[100] that PI provides the negative-charged lipid requirement in the dogfish oxidase.

β-Hydroxybutyrate dehydrogenase is the other, well-investigated example of a lipid-requiring enzyme which has an absolute requirement for a particular phospholipid head group. This enzyme is tightly associated with the inner mitochondrial membrane and exhibits a specific requirement for PC. The purified apoenzyme is completely inactive but can be reactivated by interaction with PC, presented in microdispersions,[74] liposomes,[75] or biological membranes.[181] The specificity of interaction of this enzyme is primarily determined by the polar head group of PC.[171] The hydrophobic fatty acyl chains, although necessary for efficient binding of the phospholipid to the apoenzyme, can be varied in chain length and degree of unsaturation, within limits, without affecting activity.[75] The diacylglycerol region of the phospholipid has also been shown not to be essential by the use of a branched alkyl phosphorylcholine in reactivation studies. Activation of the enzyme does not occur, however, when a phosphinate analog, in which there is a decreased separation of the phosphoryl and quanternary ammonium groups, is used.[171] Similarly, a quanternary ammonium group of defined size seems to be essential.

With such detail known about the specific interaction of PC with β-hydroxybutyrate dehydrogenase, it is possible to speculate on the role of the phospholipid in the kinetic mechanism. It has been shown[75] that the binding of NADH to the dehydrogenase is dependent upon the formation of an enzyme PC complex. The NAD(H) binding step in the reaction sequence may, therefore, be the one which confers a specific phospholipid requirement to the enzyme. The titration curves of enzyme activity as a function of phospholipid concentration are consistent with a model in which the enzyme contains two identical, noninteracting, lecithin binding sites.[182] Both have to be occupied for activation of the apoenzyme. In a kinetic analysis of the phospholipid-protein interaction for this enzyme, Cortese et al.[182] and Cortese and Vidal[183] reject the necessity to invoke cooperativity among sites to explain the sigmoidicity of the reactivation curves. The lipid activation curves of many membrane enzymes are sigmoidal, but most lipid binding studies carried out so far have yielded no indication that the sites are interacting.[184] This may turn out to be a common feature of lipid interactions with membrane-bound lipid-requiring enzymes.[185]

B. Type 2 Interactions — Lipid Solvation

After isolation and purification, many membrane-associated enzymes show a partial, or absolute, functional requirement for a boundary shell of amphipathic molecules. In the native membrane this requirement is usually met by phospholipid, but generally a high degree of specificity is not required. As long as the interacting molecules fulfill certain requirements of appropriate hydrophilic/hydrophobic balance, then a protein conformation for optimum activity can be maintained. For example, Dean and Tanford[97,108] report the full restoration of activity to a delipidated inactive Ca^{2+}-ATPase by the addition of a nonionic detergent in the absence of phospholipid.

One of the major factors involved in determining the activity of membrane enzymes via type 2 interactions appears to be the length of the hydrophobic region of the interacting amphiphile. Fatty acyl chain length has been shown to be important for reactivation of several purified membrane enzymes, including (Ca^{2+} + Mg^{2+})-ATPase,[186] (Na^{+} + K^{+})-ATPase,[187] and pyruvate oxidase.[188] With pyruvate oxidase, the binding of amphiphiles to the enzyme becomes independent of the length of the hydrocarbon chain beyond chain lengths of 14 to 16 carbon atoms.[188] Up to this length, the absolute values of the binding free energy are found to increase with increasing chain length, with each additional methylene group contributing approximately −700 cal/mol to the binding free energy. This is of the same order of magnitude as the change in free energy upon the transfer of the same groups from water to organic solvents.[11] Blake et al.[188] also report that the effect of the degree of unsaturation of the hydrocarbon chain on the interaction between pyruvate oxidase and amphiphile can be explained by the effect of unsaturation on overall chain length. The presence of an internal *cis* double bond was found to have a binding energy equivalent to the removal of one half of a methylene group, while the introduction of an internal *trans* double bond had little effect. This correlates with the effect on chain length of *cis* and *trans* double bonds, the former shortening the hydrocarbon chain by approximately one half of a bond length, and the latter having little effect.

These general findings on chain length are consistent with studies on Ca^{2+}-ATPase in-

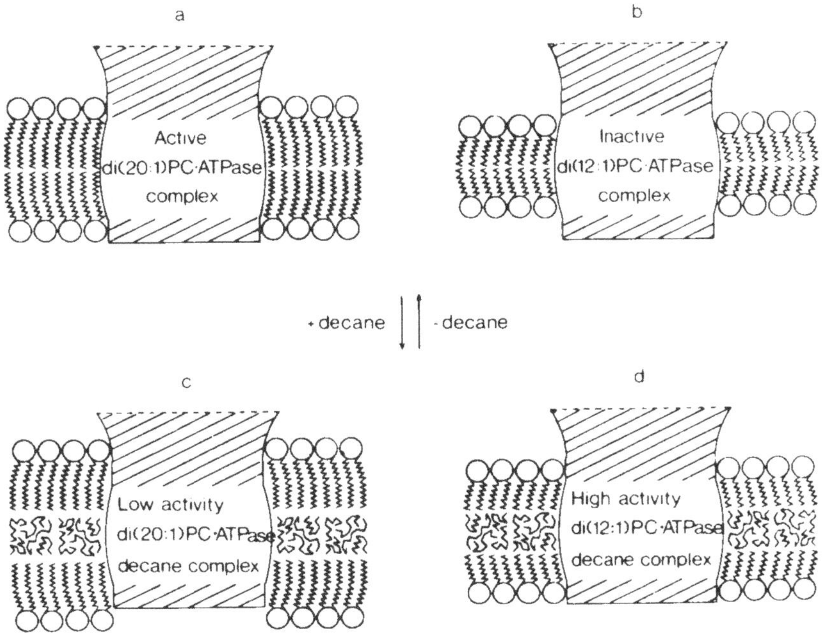

FIGURE 2. Models of Ca^{2+}-ATPase in phospholipid bilayers of different thicknesses. Ca^{2+}-ATPase was incorporated into bilayers of PC of different chain lengths. The enzyme showed high activity when incorporated into di(20:1)PC (a), but was inactive in di(12:1) PC membranes (b). When membrane thickness was increased by decane, the activity of the enzyme incorporated into di(20:1)PC membranes dropped (c), but the opposite effect was seen on addition of decane to the di(12:1)PC membranes (d). The experiments illustrate the dependence of a membrane enzyme on membrane thickness. (Reproduced from Johannsson, A., Keightley, C. A., Smith, G. A., Richards, C. D., Hesketh, T. R., and Metcalfe, J. C., *J. Biol. Chem.*, 256, 1643, 1981. With permission.)

corporated into phospholipid bilayers of different thicknesses.[186] Johannsson et al.[129] found that the activity of this enzyme in a reconstituted system was sensitive mainly to the thickness of the bilayer, which was effectively determined by the fatty acyl chain length of the phospholipid, and was relatively insensitive to details of unsaturation. Reconstituted phospholipid vesicles incorporating Ca^{2+}-ATPase in the bilayer were treated with decane to increase bilayer thickness.[129] The activity of enzyme incorporated into bilayers of phospholipid with fatty acyl chain length less than $n = 18$ first increased and then, subsequently, decreased as the proportion of decane increased. For phospholipid bilayers with $n = 20$ or above, decane only caused a progressive decrease in activity with increasing concentration. The results indicate an optimal bilayer thickness equivalent to that of di(20:1) PC which could be matched by mixtures of decane with PCs of lower chain lengths (Figure 2). Similar results are reported for the effect of bilayer thickness on the activity of ($Na^+ + K^+$)-ATPase.[187]

The general lack of specificity for the "solvation" effect of lipid on membrane enzymes raises the question of how far the properties of the boundary lipid differ from those of the bulk lipid. This is the subject of other chapters in this volume but one aspect is of direct relevance to a consideration of type 2 interactions. There have been many suggestions that the residence time of boundary lipids is longer than the residence time of free bilayer phospholipid. This is certainly the case for specific (type 1) interactions and may also be true for phospholipids involved in interfacial regulation (type 3 interactions). However, recent measurements on the lifetime of neutral phospholipids at protein boundaries are indicative of relatively weak and nonspecific phospholipid/protein interactions.[94,189-192] A discrete and tightly bound lipid annulus does not appear to be necessary for enzyme activation via the lipid solvation effect. A double-labeling scheme has been used by Davoust et al.[190] to assess the collision frequency between nitrogen-14 spin-labeled phospholipid and nitrogen-15 spin-labeled fatty acid, covalently linked to rhodopsin. ESR spin-spin interactions were used to monitor lipid/protein contacts. Little difference was found between exchange frequencies measured with spin-labeled PC, PE, or PS, indicating that rhodopsin was not selecting a specific phospholipid for its surroundings. Moreover, the results clearly indicated that lipid chain collision rates at the protein boundary were of the same order of magnitude as in the bulk lipid phase.

Similar results are reported by Tamm and Seelig[192] from ^{2}H, ^{14}N, and ^{31}P NMR studies on the interaction of the polar group and *cis* double bonds of unsaturated PC bilayers with cytochrome *c* oxidase. No strong interactions were found between the enzyme and the lipid, either between the phosphocholine head group or the nonpolar interactions in the interior of the bilayer. It should be remembered that no attempt was made in this study to remove the tightly bound DPG and the conclusions do not apply to the specific (type 1) interaction for the enzyme. Neither would they apply to interfacial regulation resulting from the preferential segregation of particular charged head groups. (Na^{+} + K^{+})-ATPase exhibits a detectably larger than average binding constant for the negatively charged PS than for the corresponding PC.[131]

Brotherus et al.[131] have pointed out that the modest values of lipid binding constants to many membrane enzymes could have the same effect as the much larger values of binding constants found for aqueous ligands in aqueous solution because of the concentrated nature of the membrane system. For example, to achieve 90% occupancy of a specific lipid binding site on a protein in a bilayer containing 10 mol% of the specific lipid requires a binding constant of 100 mol fraction units. For comparison, the same occupancy in dilute solution with an aqueous ligand concentration of 10^{-5} *M* needs 5×10^{7} in comparable units.

The necessity for almost complete lipid solvation before reactivation of most purified membrane enzymes is apparent from the sigmoidal nature of the reactivation curves. Sandermann[184,185] presents an analysis of the dependence of integral membrane enzymes on lipid activation in terms of multiple binding site kinetics. The model which gives greatest consistency with existing experimental data assumes the enzyme to have a number (n) of noninteracting and identical binding sites. If only fully substituted enzyme (EL_n) and the next most highly substituted forms (such as EL_{n-1} and EL_{n-2}) are assumed to possess enzyme activity, sigmoidal activation can be predicted with varying degrees of apparent kinetic cooperativity. Cooperativity reaches a maximum when enzyme activity starts to appear with about 80% of the full lipid substitution.

C. Type 3 Interactions — Interfacial Regulation

The charge and hydration state of the polar head groups of phospholipids has a strong influence on the activity of lipid-dependent enzymes. The term "interfacial regulation" has been used by Sandermann[193] to designate the dependence of membrane functions on the amount or structure of bound interfacial water. The term has been extended[5] to include the effects of lipid surface charge. Examples of the importance of lipid hydration and surface charge are shown in Table 3.

Most biological membranes are electrostatically charged owing to the ionized groups of membrane proteins, phospholipids, and glycolipids. For negatively charged surfaces, the electrostatic potential of the surface will decrease the local concentration of negatively charged substrates in the microenvironment of the bound enzyme molecules.[204] This has the effect of shifting the K_m for the substrate to higher values.[205] For example, adsorption of glyceraldehyde 3-phosphate dehydrogenase to negatively charged liposomes of PA and PC increases the K_m value for the negatively charged substrate glyceraldehyde 3-phosphate.[206]

Table 3
SELECTED EXAMPLES OF THE EFFECT OF LIPID HEAD-GROUP PROPERTIES ON THE FUNCTIONAL ACTIVITIES OF MEMBRANE ENZYMES

Enzyme	Effecting parameter	Ref.
Ca^{2+}-ATPase	Packing of head groups	194, 195
	Dipole orientation	196
$(Na^+ + K^+)$-ATPase	Negative surface charge	28, 58
	Packing of head groups	197
H^+-ATPase	Charge and size of head groups	202
C_{55}-isoprenoid alcohol kinase	Polar group hydration	193
Cytochrome P-450	Negative surface charge	198—201
Cytochrome *c*-oxidase	Negative surface charge	111, 112, 121
Acetylcholinesterase	Negative surface charge	168
β-Hydroxybutyrate dehydrogenase	Size of polar head group	171
5′-Nucleotide	Size of polar head group	170
Band 3 anion transporter	Inhibited by negative surface charge	203

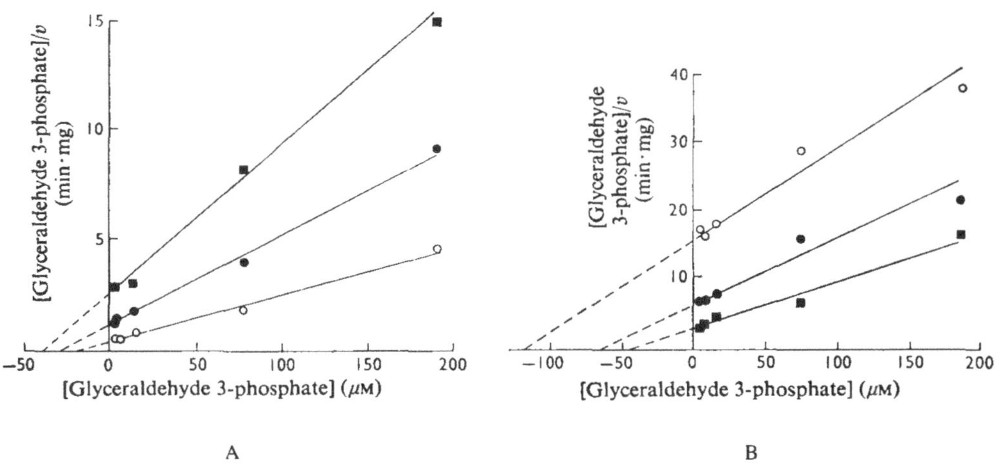

FIGURE 3. Effects on the kinetic properties of glyceraldehyde 3-phosphate dehydrogenase by adsorption to (A) negatively and (B) positively charged liposomes. (A) Plots of (S)/v against (S) using glyceraldehyde 3-phosphate in the absence (○) and presence (●, ■) of negatively charged liposomes; (B) as in (A) but using positively charged liposomes. In the first case, the apparent K_m for the negatively charged substrate is shifted to higher values, while in the second case the apparent K_m is decreased. (Reproduced from Wooster, M. S. and Wrigglesworth, J. M., *Biochem. J.*, 159, 627, 1976. With permission.)

A decreased K_m is found in the presence of positively charged phospholipid (Figure 3). The local electrostatic field in the microenvironment of a membrane-associated enzyme will also affect the local concentration of H^+.[207] All these effects could be mediated by phospholipid preferentially segregated by membrane enzymes. The magnitude of the effects is difficult to assess for any particular case since preferential segregation, presumably, involves charge interaction between ionized groups on the protein with oppositely charged groups on the lipid. The final net electrostatic contribution to the surface potential will, therefore, be influenced by the protein as well as the lipid.

An example where the interaction of charged phospholipids with a membrane enzyme strongly influences enzyme-substrate and also protein-protein interactions is the reconstituted monooxygenase system, NADPH cytochrome P-450 reductase and cytochrome P-450. The endoplasmic reticulum, from which the system can be purified, contains high amounts of negatively charged lipids;[208] PS and PI are present up to a molar content of 20%. It might be expected, therefore, that the activity of the reconstituted system would strongly depend on surface charge. This is found to be the case. Cytochrome P-450 itself strongly interacts with negatively charged lipids,[200,201] and the reduction of cytochrome P-450 by NADPH in reconstituted phospholipid vesicles is dependent on membrane charge.[198] The reaction is mediated by NADPH cytochrome P-450 reductase, and negative membranes appear to favor the formation of protein/protein complexes in the membrane.[199,209,210] A similar relationship between the reducibility of cytochrome b_5 and the negative charge of the liposomes is also reported for membranes containing NADPH cytochrome P-450 reductase and cytochrome b_5.[198]

REFERENCES

1. **Coleman, R.,** Membrane-bound enzymes and membrane ultrastructure, *Biochim. Biophys. Acta,* 300, 1, 1973.
2. **Fourcans, B. and Jain, M. K.,** Role of phospholipids in transport and enzymic reactions, in *Advances in Lipid Research,* Vol. 12, Paoletti, R. and Kritchevsky, D., Eds., Academic Press, New York, 1974, 147.
3. **Martonosi, A., Ed.,** *The Enzymes of Biological Membranes,* Vol. 1 to 4, Plenum Press, New York, 1976.
4. **Gazzotti, P. and Peterson, S. W.,** Lipid requirement of membrane-bound enzymes, *J. Bioenerg. Biomembr.,* 9, 373, 1977.
5. **Sandermann, H.,** Regulation of membrane enzymes by lipids, *Biochim. Biophys. Acta,* 515, 209, 1978.
6. **Freedman, R. B.,** Membrane-bound enzymes, in *New Comprehensive Biochemistry,* Vol. 1, Finean, J. B. and Michell, R. H., Eds., Elsevier/North-Holland, Amsterdam, 1981, 161.
7. **Warren, G. B.,** Membrane proteins: structure and assembly, in *New Comprehensive Biochemistry,* Vol. 1, Finean, J. B. and Michell, R. H., Eds., Elsevier/North-Holland, Amsterdam, 1981, 215.
8. **McElhaney, R. N.,** Effects of membrane lipids on transport and enzymic activities, *Curr. Top. Membr. Transp.,* 17, 317, 1982.
9. **Salton, N. R. J.,** Membrane associated enzymes in bacteria, *Adv. Microb. Physiol.,* 11, 213, 1974.
10. **Triggle, D. J.,** Some aspects of the role of lipids in lipid-protein interactions and cell membrane structure and function, in *Recent Progress in Surface Science,* Vol. 3, Danielli, J. F., Riddiford, A. C., and Rosenberg, M. D., Eds., Academic Press, New York, 1970, 273.
11. **Tanford, C.,** *The Hydrophobic Effect: Formation of Micelles and Biological Membranes,* John Wiley & Sons, New York, 1973.
12. **Glennis, R. B. and Jonas, A.,** Protein-lipid interactions, *Annu. Rev. Biophys. Bioenerg.,* 6, 195, 1977.
13. **Parsegian, A., Ed.,** Biophysical discussions: protein-lipid interactions in membranes, *Biophys. J.,* 37 (1), 1982.
14. **Razin, S.,** Reconstitution of biological membranes, *Biochim. Biophys. Acta,* 265, 241, 1972.
15. **Kagawa, Y.,** Reconstitution of oxidative phosphorylation, *Biochim. Biophys. Acta,* 265, 297, 1972.
16. **Montal, M.,** Experimental membranes and mechanisms of bioenergy transductions, *Annu. Rev. Biophys. Bioenerg.,* 5, 119, 1976.
17. **Racker, E.,** *A New Look at Mechanisms in Bioenergetics,* Academic Press, New York, 1976.
18. **De Pierre, J. W. and Ernster, L.,** Enzyme topology of intracellular membranes, *Annu. Rev. Biochem.,* 46, 201, 1977.
19. **Korenbrot, J. I.,** Ion transport in membranes: incorporation of biological ion-translocating proteins in model membrane systems, *Annu. Rev. Physiol.,* 39, 19, 1977.
20. **Racker, E.,** Reconstitution of membrane processes, *Methods Enzymol.,* 55, 699, 1979.
21. **Hokin, L. E.,** Reconstitution of 'carriers' in artificial membranes, *J. Membrane Biol.,* 60, 77, 1981.
22. **Hellingwert, K. J., Scholte, B. J., and Van Dam, K.,** Bacteriorhodopsin vesicles: an outline of the requirements for light-dependent H^+ pumping, *Biochim. Biophys. Acta,* 513, 66, 1978.
23. **Bennett, J. P., McGill, K. A., and Warren, G. B.,** The role of lipids in the functioning of a membrane protein: the sarcoplasmic reticulum calcium pump, *Curr. Top. Membr. Transp.,* 14, 127, 1980.

24. **Roelofsen, B.,** The (non)specificity in the lipid-requirement of calcium — and (sodium plus potassium) — transporting adenosine triphosphatases, *Life Sci.*, 29, 2235, 1981.
25. **Loh, H. H. and Law, P. Y.,** The role of lipids in receptor mechanisms, *Annu. Rev. Pharmacol. Toxicol.*, 20, 201, 1980.
26. **Houslay, M. D. and Gordon, L. M.,** The activity of adenylate cyclase is regulated by the nature of its lipid environment, *Curr. Top. Memb. Transp.*, 18, 179, 1983.
27. **Kimelberg, H. K.,** Protein-liposome interactions and their relevance to the structure and function of cell membranes, *Mol. Cell. Biochem.*, 10, 171, 1976.
28. **Ahrens, M.-L.,** Electrostatic control by lipids upon the membrane-bound (Na^+ + K^+)-ATPase, *Biochim. Biophys. Acta*, 642, 252, 1981.
29. **Jorgensen, P. L.,** Isolation and characterization of the components of the sodium pump, *Q. Rev. Biophys.*, 7, 239, 1975.
30. **Robinson, J. D. and Flashner, M. S.,** The (Na^+ + K^+)-activated ATPase: enzymic and transport properties, *Biochim. Biophys. Acta*, 549, 145, 1979.
31. **Mandersloot, J. G., Roelofsen, B., and DeGier, J.,** Phosphatidylinositol as the endogenous activator of the (Na^+ + K^+)-ATPase in microsomes of rabbit kidney, *Biochim. Biophys. Acta*, 508, 478, 1978.
32. **Ottolenghi, P.,** The relipidation of delipidated Na, K-ATPase, *Eur. J. Biochem.*, 99, 113, 1979.
33. **Hilden, S. and Hokin, L.,** Coupled Na^+-K^+ transport in vesicles containing a purified (Na K)-ATPase and only phosphatidyl choline, *Biochem. Biophys. Res. Commun.*, 69, 521, 1976.
34. **Roelofsen, B. and van Deenen, L. L. M.,** Lipid requirement of membrane-bound ATPase: studies on human erythrocyte ghosts, *Eur. J. Biochem.*, 40, 245, 1973.
35. **Roelofsen, B. and Schatzmann, H. J.,** The lipid requirement of the (Ca^{2+} + Mg^{2+})-ATPase in the human erythrocyte membrane, as studied by various highly purified phospholipases, *Biochim. Biophys. Acta*, 464, 17, 1977.
36. **Yeagle, P. L.,** Cholesterol modulation of (Na^+ + K^+)-ATPase hydrolyzing activity in the human erythrocyte, *Biochim. Biophys. Acta*, 727, 39, 1983.
37. **Giraud, F., Claret, M., Bruckdorfer, K. R., and Chailley, B.,** The effects of membrane lipid order and cholesterol on the internal and external cationic sites of the Na^+-K^+ pump in erythrocytes, *Biochim. Biophys. Acta*, 647, 249, 1981.
38. **Zambrano, F., Morales, M., Fuentes, N., and Rojas, M.,** Sulfatide role in the sodium pump, *J. Membrane Biol.*, 63, 71, 1981.
39. **Tanaka, R. and Strickland, K. P.,** Role of phospholipid in the activation of Na^+, K^+-activated adenosine triphosphatase of beef brain, *Arch. Biochem. Biophys.*, 111, 583, 1965.
40. **Wheeler, K. P. and Whittam, R.,** The involvement of phosphatidylserine in adenosine triphosphatase activity of the sodium pump, *J. Physiol.*, 207, 303, 1970.
41. **Tanaka, R., Sakamoto, T., and Sakamoto, Y.,** Mechanism of lipid activation of Na^+, K^+, Mg^{2+}-activated adenosine triphosphatase and K^+, Mg^{2+}-activated phosphatase of bovine cerebral cortex, *J. Membr. Biol.*, 4, 42, 1971.
42. **Noguchi, T. and Freed, S.,** Dissociation of lipid components and reconstitution at $-75°C$ of Mg^{2+} dependent, Na^+ and K^+ stimulated, adenosine triphosphatase in rat brain, *Nature (London), (New Biol.)*, 230, 148, 1971.
43. **De Pont, J. J. H. H. M., Van Prooijen-van Eeden, A., and Bonting, S. L.,** Studies on (Na^+-K^+)-activated ATPase: phosphatidylserine not essential for (Na^+-K^+)-ATPase activity, *Biochim. Biophys. Acta*, 323, 487, 1973.
44. **Abeywardena, M. Y. and Charnock, J. S.,** Modulation of cardiac glycoside inhibition of (Na^+-K^+)-ATPase by membrane lipids, *Biochim. Biophys. Acta*, 729, 75, 1983.
45. **Fenster, L. J. and Copenhaver, J. H., Jr.,** Phosphatidyl serine requirement of (Na^+-K^+)-activated adenosine triphosphatase from rat kidney and brain, *Biochim. Biophys. Acta*, 137, 406, 1967.
46. **Stahl, W. L.,** Role of phospholipids in the Na^+, K^+-stimulated adenosine triphosphatase system of brain microsomes, *Arch. Biochem. Biophys.*, 154, 56, 1973.
47. **Kimelberg, H. K. and Papahadjopoulos, D.,** Phospholipid requirements for (Na^+ + K^+)-ATPase activity: head-group specificity and fatty acid fluidity, *Biochim. Biophys. Acta*, 282, 277, 1972.
48. **Wheeler, K. P., Walker, J. A., and Barker, D. M.,** Lipid requirement of the membrane sodium-plus-potassium ion-dependent adenosine triphosphatase system, *Biochem. J.*, 146, 713, 1975.
49. **Peters, W. H. M., Fleuren-Jakobs, A. M. M., De Pont, J. J. H. H. M., and Bonting, S. L.,** Studies on (Na^+ + K^+)-activated ATPase. XLIX. Content and role of cholesterol and other neutral lipids in highly purified rabbit kidney enzyme preparation, *Biochim. Biophys. Acta*, 649, 541, 1981.
50. **De Pont, J. J. H. H. M., Van Prooijen-van Eeden, A., and Bonting, S. L.,** Role of negatively charged phospholipids in highly purified (Na^+ + K^+)-ATPase from rabbit kidney outer medulla, *Biochim. Biophys. Acta*, 508, 464, 1978.
51. **Walker, J. A. and Wheeler, K. P.,** Polar head-groups and acyl side-chain requirements for phospholipid-dependent (Na^+ + K^+)-ATPase, *Biochim. Biophys. Acta*, 394, 135, 1975.

52. **Kimelberg, H. K. and Papahadjopoulos, D.,** Effects of phospholipid acyl chain fluidity, phase transitions and cholesterol on (Na^+ + K^+)-stimulated adenosine triphosphatase, *J. Biol. Chem.*, 249, 1071, 1974.
53. **Palatini, P., Dabbeni-Sala, F., Pitotti, A., Bruni, A., and Mandersloot, J. C.,** Activation of (Na^+ + K^+)-dependent ATPase by lipid vesicles of negative phospholipids, *Biochim. Biophys. Acta*, 466, 1, 1977.
54. **Foussard-Guilbert, F., Ermias, A., Laget, P., Tanguy, G., Girault, M., and Jallet, P.,** Detergent effects of kinetic properties of (Na^+ + K^+)-ATPase from kidney membranes, *Biochim. Biophys. Acta*, 692, 296, 1982.
55. **Palatini, P., Dabbeni-Sala, F., and Bruni, A.,** Reactivation of a phospholipid-depleted sodium, potassium-stimulated ATPase, *Biochim. Biophys. Acta*, 288, 413, 1972.
56. **Jarnefelt, J.,** Lipid requirements of a functional membrane structures as indicated by the reversible inactivation of (Na^+-K^+)-ATPase, *Biochim. Biophys. Acta*, 266, 91, 1972.
57. **Goldman, S. S. and Albers, R. W.,** Sodium-potassium-activated adenosine triphosphatase: the role of phospholipids, *J. Biol. Chem.*, 248, 867, 1973.
58. **Brotherus, J. R., Jost, P. C., Griffith, O. H., Keana, J. F. W., and Hokin, L. E.,** Charge specificity at the lipid-protein interface of membranous Na, K-ATPase, *Proc. Natl. Acad. Sci. U.S.A.*, 77, 272, 1980.
59. **Hansson, G. C., Heilbronn, E., Karlsson, K.-A., and Samuelsson, B. E.,** The lipid composition of the electric organ of the ray, *Torpedo marmorata*, with specific reference to sulfatides and Na^+-K^+-ATPase, *J. Lipid Res.*, 20, 509, 1979.
60. **Poon, R., Richards, J. M., and Clark, W. R.,** The relationship between plasma membrane lipid composition and physical-chemical properties. II. Effect of phospholipid fatty acid modulation on plasma membrane physical properties and enzymic activities, *Biochim. Biophys. Acta*, 649, 58, 1981.
61. **Singer, S. J. and Nicholson, G. L.,** The fluid mosaic model of the structure of cell membranes, *Science*, 175, 720, 1972.
62. **Helenius, A. and Simons, K.,** Solubilization of membranes by detergents, *Biochim. Biophys. Acta*, 415, 29, 1975.
63. **Tanford, C. and Reynolds, J. A.,** Characterisation of membrane proteins in detergent solutions, *Biochim. Biophys. Acta*, 457, 133, 1976.
64. **Fong, S.-L., Tsin, A. T. C., Bridges, C. D. B., and Liou, G. I.,** Detergents for extraction of visual pigments: types, solubilization and stability, *Methods Enzymol.*, 81, 133, 1982.
65. **Zahler, P. and Niggli, V.,** The use of organic solvents in membrane research, *Methods Membr. Biol.*, 8, 1, 1977.
66. **Kuboyama, M., Yong, F. C., and King, T. E.,** Studies on cytochrome oxidase. VIII. Preparation and some properties of cardiac cytochrome oxidase, *J. Biol. Chem.*, 247, 6375, 1972.
67. **Capaldi, R. A. and Hayashi, H.,** The polypeptide composition of cytochrome oxidase from beef heart mitochondria, *FEBS Lett.*, 26, 261, 1972.
68. **Pentilla, T., Saraste, M., and Wikström M.,** The number of subunits in bovine cytochrome *c* oxidase, *FEBS Lett.*, 101, 295, 1979.
69. **Downer, N. W., Robinson, N. C., and Capaldi, R. A.,** Characterization of a seventh different subunit of beef heart cytochrome c oxidase: similarities between the beef heart enzyme and that from other species, *Biochemistry*, 15, 2930, 1976.
70. **Merle, P. and Kadenbach, B.,** The subunit composition of mammalian cytochrome c oxidase, *Eur. J. Biochem.*, 105, 499, 1980.
71. **Wikström, M., Krab, K., and Saraste, M.,** *Cytochrome Oxidase, a Synthesis*, Academic Press, London, 1981.
72. **Robinson, N. C. and Capaldi, R. A.,** Interaction of detergents with cytochrome c oxidase, *Biochemistry*, 16, 375, 1977.
73. **Saraste, M., Penttila, T., and Wikström, M.,** Quaternary structure of bovine cytochrome oxidase, *Eur. J. Biochem.*, 115, 261, 1981.
74. **Bock, H. G. and Fleischer, S.,** Preparation of a homogenous soluble D-β-hydroxybutyrate apodehydrogenase from mitochondria, *J. Biol. Chem.*, 250, 5774, 1975.
75. **Gazzotti, P., Bock, H.-G., and Fleischer, S.,** Interaction of D-β-hydroxybutyrate apodehydrogenase with phospholipids, *J. Biol. Chem.*, 250, 5782, 1975.
76. **Small, D. M., Penkett, S. A., and Chapman, D.,** Studies on simple and mixed bile salt micelles by nuclear magnetic resonance spectroscopy, *Biochim. Biophys. Acta*, 176, 178, 1969.
77. **Yedgar, S., Barenholz, Y., and Cooper, V. G.,** Molecular weight, shape and structure of mixed micelles of Triton X-100 and sphingomyelin, *Biochim. Biophys. Acta*, 363, 98, 1974.
78. **Collins, M. L. P. and Salton, M. R. J.,** Solubility characteristics of *Micrococcus lysodeikticus* membrane components in detergents and chaotropic salts analysed by immunoelectrophoresis, *Biochim. Biophys. Acta*, 553, 40, 1979.
79. **Hjelmeland, L. M., Nebert, D. W., and Osborne, J. C.,** Sulfobetaine derivatives of bile acids: nondenaturing surfactants for membrane biochemistry, *Anal. Biochem.*, 130, 72, 1983.

80. **McMurchie, E. J., Abeywardena, M. Y., Charnock, J. S., and Gibson, R. A.,** Differential modulation of rat heart mitochondrial membrane-associated enzymes by dietary lipid, *Biochim. Biophys. Acta,* 760, 13, 1983.
81. **Helenius, A., McCaslin, E. F., and Tanford, C.,** Properties of detergents, *Methods Enzymol.,* 56, 734, 1979.
82. **Furth, A. J.,** Removing unbound detergent from hydrophobic proteins, *Anal. Biochem.,* 109, 207, 1980.
83. **Houslay, M. D. and Stanley, K. K.,** *Dynamics of Biological Membranes,* John Wiley & Sons, Chichester, 1982.
84. **Andersen, J. P., Fellmann, P., Moller, J. V., and Devaux, P. F.,** Immobilization of a spin-labeled fatty acid chain covalently attached to Ca^{2+}-ATPase from sarcoplasmic reticulum suggests an oligomeric structure, *Biochemistry,* 20, 4928, 1981.
85. **Andersen, J. P., Le Maire, M., Kragh-Hansen, U., Champeil, P., and Moller, J. V.,** Perturbation of the structure and function of membranous Ca^{2+}-ATPase by non-solubilizing concentrations of a non-ionic detergent, *Eur. J. Biochem.,* 134, 205, 1983.
86. **Fellmann, P., Andersen, J., Devaux, P. F., le Maire, M., and Bienvenue, A.,** Photoaffinity spin-labeling of the Ca^{2+}-ATPase in sarcoplasmic reticulum: evidence for oligomeric structure, *Biochem. Biophys. Res. Commun.,* 95, 289, 1980.
87. **Lichtenberg, D., Robson, R., and Dennis, E. A.,** Solubilization of phospholipids by detergents. Structural and kinetic aspects, *Biochim. Biophys. Acta,* 737, 285, 1983.
88. **Jost, P. C., Griffith, O. H., Capaldi, R. A., and Vanderkooi, G.,** Evidence for boundary lipid in membranes, *Proc. Natl. Acad. Sci. U.S.A.,* 70, 480, 1973.
89. **Jost, P., Griffith, O. H., Capaldi, R. A., and Vanderkooi, G.,** Identification and extent of fluid bilayer regions in membranous cytochrome oxidase, *Biochim. Biophys. Acta,* 311, 141, 1973.
90. **Jost, P. C., Nadakavukaren, K. K., and Griffith, O. H.,** Phosphatidylcholine exchange between the boundary lipid and bilayer domains in cytochrome oxidase containing membranes, *Biochemistry,* 16, 3110, 1977.
91. **Warren, G. B., Toon, P. A., Birdsall, N. J. M., Lee, A. G., and Metcalfe, J. C.,** Reversible lipid titrations of the activity of pure adenosine triphosphatase-lipid complexes, *Biochemistry,* 13, 5501, 1974.
92. **Warren, G. B., Houslay, M. D., Metcalfe, J. C., and Birdsall, N. J. M.,** Cholesterol is excluded from the phospholipid annulus surrounding an active calcium transport protein, *Nature (London),* 255, 684, 1975.
93. **Hesketh, T. R., Smith, G. A., Houslay, M. D., McGill, K. A., Birdsall, N. J. M., Metcalfe, J. C., and Warren, G. B.,** Annular lipids determine the ATPase activity of a calcium transport protein complexed with dipalmitoyllecithin, *Biochemistry,* 15, 4145, 1976.
94. **Kang, S. Y., Gutowsky, H. S., Hsung, J. C., Jacobs, R., King, T. E., Rice, D., and Oldfield, E.,** Nuclear magnetic resonance investigation of the cytochrome oxidase-phospholipid interaction: a new model for boundary lipid, *Biochemistry,* 18, 3257, 1979.
95. **Kinosita, K., Kawato, S., Ikegami, A., Yoshida, S., and Orii, Y.,** The effect of cytochrome oxidase on lipid chain dynamics: a nanosecond fluorescence depolarization study, *Biochim. Biophys. Acta,* 647, 7, 1981.
96. **MacLennan, D. H., Seeman, P., Iles, G. H., and Yip, C. C.,** Membrane formation by the adenosine triphosphatase of sarcoplasmic reticulum, *J. Biol. Chem.,* 246, 2702, 1971.
97. **Dean, W. L. and Tanford, C.,** Properties of a delipidated, detergent-activated Ca^{2+}-ATPase, *Biochemistry,* 17, 1683, 1978.
98. **Cable, M. B. and Powell, G. L.,** Spin-labeled cardiolipin: preferential segregation in the boundary layer of cytochrome *c* oxidase, *Biochemistry,* 19, 5679, 1980.
99. **Awashi, Y. C., Chuang, T. F., Keenan, T. W., and Crane, F. L.,** Tightly bound cardiolipin in cytochrome oxidase, *Biochim. Biophys. Acta,* 226, 42, 1971.
100. **Al-Fai, W., Jones, M. G., Rashid, K., and Wilson, M. T.,** An active cytochrome *c* oxidase that has no tightly bound cardiolipin, *Biochem. J.,* 209, 901, 1983.
101. **Knowles, P. F., Watts, A., and Marsh, D.,** Spin-labelled studies of head-group specificity in the interaction of phospholipids with yeast cytochrome oxidase, *Biochemistry,* 20, 5888, 1981.
102. **Brierley, G. P. and Merola, A. J.,** Studies on the electron-transfer system. XLVIII. Phospholipid requirements in cytochrome oxidase, *Biochim. Biophys. Acta,* 64, 205, 1962.
103. **Fleischer, S., Brierly, G., Klouwen, H., and Slautterback, D. B.,** Studies of the electron transfer system. XLVII. The role of phospholipids in electron transfer, *J. Biol. Chem.,* 237, 3264, 1962.
104. **Benga, G., Porumb, T., and Wrigglesworth, J. M.,** Estimation of lipid regions in a cytochrome oxidase-lipid complex using spin labeling electron spin resonance: distribution effects on the spin label, *J. Bioenerg. Biomembr.,* 13, 269, 1981.
105. **Yu, C.-A., Yu, L., and King, T. E.,** Studies on cytochrome oxidase: interactions of the cytochrome oxidase protein with phospholipids and cytochrome c, *J. Biol. Chem.,* 250, 1383, 1975.
106. **Knowles, A. F., Eytan, E., and Racker, E.,** Phospholipid-protein interactions in the Ca^{2+}-adenosine triphosphatase of sarcoplasmic reticulum, *J. Biol. Chem.,* 251, 5161, 1976.

107. **Helenius, A., Fries, E., Garoff, H., and Simons, K.,** Solubilization of the semliki forest virus membrane with sodium deoxycholate, *Biochim. Biophys. Acta,* 436, 319, 1976.
108. **Dean, W. L. and Tanford, C.,** Reactivation of lipid-depleted Ca^{2+}-ATPase by a nonionic detergent, *J. Biol. Chem.,* 252, 3551, 1977.
109. **Robinson, N. C., Strey, F., and Talbert, L.,** Investigation of the essential boundary layer phospholipids of cytochrome *c* oxidase using Triton X-100 delipidation, *Biochemistry,* 19, 3656, 1980.
110. **Vik, S. B. and Capaldi, R. A.,** Lipid requirements for cytochrome *c* oxidase activity, *Biochemistry,* 16, 5755, 1977.
111. **Robinson, N. C.,** Specificity and binding affinity of phospholipids to the high-affinity cardiolipin sites of beef heart cytochrome c oxidase, *Biochemistry,* 21, 184, 1982.
112. **Robinson, N. C.,** The specificity and affinity of phospholipids for cytochrome *c* oxidase, *Biophys. J.,* 37, 65, 1982.
113. **Folch, J., Lees, M., and Sloane Stanley, H.,** A simple method for the isolation and purification of total lipids from animal tissues, *J. Biol. Chem.,* 226, 497, 1957.
114. **Bligh, E. G. and Dyer, W. J.,** A rapid method of total lipid extraction and purification, *Can. J. Biochem. Physiol.,* 37, 911, 1959.
115. **Rose, H. G. and Oklander, M.,** Improved procedure for the extraction of lipids from human erythrocytes, *J. Lipid Res.,* 6, 428, 1965.
116. **Lenaz, G., Parenti Castelli, G., Sechi, A. M., and Masotti, L.,** Lipid-protein interactions in mitochondria, *Arch. Biochem. Biophys.,* 148, 391, 1972.
117. **Hamaguchi, H. and Cleve, H.,** Solubilization and comparative analysis of mammalian erythrocyte membrane glycoproteins, *Biochem. Biophys. Res. Commun.,* 47, 459, 1972.
118. **Agrawal, H. C., Burton, R. M., Fishman, M. A., Mitchell, R. F., and Prensky, A. L.,** Partial characterization of a new myelin protein component, *J. Neurochem.,* 19, 2083, 1972.
119. **Roelofsen, B., de Gier, J., and van Deenen, L. L. M.,** Binding of lipids in the red cell membrane, *J. Cell. Comp. Physiol.,* 63, 233, 1964.
120. **Zamudio, I., Cellino, M., and Canessa-Fischer, M.,** The relation between membrane structure and NADH: (acceptor) oxidoreductase activity of erythrocyte ghosts, *Arch. Biochem. Biophys.,* 129, 336, 1969.
121. **Vik, S. B., Georgevich, G., and Capaldi, R. A.,** Diphosphatidylglycerol is required for optimal activity of beef heart cytochrome *c* oxidase, *Proc. Natl. Acad. Sci. U.S.A.,* 78, 1456, 1981.
122. **Bielski, B. H. J. and Freed, S.,** Enzymatic reactions below 0° of α-chymotrypsin in methanol-water solvents, *Biochim. Biophys. Acta,* 89, 314, 1964.
123. **Freed, S.,** Chemical-biochemical signal and noise, *Science,* 150, 576, 1965.
124. **Magee, A. I. and Schlesinger, M. J.,** Fatty acid acylation of eucaryotic cell membrane proteins, *Biochim. Biophys. Acta,* 694, 279, 1982.
125. **Di Rienzo, J. M., Nakamura, K., and Inouye, M.,** The outer membrane proteins of gram-negative bacteria: biosynthesis, assembly and functions, *Annu. Rev. Biochem.,* 47, 481, 1978.
126. **Nielsen, J. B. K., Caulfield, M. P., and Lampen, J. O.,** Lipoprotein nature of *B. licheniformis* membrane penicillinase, *Proc. Natl. Acad. Sci. U.S.A.,* 78, 3511, 1981.
127. **Yamamoto, S. and Lampen, J. O.,** The hydrophobic membrane penicillinase of *B. licheniformis, J. Biol. Chem.,* 251, 4102, 1976.
128. **Warren, G. B., Toon, P. A., Birdsall, N. J. M., Lee, A. G., and Metcalfe, J. C.,** Reconstitution of a calcium pump using defined membrane components, *Proc. Natl. Acad. Sci. U.S.A.,* 71, 622, 1974.
129. **Johannsson, A., Keightley, C. A., Smith, G. A., Richards, C. D., Hesketh, T. R., and Metcalfe, J. C.,** The effect of bilayer thickness and n-alkanes on the activity of the (Ca^{2+} + Mg^{2+})-dependent ATPase of sarcoplasmic reticulum, *J. Biol. Chem.,* 256, 1643, 1981.
130. **Johannsson, A., Keightley, C. A., Smith, G. A., and Metcalfe, J. C.,** Cholesterol in sarcoplasmic reticulum and the physiological significance of membrane fluidity, *Biochem. J.,* 196, 505, 1981.
131. **Brotherus, J. R., Griffith, O. H., Brotherus, M. O., Jost, P. C., Silvius, J. R., and Hokin, L. E.,** Lipid-protein multiple binding equilibria in membranes, *Biochemistry,* 20, 5261, 1981.
132. **Dreesen, T. D. and Koch, R. B.,** Odorous chemical perturbations of (Na^+ + K^+)-dependent ATPase activities, *Biochem. J.,* 203, 69, 1982.
133. **Bisson, R. and Montecucco, C.,** Photolabelling of membrane proteins with photoactive phospholipids, *Biochem. J.,* 193, 757, 1981.
134. **Moore, B. M., Lentz, B. R., Hoechli, M., and Meissner, G.,** Effect of lipid membrane structure on the adenosine 5′-triphosphate hydrolysing activity of the calcium stimulated adenosine triphosphatase of sarcoplasmic reticulum, *Biochemistry,* 20, 6810, 1981.
135. **Kamp, H. H., Wirtz, K. W. A., and van Deenen, L. L. M.,** Some properties of phosphatidylcholine exchange protein purified from beef liver, *Biochim. Biophys. Acta,* 318, 313, 1973.
136. **Johnson, L. W. and Zilversmit, D. B.,** Catalytic properties of phospholipid exchange protein from bovine heart, *Biochim. Biophys. Acta,* 375, 165, 1975.

137. **Poorthuis, B. J. H. M., van de Krift, T. P., Teerlink, T., Akeroyd, R., Hostetler, K. Y., and Wirtz, K. W. A.,** Phospholipid transfer activities in Morris hepatomas and the specific contribution of the phosphatidylcholine exchange protein, *Biochim. Biophys. Acta*, 600, 376, 1980.
138. **Douady, D., Grosbois, M., Guerbette, F., and Kader, J.-C.,** Purification of a basic phospholipid transfer protein from maize seedlings, *Biochim. Biophys. Acta*, 710, 143, 1982.
139. **Badger, C. R. and Helmkamp, G. M.,** Modulation of phospholipid transfer protein activity: inhibition by local anaesthetics, *Biochim. Biophys. Acta*, 692, 33, 1982.
140. **Dyatovitskaya, E. V., Timofeeva, N. G., Yakimenko, E. F., Barsukov, L. I., Muzya, G. I., and Bergelson, L. D.,** A sphingomyelin transfer protein in rat tumors and fetal liver, *Eur. J. Biochem.*, 123, 311, 1982.
141. **Yamada, K. and Sasaki, T.,** A rat brain cytosol protein which accelerates the translocation of galactosylceramide, lactosylceramide and glucosylceramide between membranes, *Biochim. Biophys. Acta*, 687, 195, 1982.
142. **Bruckdorfer, K. R., Graham, J. M., and Green, C.,** The incorporation of steroid molecules into lecithin sols, β-lipoproteins and cellular membranes, *Eur. J. Biochem.*, 4, 512, 1968.
143. **Whetton, A. D., Gordon, L. M., and Houslay, M. D.,** Adenylate cyclase is inhibited on depletion of plasma membrane cholesterol, *Biochem. J.*, 212, 331, 1983.
144. **Madden, T. D., Chapman, D., and Quinn, P. J.,** Cholesterol modulates activity of calcium-dependent ATPase of the sarcoplasmic reticulum, *Nature (London)*, 279, 538, 1979.
145. **Drenthe, E.H. S. and Daemen, F. J. M.,** Phospholipases as tools for studying structure and function of photoreceptor membranes, *Methods Enzymol.*, 81, 320, 1982.
146. **Op den Kamp, J. A. F.,** Lipid asymmetry in membranes, *Annu. Rev. Biochem.*, 48, 47, 1979.
147. **Nestruck-Goyke, A. C. and Hasselbach, W.,** Preparative isolation of Apo (Ca^{2+}-ATPase) from sarcoplasmic reticulum and the reactivation by lysophosphatidylcholine of Ca^{2+}-dependent ATP hydrolysis and partial-reaction steps of the enzyme, *Eur. J. Biochem.*, 114, 339, 1981.
148. **Swoboda, G., Fritsche, J., and Hasselbach, W.,** Effects of phospholipase A_2 and albumin on the calcium-dependent ATPase and the lipid composition of sarcoplasmic membranes, *Eur. J. Biochem.*, 95, 77, 1979.
149. **Coleman, R. and Bromley, T. A.,** Hydrolysis of erythrocyte membrane phospholipids by a preparation of phospholipase C from *Clostridium Welchii:* deactivation of (Ca^{2+}, Mg^{2+})-ATPase and its reactivation by added lipids, *Biochim. Biophys. Acta*, 382, 565, 1975.
150. **Warner, T. G. and Dennis, E. A.,** Action of the highly purified, membrane-bound enzyme phosphatidylserine decarboxylase *Escherichia coli* toward phosphatidylserine in mixed micelles and erythrocyte ghosts in the presence of surfactant, *J. Biol. Chem.*, 250, 8004, 1975.
151. **Rizzolo, L. J.,** Kinetics and protein subunit interactions of *Escherichia coli* phosphatidylserine decarboxylase in detergent solution, *Biochemistry*, 20, 868, 1981.
152. **Stadtlander, K., Rade, S., and Ahlers, J.,** Influence of growth conditions on the composition of the plasma membrane from yeast and on kinetic properties of two membrane functions, *J. Cell. Biochem.*, 20, 369, 1982.
153. **Chapman, D. J., De-Felice, J., and Barber, J.,** Influence of winter and summer growth conditions on leaf membrane lipids of *Pisum sativum* L., *Planta*, 157, 218, 1983.
154. **Martin, C. E. and Johnston, A. M.,** Changes in fatty acid distribution and thermotropic properties of phospholipids following phosphatidylcholine depletion in a choline-requiring mutant of *Neurospora crassa*, *Biochim. Biophys. Acta*, 730, 10, 1983.
155. **Denning, G. M., Figard, P. H., and Spector, A. A.,** Effect of fatty acid modification on prostaglandin production by cultured 3T3 cells, *J. Lipid Res.*, 23, 584, 1982.
156. **Seiler, D. and Hasselbach, W.,** Essential fatty acid deficiency and the activity of the sarcoplasmic calcium pump, *Eur. J. Biochem.*, 21, 385, 1971.
157. **Vajreswari, A., Rao, P. S., Kaplay, S. S., and Tulpule, P. G.,** Erythrocyte membrane in rats fed high erucic-acid containing mustard oil: osmotic fragility, lipid composition and sodium potassium ATPase and calcium, magnesium ATPase, *Biochem. Med.*, 29, 74, 1983.
158. **Englehard, V. H., Glaser, M., and Storm, D. R.,** Effect of membrane phospholipid changes on adenylate cyclase in LM cells, *Biochemistry*, 17, 3191, 1978.
159. **Gabrielides, C., Zrike, J., and Scott, W. A.,** Cyclic AMP levels in relation to membrane phospholipid variations in *Neurospora crassa*, *Arch. Microbiol.*, 134, 108, 1983.
160. **Holloway, C. T. and Holloway, P. W.,** The dietary regulation of stearyl coenzyme A desaturase activity and membrane fluidity in rat aorta, *Lipids*, 12, 1025, 1977.
161. **Smith, J. D.,** Effect of modification of membrane phospholipid composition on the activity of phosphatidylethanolamine N-methyl-transferase of *Tetrahymena*, *Arch. Biochem. Biophys.*, 223, 193, 1983.
162. **Brenneman, D. E., Kaduce, T., and Spector, A. A.,** Effect of dietary fat saturation on acyl coenzyme A: cholesterol acyltransferase activity of Ehrlich cell microsomes, *J. Lipid Res.*, 18, 582, 1977.
163. **Spector, A. A., Kaduce, T. L., and Dane, R. W.,** Effects of dietary fat saturation on acylcoenzyme A: cholesterol acyltransferase activity of rat liver microsomes, *J. Lipid Res.*, 21, 169, 1980.

164. **Farias, R. N., Bloj, B., Morero, R. D., Sineriz, F., and Trucco, R. E.,** Regulation of allosteric membrane bound enzymes through changes in membrane lipid composition, *Biochim. Biophys. Acta,* 415, 231, 1975.
165. **Abeywardena, M. Y., Allen, T. M., and Charnock, J. S.,** Lipid-protein interactions of reconstituted membrane-associated adenosinetriphosphatases, *Biochim. Biophys. Acta,* 729, 62, 1983.
166. **Dyatlovitskaya, E. V., Lemenovskaya, A. F., and Bergelson, L. D.,** Use of protein-mediated lipid exchange in the study of membrane-bound enzymes: the lipid dependence of glucose-6-phosphatase, *Eur. J. Biochem.,* 99, 605, 1979.
167. **Roelofsen, B. and van Linde-Sibenius Trip, W.,** The fraction of phosphatidylinositol that activates the $(Na^+ + K^+)$-ATPase in rabbit kidney microsomes is closely associated with the enzyme protein, *Biochim. Biophys. Acta,* 647, 302, 1981.
168. **Beauregard, G. and Roufogalis, B. D.,** The role of tightly bound phospholipid in the activity of erythrocyte acetylcholinesterase, *Biochem. Biophys. Res. Commun.,* 77, 211, 1977.
169. **Heron, C., Corina, D., and Ragan, C. I.,** The phospholipid annulus of mitochondrial NADH-ubiquinone reductase: a dual phospholipid requirement for enzyme activity, *FEBS Lett.,* 79, 399, 1977.
170. **Merisko, E. M., Ojakian, G. K., and Widnell, C. C.,** The effects of phospholipids on the properties of hepatic 5′-nucleotidase, *J. Biol. Chem.,* 256, 1983, 1981.
171. **Isaacson, Y. A., Deroo, P. W., Rosenthal, A. F., Bittman, R., McIntyre, J. O., Bock, H. G., Gazzotti, P., and Fleischer, S.,** The structural specificity of lecithin for activation of purified D-β-hydroxybutyrate apodehydrogenase, *J. Biol. Chem.,* 254, 117, 1979.
172. **Yeagle, P. L.,** ^{31}P nuclear magnetic resonance studies of the phospholipid-protein interface in cell membranes, *Biophys. J.,* 37, 227, 1982.
173. **Ferguson-Miller, S., Brautigan, D. L., and Margoliash, E.,** Correlation of the kinetics of electron transfer activity of various eukaryotic cytochromes c with binding to mitochondrial cytochrome *c* oxidase, *J. Biol. Chem.,* 251, 1104, 1976.
174. **Ferguson-Miller, S., Brautigan, D., and Margoliash, E.,** Definition of cytochrome *c* binding domains by chemical modification. III. Kinetics of reaction of carboxydinitrophenyl cytochromes *c* with cytochrome *c* oxidase, *J. Biol. Chem.,* 253, 149, 1978.
175. **Speck, S. H., Ferguson-Miller, S., Osheroff, N., and Margoliash, E.,** Definition of cytochrome *c* binding domains by chemical modification: kinetics of reaction with beef mitochondrial reductase and functional organization of the respiratory chain, *Proc. Natl. Acad. Sci. U.S.A.,* 76, 155, 1979.
176. **Smith, L., Davies, H. C., and Nava, M. E.,** Studies of the kinetics of oxidation of cytochrome c by cytochrome *c* oxidase: comparison of spectrophotometric and polarographic assays, *Biochemistry,* 18, 3140, 1979.
177. **Nicholls, P., Hildebrandt, V., Hill, B. C., Nicholls, F., and Wrigglesworth, J. M.,** Pathways of cytochrome *c* oxidation by soluble and membrane-bound cytochrome aa_3, *Can. J. Biochem.,* 58, 969, 1980.
178. **Bisson, R., Jacobs, B., and Capaldi, R. A.,** Binding of arylazidocytochrome c derivatives to beef heart cytochrome *c* oxidase: cross-linking in the high- and low-affinity binding site, *Biochemistry,* 19, 4173, 1980.
179. **Watts, A., Marsh, D., and Knowles, P. F.,** Lipid-substituted cytochrome oxidase: no absolute requirement of cardiolipin for activity, *Biochem. Biophys. Res. Commun.,* 81, 403, 1978.
180. **Fukumori, Y. and Yamanaka, T.,** Effect of cardiolipin on the enzymatic activity of *Nitrobacter agilis* cytochrome *c* oxidase, *Biochim. Biophys. Acta,* 681, 305, 1982.
181. **McIntyre, J. O., Wang, C., and Fleischer, S.,** The insertion of purified D-β-hydroxybutyrate apodehydrogenase into membranes, *J. Biol. Chem.,* 254, 5199, 1979.
182. **Cortese, J. D., Vidal, J. C., Churchill, P., McIntyre, J. O., and Fleischer, S.,** Reactivation of D-β-hydroxybutyrate dehydrogenase with short-chain lecithins: stoichiometry and kinetic mechanism, *Biochemistry,* 21, 3899, 1982.
183. **Cortese, J. D. and Vidal, J. C.,** Kinetic studies on the reactivation of D-β-hydroxybutyrate dehydrogenase with mixtures of short-chain lecithins, *Arch. Biochem. Biophys.,* 224, 351, 1983.
184. **Sandermann, H.,** Lipid-dependent membrane enzymes: a kinetic model for cooperative activation in the absence of cooperativity in lipid binding, *Eur. J. Biochem.,* 127, 123, 1982.
185. **Sandermann, H.,** Lipid solvation and kinetic cooperativity of functional membrane proteins, *TIBS,* 8, 408, 1983.
186. **Caffrey, M. and Feigenson, G. W.,** Fluorescence quenching in model membranes: relationship between Ca^{2+}-ATPase activity and the affinity of the protein for phosphatidylcholines of different acyl chain characteristics, *Biochemistry,* 20, 1949, 1981.
187. **Johannsson, A., Smith, G. A., and Metcalfe, J. C.,** The effect of bilayer thickness on the activity of $(Na^+ + K^+)$-ATPase, *Biochim. Biophys. Acta,* 641, 416, 1981.
188. **Blake, R. and Hager, L. P.,** Activation of pyruvate oxidase by monomeric and micellar amphiphiles, *J. Biol. Chem.,* 253, 1963, 1978.

189. **Seelig, J., Tamm, L., Hymel, L., and Fleischer, S.,** Deuterium and phosphorus nuclear magnetic resonance and fluorescence depolarization studies of functional reconstituted sarcoplasmic reticulum membrane vesicles, *Biochemistry*, 20, 3922, 1981.
190. **Davoust, J., Seigneuret, M., Herve, P., and Devaux, P. P.,** Collisions between nitrogen-14 and nitrogen-15 spin labels. II. Investigations on the specificity of the lipid environment of rhodopsin, *Biochemistry*, 22, 3146, 1983.
191. **Banaszak, L. J. and Seelig, J.,** Lipid domains in the crystalline lipovitellin/phosvitin complex: a phosphorus-31 and deuterium nuclear magnetic resonance study, *Biochemistry*, 21, 2436, 1982.
192. **Tamm, L. K. and Seelig, J.,** Lipid solvation of cytochrome c oxidase. Deuterium nitrogen-14, and phosphorus-31 nuclear magnetic resonance studies on the phosphocholine head group and on cis-unsaturated fatty acyl chains, *Biochemistry*, 22, 1474, 1983.
193. **Sandermann, H.,** A possible correlation between lipid hydration and lipid activation of the C_{55}-isoprenoid alcohol phosphokinase apoprotein, *Eur. J. Biochem.*, 62, 479, 1976.
194. **East, J. M. and Lee, A. G.,** Lipid selectivity of the calcium and magnesium ion dependent adenosinetriphosphatase, studied with fluorescence quenching by a brominated phospholipid, *Biochemistry*, 21, 4144, 1982.
195. **London, E. and Feigenson, G. W.,** Fluorescence quenching in model membranes. II. Determination of the local lipid environment of the calcium adenosine-triphosphatase from sarcoplasmic reticulum, *Biochemistry*, 20, 1939, 1981.
196. **Davis, D. G., Inesi, G., and Gulik-Krzywicki, T.,** Lipid molecular motion and enzyme activity in sarcoplasmic reticulum membrane, *Biochemistry*, 15, 1271, 1976.
197. **Boldyrev, A., Ruuge, E., Smirnova, I., and Tabak, M.,** Na^+, K^+-ATPase: the role of state of lipids and Mg ions in activity regulation, *FEBS Lett.*, 80, 303, 1977.
198. **Ingelman-Sundberg, M., Haaparanta, T., and Rydström, J.,** Membrane charge as effector of cytochrome P-450 catalyzed reactions in reconstituted liposomes, *Biochemistry*, 20, 4100, 1981.
199. **Ingelman-Sundberg, M., Blanck, J., Smettan, G., and Ruckpaul, K.,** Reduction of cytochrome P-450 LM_2 by NADPH in reconstituted phospholipid vesicles is dependent on membrane charge, *Eur. J. Biochem.*, 134, 157, 1983.
200. **Bösterling, B., Trudell, J. R., and Galla, H. J.,** Phospholipid interactions with cytochrome P-450 in reconstituted vesicles: preference for negatively-charged phosphatidic acid, *Biochim. Biophys. Acta*, 643, 547, 1981.
201. **Bösterling, B. and Stier, A.,** Specificity in the interaction of phospholipids and fatty acids with vesicle reconstituted cytochrome P-450: a spin label study, *Biochim. Biophys. Acta*, 729, 258, 1983.
202. **Knowles, A. F., Kandrach, A., Racker, E., and Khorana, H. G.,** Acetyl phosphatidylethanolamine in the reconstitution of ion pumps, *J. Biol. Chem.*, 250, 1809, 1975.
203. **Köhne, W., Deuticke, B., and Hoest, C. W. M.,** Phospholipid dependence of the anion transport system of the human erythrocyte membrane, *Biochim. Biophys. Acta*, 730, 139, 1983.
204. **Katchalski, E., Silman, I., and Goldman, R.,** Effect of microenvironment on the mode of action of immobilized enzymes, *Adv. Enzymol.*, 34, 445, 1971.
205. **Goldstein, L., Levin, Y., and Katchalski, E.,** A water-insoluble polyanionic derivative of trypsin. II. Effect of the polyelectrolyte carrier on the kinetic behaviour of the bound trypsin, *Biochemistry*, 3, 1913, 1964.
206. **Wooster, M. S. and Wrigglesworth, J. M.,** Modification of glyceraldehyde 3-phosphate dehydrogenase activity by adsorption on phospholipid vesicles, *Biochem. J.*, 159, 627, 1976.
207. **Goldman, R., Kedem, O., and Katachalski, E.,** Kinetic behaviour of alkaline phosphatase-collodion membranes, *Biochemstry*, 10, 165, 1971.
208. **Depierre, J. W. and Dallner, G.,** Structural aspects of the membrane of the endoplasmic reticulum, *Biochim. Biophys. Acta*, 415, 411, 1975.
209. **Nisimoto, Y., Kinosita, K., Ikegami, A., Kawai, N., Ichihara, I., and Shibata, Y.,** Possible association of NADPH-cytochrome P-450 reductase and cytochrome P-450 in reconstituted phospholipid vesicles, *Biochemistry*, 22, 3586, 1983.
210. **Kawato, S., Gut, J., Cherry, R. J., Winterhalter, K. H., and Richter, C.,** Rotation of cytochrome P-450. I. Investigations of protein-protein interactions of cytochrome P-450 in phospholipid vesicles and liver microsomes, *J. Biol. Chem.*, 257, 7023, 1982.

Chapter 8

PROTEIN-LIPID INTERACTIONS IN BIOLOGICAL MEMBRANES

Gheorghe Benga

TABLE OF CONTENTS

I. INTRODUCTION

The interaction between protein and lipid is a fundamental process of biological systems that has received increasing attention in recent years.[1] Among biological systems in which protein-lipid interactions are considered to be of importance, one can include certain enzyme and receptor systems, serum lipoproteins, blood coagulation processes, lung surfactant, and many processes involving cell membranes (cell recognition, membrane biogenesis, ion transport, photosynthesis, and oxidative phosphorylation).

The ways in which the membrane lipids and proteins interact within cell membranes are of paramount importance for determining the organization of membrane structure as described in several chapters of this book. It is clear that proteins, by their interrelation with the lipids, carry out most of the dynamic activities of membranes.

There are many aspects of protein-lipid interactions in biological membranes, from theoretical treatments of the interaction forces involved to studies on model and reconstituted systems using components purified from membranes, studies of the influence of proteins on membrane fluidity, or on the lipid requirement of enzymes. Since many of these aspects are described in various reviews, including several chapters of this book, I shall concentrate here mainly on the question of molecular aspects of interactions between proteins and lipids in biological membranes, particularly on interactions between integral proteins and lipids.

A discussion of molecular interactions in biological membranes usually begins with a description of composition of cell membranes. However, there are already many books and reviews presenting the lipid and protein molecules found in cell membranes which are available to an interested reader.[2-7] Therefore, I shall only mention the main classes of lipid and protein molecules occurring in cell membranes. Membrane lipids consist of *polar lipids* (such as phospholipids and glycolipids) and *nonpolar lipids* (such as mono-, di-, or triacylglycerols, sterols, and steryl esters of long-chain fatty acids). Membrane proteins can be divided into two groups: peripheral and integral. The peripheral (extrinsic) proteins can be removed from membranes by mild treatments (such as the addition of salts) and are soluble in aqueous buffers. The integral (intrinsic) proteins, because of their hydrophobic properties, require a detergent to solubilize them out of the membrane. In addition, upon extraction of the lipids they may form aggregates and lose their activity. Many such proteins span the membrane lipid bilayer.

The interactions between proteins and lipids in biological membranes can be divided into three types: (1) the nonspecific association of soluble proteins with membranes; (2) the association of peripheral proteins with membranes, and (3) the interactions between integral membrane proteins and lipids.

II. ASSOCIATION OF SOLUBLE PROTEINS WITH MEMBRANES

Before well-characterized membrane proteins became available, a good deal of work had been done using pure soluble proteins, most of which are not normally associated with membranes (such as albumin, ribonuclease, lysozyme, etc.). The interaction of these proteins with well-characterized lipids has been studied in a variety of physical situations, using several analytical techniques (see the review of Chapman and Quinn[8]). Although the relevance of such an approach to the biological membrane structure has been questioned, this had the advantage of a well-defined system for studying protein-lipid interactions. Recent data show that studies of association of soluble proteins with membranes do have physiological significance.

Many of the earlier studies of protein-lipid interactions in *monolayers* are difficult to interpret because of the inadequate techniques used and because the proteins and/or phospholipids employed were often impure. Doty and Schulman[9] and Eley and Hedge[10] have studied the adsorption of serum proteins on monolayers of charged lipids, suggesting that the primary association which occurs is almost entirely between the charged ionic groups of the interface and the proteins; this was in agreement with the dominant idea of that time of a purely ionic interaction between lipids and proteins.[11] Eley and Hedge[12] measured the adsorption of various proteins (bovine plasma albumin, insulin, lysozyme) onto monolayers of cholesterol and stearic acid (or distearin) at initial pressures of 10 and 2 dyn/cm, respectively. None of these lipid films would contain charged groups orientated towards the subphase since carboxylic acid groups of stearic acid are apparently only ionized at an interface above a bulk pH of 9. The increase in surface pressure on injecting proteins into the subphase has been attributed to penetration of the film by the larger nonpolar side chains of the first adsorbed layer of protein. This first layer of protein, considered completely unfolded, would be followed by a sublayer of "native" protein with an area/molecule of protein which is many times less than the denatured protein layer. Such an interpretation was favored by Colacicco et al.[13] in their investigations of the interaction of various proteins (rabbit γ-globulin, rabbit serum albumin, ribonuclease) with monolayers of dihydroceramide lactosides. Because of their amphipathic nature protein molecules are surface active. When a protein is adsorbed at an interface, an alternative way of folding the molecule to minimize the hydrophobic free energy becomes possible; thus, besides folding to the interior of a globule in the aqueous phase, apolar side chains can be located in either the air or oil phase.[14]

The studies on lipid monolayer penetration by proteins have been greatly extended and refined in recent years (see Chapter 2, Volume III). Attention has been devoted to various classes of proteins: toxins, proteins involved in blood clotting, phospholipid exchange proteins, etc.[15] Monolayers have intensively been used as substrates for lipolytic enzymes. Dawson[16] has reviewed his studies of the digestion of monomolecular films of phospholipids by phospholipases and has also presented a detailed description of the physical chemistry of the phospholipid-water interface.

The monolayer studies have shown that proteins interact with lipid films at low surface pressure (2 to 10 dyn/cm) by a process of "free penetration", where whole protein molecules occupy the interstitial spaces at the air-water interphase between lipids; the process is characterized by large increases in surface pressure. Charge-charge interactions between phospholipids and proteins promote both binding and penetration as observed by an increase in surface pressure or film expansion, even at high initial film pressures (12 to 35 dyn/cm). Finally, all above interactions depend on the chemistry and configuration of both lipids and proteins, and they appear to involve both polar and nonpolar (hydrophobic) associations.

The effects of soluble proteins on *permeability of lipid vesicles* have been studied with the aim to establish a functional aspect of protein-lipid interactions. Sweet and Zull[17] found that albumin increases the ^{14}C-glucose permeability of lecithin-cholesterol-dicetyl phosphate liposomes at pH values below the isoelectric point of albumin. The activation of diffusion takes place with negatively charged micelles and is not disrupted by high salt concentrations. It was suggested that albumin binds through electrostatic interactions initially, but that subsequent formation of apolar bonds brings about the activation of glucose diffusion. However, in a recent paper, Sogor and Zull[18] have found a much smaller effect of albumin on glucose permeability, suggesting that the transitory disruption of the vesicles by the albumin as it becomes associated with the lipid explain their previous results. Lysozyme and cytochrome *c*, but not ribonuclease, have been reported to increase the $^{22}Na^+$ permeability of sonicated vesicles formed from phosphatidylserine (PS) by several orders of magnitude, at neutral pH and low ionic strength.[19] It has been found that hemoglobin at neutral pH, high ionic strength, and very low concentrations caused large increases in the permeability of PS-PC (phosphatidylcholine) vesicles to $^{86}Rb^+$. No effect was seen on glucose diffusion and other proteins had little or no effect at the same concentrations and under the same conditions. Apart from the case of hemoglobin, most of the above-mentioned results

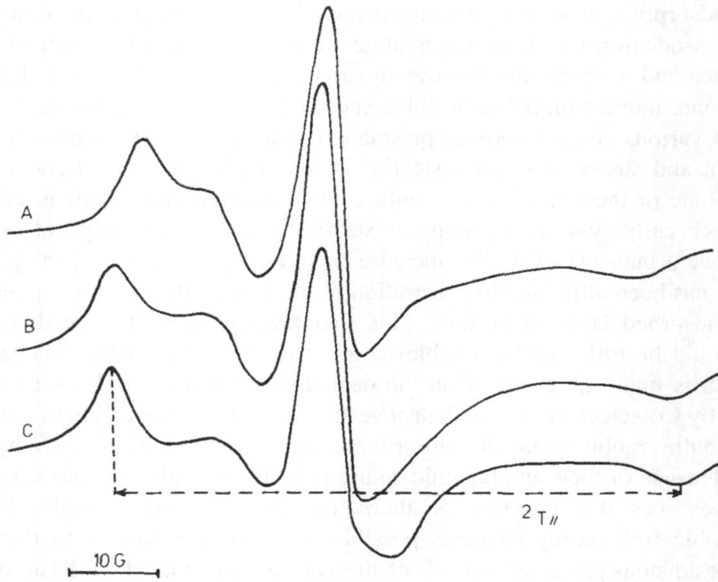

FIGURE 1. ESR spectra of M 12 NS (the methyl ester of N-oxyl-4',4',dimethyloxazolidine derivative of 12 ketostearic acid) in liposomes (A) formed from DPL, in the mixture of liposomes and albumin, (B) 10:1 molar ratio DPL/albumin, and in albumin alone (C). (From Benga, Gh. and Chapman, D., *Rev. Roum. Biochim.*, 13, 251, 1976.)

have been interpreted[19] as an initial electrostatic interaction that could stabilize the protein at the lipid interface, allowing subsequent relatively weak hydrophobic forces to be formed. These, in turn, result in distortion of the bilayer interior and subsequent permeability increases. If the number of such hydrophobic interactions is large, they can become the *dominant* attractive forces after the initial electrostatic interaction. It has also been suggested[19] that initial electrostatic interactions might induce conformational changes in the protein exposing nonpolar sites, thus increasing the potential for hydrophobic interactions. The effects of proteins in increasing vesicle permeability appear to require fluid fatty acyl chains.

There have been a number of studies of soluble protein-lipid interactions using the technique of *spin labeling electron spin resonance* (ESR). Barrat et al.[20] have studied the interaction of liposomes with spin-labeled proteins, while others have introduced spin-labeled lipid probes into liposomes interacting with proteins. Butler et al.[21] have studied the interaction between brain multibilayer films and a series of proteins and peptides. A predominantly electrostatic interaction has been postulated for some proteins (ovalbumin, lactoglobulin, pepsin, cytochrome *c*). Disordering effects like those observed with poly Lys-Phe were attributed to intercalation of hydrophobic amino acid side chains into the lipid bilayers.

Benga and Chapman[22] studied the ESR spectra of spin-labeled derivatives of fatty acids, phospholipid, and cholestane in liposomes formed from various phospholipids and in the presence of albumin. With the albumin-dipalmitoyllecithin (DPL) liposome system a distribution of the spin-labeled fatty acids takes place between the protein and lipid giving a spectrum showing an immobilization of the label (Figure 1). With spin-labeled phospholipid, a much smaller immobilization was noted as indicated by the values of the splitting between the low and high field extremes, $2T_{\parallel}$, which were considered as parameters for immobilization (Table 1).

The distribution of spin-labeled fatty acid was influenced by the binding abilities of protein (e.g., the values of $2T_{\parallel}$ decreased below pH 4.0 when binding of fatty acid to albumin is

Table 1
THE VALUES OF 2 T_{II} FOR THE FATTY ACID AND PHOSPHOLIPID SPIN-LABELED DPL LIPOSOMES AND IN THE PRESENCE OF ALBUMIN[22]

Sample		2 T_{II}(gauss) Methyl ester of 12 doxyl stearic acid	Spin-labeled phospholipid
DPL		55.1	55.0
DLP/albumin			
Molar ratio			
500/1		56.2	55.5
100/1		59.3	57.5
50/1		60.5	57.8
10/1		61.8	57.8
Albumin	29%	62.3	63.2

also decreased) and the fluidity of the lipid. With egg yolk lecithin there was a similar decrease in the mobility of the spin label, using fatty acid derivatives and also a spin-labeled phospholipid.

The study showed that the use of fatty acid spin labels may not give an accurate picture of protein-phospholipid interactions. This is because the protein itself can bind fatty acids. If a distribution of the spin label between phospholipid and protein occurs, the combination of spectra due to the spin label bound to the protein and free in the lipid phase may give an overall impression of immobilization, but this need not be due to the interaction between protein and phospholipids. Similar problems with regard to studies of protein-lipid interactions may occur with any probes (e.g., fluorescent probes) which can distribute themselves between protein and lipid.

On the other hand, it was interesting to note that the same values for $2T_{\parallel}$ were obtained when the spin label was added before or after the liposomes were prepared. This can be best explained by considering that the partition of the spin label between protein and various bilayers of lipid may take place by diffusion through the aqueous phase, across a bilayer, and within a bilayer.[22] The transfer of fatty acids through different bilayers is an interesting finding, because it could give some clues to the understanding of the mechanisms of absorption of free fatty acids through membranes of microvili or the uptake of fatty acids from plasma into hepatocytes.

Defatted serum albumin has been shown to induce morphological changes (e.g., a cup shape) in erythrocytes. The membrane integral proteins seem to be involved in mediating the shape modulating effects of albumin.[23] However, serum albumin also removes lyso-PC and free fatty acids from the erythrocyte membrane, particularly from the outer monolayer.[24]

Recent studies of fatty acid, sulfobromphthalein, and bilirubin uptake in perfused rat liver and isolated rat hepatocytes have been interpreted as indicating the existence on the liver cell plasma membrane of a specific albumin receptor, which might speed dissociation of the albumin-ligand complex by a variety of mechanisms. Studies of albumin binding to isolated liver plasma membranes failed to support the albumin receptor hypothesis; it, however, provided evidence for the existence of a fatty acid-binding membrane protein.[25] It is thus obvious that studies of interactions between soluble proteins and lipids are relevant for physiological conditions.

The binding of lipoproteins to phospholipids has been studied extensively and the structural features of this interaction are relatively well understood.[26] The particular structural feature of apolipoproteins involved in binding is the amphipathic α-helix. This α-helix is located in the lipid/water interface with polar amino acid residues mainly located on one side and exposed to the aqueous phase and with apolar residues on the opposite side embedded among the phospholipid hydrocarbon chains.[27] Insertion of apolipoproteins into phospholipid bilayers to form lipid-protein complexes brings about considerable changes in the physical properties of the phospholipids, such as changes in hydrocarbon chain packing and motion and a loss of cooperativity in chain melting of the phospholipids. Recent studies of Reijngoud et al.[28] have shown that the presence of apolipoprotein at a phospholipid water interface does not perturb the PC polar group conformation which is determined by intramolecular effects. This suggests that electrostatic forces do not play a major role; while, on the contrary, hydrophobic interactions are the dominant stabilizing force.

The primary effect of amphipathic α-helix-forming apoliproteins at the surface of lipoprotein particles is spacing out of the phospolipid molecules.

The adsorption to cell membranes of certain cytosol proteins that exist at high concentrations within the cell is another situation which may have physiological relevance. There is good evidence in model systems that soluble enzymes such as lactate dehydrogenase, malate dehydrogenase, glutamate dehydrogenase, and fructose 1,6-diphosphatase can adsorb, to varying extents, to lipid bilayers at high protein concentrations. This results in changes in their thermal stability, sensitivity to proteases, and kinetic properties. During subcellular fractionation which involves considerable dilution the cytosol proteins will be desorbed and will be found in the soluble fractions. However, since they exist at high concentrations in the cell it is possible for an association with the cell membrane to happen in vivo.[7]

III. INTERACTION OF PERIPHERAL PROTEINS WITH LIPIDS

The distinction between cytoplasmatic and peripheral membrane proteins is not clear-cut. Hemoglobin and tubulin are two examples of proteins primarily found in the cytoplasm, but both specifically associate with membranes so they can be classified as peripheral proteins.[5] A distinction, perhaps, can be made between association of peripheral proteins with integral membrane proteins and association with membrane lipids.

The interactions between peripheral and integral proteins are discussed in Chapter 2. It should be emphasized that by such interactions peripheral proteins achieve a specific localization for structural and functional purposes, as exemplified by glyceraldehyde-3-phosphate dehydrogenase, spectrin and ankyrin from human erythrocyte membrane, and by insulin-stimulated cyclic AMP phosphodiesterase from rat liver plasma membranes. There are other peripheral proteins which interact with integral proteins through ionic interactions: calmodulin, enzymes (acetylcholinesterase), and peptide hormones or growth factors which bind to surface receptors. It is possible that their association with membrane protein does influence the properties and distribution of membrane lipids, even if no direct interaction with membrane lipids occur. An example is the stabilization of phospholipid asymmetry in erythrocyte membranes by spectrin.[29]

A great deal of work has been done to study interactions of peripheral proteins with lipids in model systems. Spectrin, when added to multilamellar liposomes containing either negative or positive charges, produced a two- to fivefold increase in the *permeability* to glucose.[30] Since the protein was equally effective in increasing the permeability of negatively or positively charged liposomes at neutral pH, hydrophobic associations were suggested. In recent years the interaction of spectrin with phospholipids has been studied in a variety of model systems (see a recent review of Haest[31]).

The preferential interaction of spectrin with negatively charged phospholipids (i.e., PS, phosphatidylionositol [PI], and cardiolipin) indicated an electrostatic interaction, besides the hydrophobic one. There is some information on the structure of the hydrophobic domains in the spectrin molecule that may interact with lipids. It is possible that spectrin binds via these multiple hydrophobic sites to phospholipids, in particular, to phosphatidylethanolamine/phosphatidylserine (PE/PS) domains in the native membrane.[31]

Papahadjopoulos et al.[32] have studies by *differential scanning calorimetry* of the interaction of a variety of proteins (polypeptides, soluble proteins, and membrane proteins) with dipalmitoyllecithin (DPL) and dipalmitoylphosphatidylglycerol (DPPG). They suggested three types of interactions: (1) purely electrostatic, i.e., simple surface binding (ribonuclease and polylysine), where there was no effect on the transition temperature (T_c) and only a small increase in the enthalpy of transition (ΔH); (2) surface binding followed by partial penetration and deformation of the bilayer (basic myelin protein and cytochrome *c*), where a drastic decrease in both T_c and ΔH occurred; (3) hydrophobic interactions (major apoprotein of myelin proteolipid and gramicidin A), where there was no appreciable effect on the T_c, but a linear decrease in the magnitude of ΔH proportional to the percentage of protein by weight was noticed.

Examples of *model systems where the details of molecular interactions* are becoming clear and are very suggestive for in vivo situations are the interactions of myelin basic protein and another basic protein, cytochrome *c*, with lipids. The interaction of myelin basic protein with different lipids has been studied at the air-water interface[33-34] and on protein-lipid complexes.[35] The highest affinity of the myelin basic protein was found for cerebroside sulfate, a lipid which is characteristic for myelin. Neutral lipids such as lecithin, cholesterol, and cerebrosides showed markedly less affinity for the basic protein. A positively charged lipid, lysylphosphatidylglycerol from *Staphylococcus aureus*, showed no interaction at all. It could be concluded that ionic forces are involved in the interaction of basic protein and lipids. On the other hand, hydrophobic forces were also very much apparent, since the interaction was affered by the fatty acid chain length and by presence of KCNS. The authors were able to conclude which parts of the protein molecule are particularly involved in the interaction. From the amino acid sequence of the basic protein it was suggested[36] that the protein consists of an open chain with a sharp bend at the Pro-Arg-Thr-Pro-Pro-Pro position 96 to 101. The N-terminal part of the molecule (positions 1 to 116) contains far more nonpolar amino acids at 8.0 than the C-terminal part of the molecule (positions 117 to 170), mainly because of the asymmetric distribution of the histidines which are neutral at pH values above 7.0, and then form hydrophobic structures. Since ionic as well as hydrophobic interactions are involved in the basic protein-phospholipid interactions, it was suggested that the N-terminal part of the molecule (positions 1 to 116) is primarily interacting with the lipid layer.

The association of basic protein with acidic lipids results in the protein producing a phase separation of acidic lipids from neutral phospholipids in artificial membranes. Boggs and Moscarello[37] have shown that portions of the protein penetrate into the lipid bilayer. This occurred particularly with phosphatidylglycerol (PG) which does not interact intermolecularly by hydrogen bonding, or with phosphatidic acid whose charge may be neutralized by interaction with charged residues on the protein. The interaction with PG is complex and is related to the phase state of the lipid.

The above-mentioned studies pointed to basic protein as the structural protein factor of the myelin membrane. On the other hand, the interaction with lipid also affects the structure of the bound protein. In aqueous solution basic myelin protein lacks ordered α-helical or β-sheet structures and behaves as a flexible linear polyelectrolyte. The interaction with lipid induces 20% α-helical and 12% β-sheet structure in the protein.[38] When the protein's two methionine residues were spin labeled, the mobility of the spin label was found to be sensitive to the degree of hydrophobic interaction of the protein with the lipid bilayer, even though the labeled methionines are not thought to be directly involved in the hydrophobic interaction. These hydrophobic and electrostatic interactions of basic myelin with acidic lipids are likely to be important in inducing and maintaining the characteristic multimamellar structure of myelin sheaths, if monomers or dimers of the protein form bridges between opposing bilayers. The myelin membrane probably consists of a lipid bilayer with little protein penetrating deep into the hydrocarbon core of the membrane.

Cytochrome *c* is a protein found exclusively at the cytosol side of the mitochondrial inner membrane, where it is responsible for transferring electrons from cytochrome c_1 to cytochrome oxidase. The ability of cytochrome *c* to interact with proteins involved in the mitochondrial electron transport chain is essential for its functional role. Cytochrome *c* binds to both cytochrome oxidase and to mitochondrial phospholipids where it is able to undergo fast lateral diffusion. The basic nature of this protein is reflected in its electrostatic interactions with phospholipids in model systems. Investigations of cytochrome *c*-phospholipid complexes began in 1958 when Widmer and Crane isolated a form of cytochrome *c* from beef heart which was soluble in lipid solvents, but not in water. This material was enzymatically active and was called lipid-cytochrome *c*. Since then a good deal of attention has been paid to the complexes formed between cytochrome *c* and phospholipids. When the protein and ultrasonically dispersed phospholipids are mixed within a certain pH range, precipitation occurs and the precipitate can be extracted into iso-octane in contrast to either of the free components of the complex. Das and Crane[39] studied the factors affecting lipid-cytochrome *c* complex formation. Inhibition of complex formation by monovalent cations was proportional to ionic strength of the solution, whereas di- and trivalent cations completely inhibit complex formation. It was concluded that the primary formation of a complex involves association between the negatively charged phospholipids and the positive free amino groups of cytochrome *c*. Purified acidic phospholipids such as cardiolipin or PI can form complexes with cytochrome *c* which contain ten molecules of phospholipid to one of cytochrome *c* and are insoluble in iso-octane.

With the less acidic PE the equivalent ratio was 24 and purified PC did not react at all. However, both PC and PE can also react with the complexes of cytochrome *c*, with the more acidic phospholipids forming highly soluble complexes in iso-octane. In such complexes the neutralization of all external basic sites of cytochrome *c* convert them from iso-octane-insoluble complexes to soluble ones. Here, presumably, hydrophobic bonding between the very acidic phospholipid, bound by salt linkages to the protein, and the lecithin or PE may occur.

Further evidence of electrostatic interactions between cytochrome *c* and lipids has been obtained from studies performed on other model systems. X-ray studies of Gulik-Krzywicky et al.[40] showed that cytochrome *c* is bound to the polar head groups of aqueous dispersions of phospholipid membranes without perturbing the structure of the phospholipid bilayers. The interaction of cytochrome *c* with black lipid films has been studied by Steinemann and Laüger,[41] who found that at low ionic strength, about 10^{13} cytochrome *c* molecules per square centimeter are bound to the lipid surface. The fast desorption of the protein after a rise in ionic strength showed that the protein did not penetrate into the lipid bilayer appreciably and that the interaction was mainly electrostatic. This is consistent with the ESR studies[42] of complexes of cytochrome *c* with a mixture of egg lecithin and cardiolipin, in which spin-labeled fatty acids or cholestan had been diffused. The effects of cytochrome *c* on the ESR spectra of lipid spin labels were small, i.e., no immobilization of spin labels was observed.

All these studies indicate that cytochrome *c* can bind electrostatically to charged groups of the phospholipids. Its ability to interact with acidic phospholipids explains why cytochrome *c* is found associated with the inner mitochondrial membrane. The localization on the cytosol side of the mitochondrial inner membrane reflects its interaction with certain specific integral membrane proteins exposed at the external surface of the inner membrane: its site of synthesis on cytosol ribosomes and its inability to penetrate the inner membrane.

The fact that both the inner mitochondrial membrane and the electron-exchange partners of cytochrome *c* are negatively charged, while cytochrome *c* itself has a strong net positive charge of possibly +8 and a dipole above 300 debye, allowed a simple electrostatic model for the kinetics of reaction of cytochrome *c* to be developed.[43]

Interactions of cytochrome *c* and myelin basic protein with membrane lipids may be considered as good examples for biological relevance of studies in model systems. Several exciting developments in cytoskeleton proteins-membrane interactions have recently been described.

Steer et al.[44] have shown that clathrin, the major protein of the coated pits and coated vesicles, interacts, with both black lipid membranes of oxidized cholesterol and small unilamellar dioleoyl (DO-) and dipalmitoylphosphatidylcholine (DPPC) vesicles. The interactions were found to be pH and calcium dependent, suggesting that perturbation of lipid bilayer may in part be associated with clathrin polymerization into organized lattice structures. It is known the basic unit of this polygonal lattice is a molecular complex of 630,000 daltons composed of these clathrin chains and three light chains of subunit molecular weight 33,000 to 36,000. These assembly units are called triskelions.[45] Coated pits and coated vesicles are fundamental structures in membrane recycling, receptor-mediated endocytosis, and secretion of glycoproteins, as well as intracellular protein translocation. It is, thus, obvious that studies of interactions between peripheral proteins and lipids in various model systems are quite promising for a better understanding of various physiological conditions.

IV. INTERACTIONS BETWEEN INTEGRAL MEMBRANE PROTEINS AND LIPIDS. THE BOUNDARY LAYER CONCEPT

The functional implications of these interactions have been revealed in a number of cases. The Ca^{2+}-ATPase of sarcoplasmic reticulum (SR) requires, for optimal function, a fluid membrane with a minimum bilayer thickness and containing unsaturated phospholipid acyl chains;[46] the coupling state of Ca^{2+}-ATPase is stabilized by cone-shaped lipid molecules (e.g., dioleoylphosphatidylethanolamine (DOPE) and monogalactosyldiglyceride).[47] It has been shown that β-hydroxybutyrate dehydrogenase is a lipid-requiring enzyme with a specific requirement of lecithin for enzymatic function,[48] and it has been proposed that cytochrome oxidase has a specific requirement for cardiolipin.[49] The modulating role of the membrane phospholipid milieu upon the adenylate cyclase in plasma membranes has been demonstrated in intact membranes and cells.[50]

A crucial question regarding the protein-lipid interactions in biological membranes is the perturbation which the proteins embedded in the lipid bilayer may cause on the neighboring lipid, and how, in molecular terms, the protein is interfaced with the fluid-lipid environment. Several techniques have been applied on certain model systems in order to obtain an answer to protein-lipid interactions. Various terms now exist in the literature to describe the perturbed lipid adjacent to intrinsic proteins which are sometimes used in a synonymous way. These include boundary layer lipid, annulus lipid, and halo lipid.

The boundary layer lipid concept has been introduced from ESR studies on the cytochrome oxidase isolated from beef heart mitochondria.[51-54] After isolation and purification procedures a complex of the protein with phospholipids was obtained and this complex has been considered as a model membrane system, because under certain circumstances the purified preparation forms vasicular structures. The approach by Jost et al.[51-53,55] was to gradually reduce the phospholipid content of the purified cytochrome oxidase complex by successive acetone extractions and to use spin-labeled lipids to study protein-lipid interactions in the complex. At low lipid levels (below 0.2 mg phospholipid per milligram of protein) the spectrum corresponded to that of a highly immobilized nitroxide label, whereas at high lipid levels (0.5 to 0.7 mg phospholipid per milligram of protein) the label had a considerable degree of motional freedom. At intermediate lipid levels, two components of the spectrum appeared; a mobile component and an immobile component (or ''bound'') (Figure 2). It has

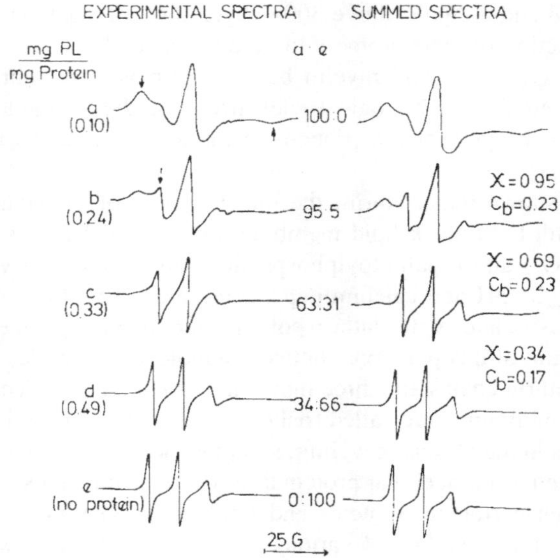

FIGURE 2. Electron spin resonance spectra (left column) of 16-doxyl-stearic acid diffused into aqueous suspensions of cytochrome oxidase with varying phospholipid content. In the right column are synthesized spectra obtained by summing various amounts of spectrum *a* (lipid/protein ratio of 0.10) and spectrum *e* (lipids extracted from membranous cytochrome oxidase) in the proportions indicated. All spectra have been normalized to the same center line height. (Adapted from Jost, P., Griffith, R. A., and Vanderkooi, G., *Biochim. Biophys. Acta*, 311, 141, 1973.)

been concluded that the two components correspond to two lipid environments: (1) lipids tightly bound to the protein and immobilized (up to 0.2 mg of phospholipid per milligram of protein) and (2) the lipids in a fluid bilayer.

Assuming the distribution of the spin label faithfully reflects the distribution of phospholipids between the two environments, the amount of bound lipid, C_b, is $C_b = \chi C_t$, the experimental value for the total phospholipid content of the sample, and χ is the fraction of the total absorption contributed by the bound component. The values of C_b calculated in different ways were similar (0.2 mg phospholipid per milligram of protein) and this led Jost et al. to conclude that the amount of phospholipid bound to the protein is independent of the extent of the fluid bilayer region. Jost et al.[51-53] presented estimates of how the lipid immobilized by cytochrome oxidase can be accounted for in molecular terms. They assumed approximate dimensions of the protein complex (in the membrane plane) of 52 × 60 Å and the diameter of one aliphatic chain of 4.8 Å. Division of the perimeter of cytochrome oxidase (224 Å) by 4.8 Å yields 47 aliphatic chains. This number must be divided by 2 to get the number of equivalent phospholipid molecules, but must also be multiplied by 2 since the bilayer arrangement is assumed, giving 47 first-layer phospholipids per protein complex. (An equivalent phospholipid molecule was defined as containing one phosphorus atom and two aliphatic chains. On this basis one molecule of cardiolipin corresponds to two equivalent phospholipid molecules.) From the molecular weights used for the protein complex (210,000) and the phospholipid molecules (775) the result was obtained that 0.17 mg of phospholipid per milligram of protein can be accommodated in the first layer around one protein complex. This amount is very close to the observed amount immobilized (0.20 mg of phospholipid per milligram of protein). Jost et al.[51-53] have summarized their hypothesis in terms on the

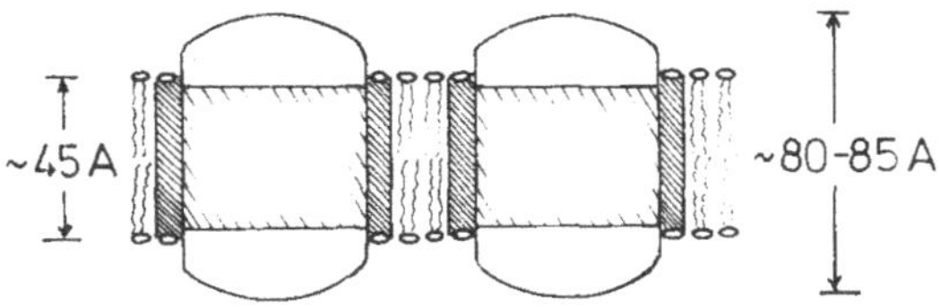

FIGURE 3. Model proposed for the boundary lipid layer in cytochrome oxidase membranes. (From Jost, P., Griffith, R. A., and Vanderkooi, G., *Biochim. Biophys. Acta*, 311, 141, 1973. With permission.)

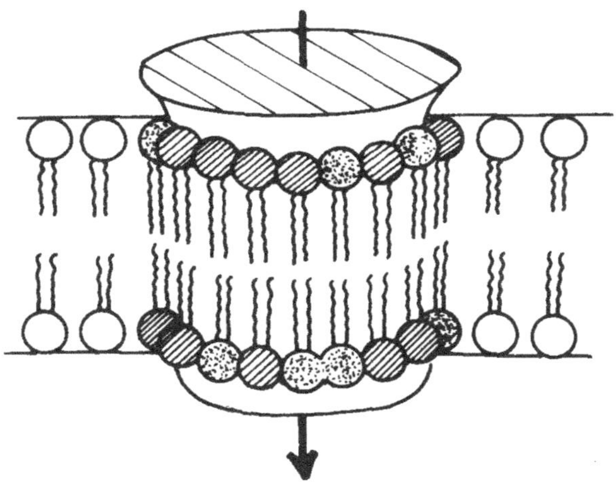

FIGURE 4. Model proposed for the organization of the anular lipids associated with the calcium transport protein from SR ATPase. (From Warren, G. B., Toon, P. A., Birdsall, N. J. M., Lee, A. G., and Metcalfe, J. C., *Proc. Natl. Acad. Sci. U.S.A.* 71, 622, 1974. With permission.)

boundary lipid model (Figure 3) where the protein complex of membrane cytochrome oxidase is shown extending through the phospholipid region. The hydrophobic surface of the protein complex tightly binds the first layer of lipid, indicated as *boundary lipid*. Beyond the boundary lipid is the more fluid phospholipid bilayer. Jost et al.[51-53] have suggested as a general feature of biological membranes the presence of these two lipid environments.

Similar arrangements of lipids have been suggested in other membrane systems. The SR Ca^{2+}-ATPase and its associated phospholipid can be dispersed in cholate without loss of activity as measured by ATP hydrolysis. Codispersal of the ATPase-lipid complex with large excess of a defined phospholipid in cholate allows equilibration of the lipid pools. In this way a complex in which one defined lipid predominates can be prepared. Excess lipid can be removed from the ATPase-lipid complex in cholate by centrifugation through detergent-free sucrose. The amount of lipid remaining in association with the protein is dependent on the amount of cholate initially present, so that after centrifugation detergent-free preparations of various lipid/protein ratios can be obtained. Regardless of the structure of lipid present it was found that maximum activity requires a minimum of 26 to 33 mol of phospholipid per mole of protein. It was considered that about 30 lipid molecules are in direct contact with the protein, forming a single bilayer shell called *annulus* (a word which is derived from the Latin, anulus, meaning a ring) around that part of the protein which penetrates the membrane[56,57] (Figure 4). Thus, in SR membranes 15 to 20% of the lipids were considered to be in a more ordered state than the bulk lipid.

In liver microsomal membranes heterogenous lipid environment has been suggested by Stier and Sackman.[58] The spin label, 5 NS, was diffused into the membrane preparation, and the rate of decay of the ESR signal was measured as a function of temperature. The rate of signal loss was greatly increased by the addition of NADPH, and this signal loss was attributed to enzymatic reduction by the cytochrome P_{450} — cytochrome P_{450} reductase hydroxylating enzyme system. There was a break in the Arrhenius plot at about 32°C. Below 32°C the signal disappeared slowly, whereas above 32°C the rate of signal disappearance increased. Control experiments ruled out phase changes in enzymatic activity and in the bulk lipids. From these data, it was suggested that a quasicrystalline environment surrounded the protein, but underwent a phase transition at 32°C, becoming more fluid above that temperature.[58] The lipid bound to the cytochrome P_{450} reductase complex appears to be around 20% of the total membrane protein; this had been referred to as *halo lipid*.[58] Träuble and Overath[59] studied lipid phase transitions in *Escherichia coli* membranes using fluorescent probes. They suggested (by comparing the amount of lipids which takes part in the phase transition with the total extracted lipids of the cell membrane) that some 20% of the lipids is removed from the phase transition process. On this basis it was suggested that one integral membrane protein is surrounded by about 600 lipid molecules with some 130 of these forming a *halo* around the protein.

Various terms — boundary layer lipid, annulus lipid, halo lipid — are, in fact, different names for the same lipid domain and thus correspond to a single concept. The main features of this concept are the following:

1. The integral membrane proteins embedded in the lipid bilayer influence in a particular way only *one layer,* i.e., the first layer of phospholipids coating the hydrophobic surface of proteins.
2. The phospholipids in this first (boundary) layer are *tightly bound* to the protein and, consequently, *immobilized* and *ordered* by the protein, while the remaining lipids in the membrane form a fluid bilayer. This has led to the development of a variety of theoretical treatments of protein-lipid interaction[60-62] which emphasized the ordering effects of protein on lipid hydrocarbon.
3. Although it was not ruled out that the lipids in the boundary layer could exchange with the bulk lipids, like water molecules in a shell of hydration, it was originally proposed that the time scale of such exchange was several orders of magnitude lower than the exchange rate between two phospholipids in a bilayer.

In recent years this concept has been critically evaluated with regard to all the above-mentioned features in various laboratories using several techniques. Alternate concepts on integral protein-lipid interactions have been developed. Some authors using *spin-labeling ESR* have reported studies concluding to the existence of an immobilized boundary layer associated with lipophiline rhodopsin cytochrome oxidase or the cholinergic receptor from *Torpedo marmorata*.[37,63] On the contrary, other authors have provided evidence in favor of a fluid environment around integral proteins in membranes.

Studies of Benga et al.[64-66] on the cytochrome oxidase-lipid complex have aimed to contribute to further understanding of spin-label data. With lipid extracted cytochrome oxidase and using spin-labeled fatty acids, the spectrum shows both a mobile and an immobile component, in agreement with data of Jost et al.[51-53] However, the assumption that these two components observed in the spectrum of spin-labeled fatty acid faithfully reflect the extent of the boundary and fluid bilayer appears not be necessarily correct. The appearance of the spectrum depended upon the amount of spin-labeled molecules added and the concentration of the dispersion. It was clear that for the mobile component there are two contributions: one from the labeled molecule in lipid; the other from the labeled molecule

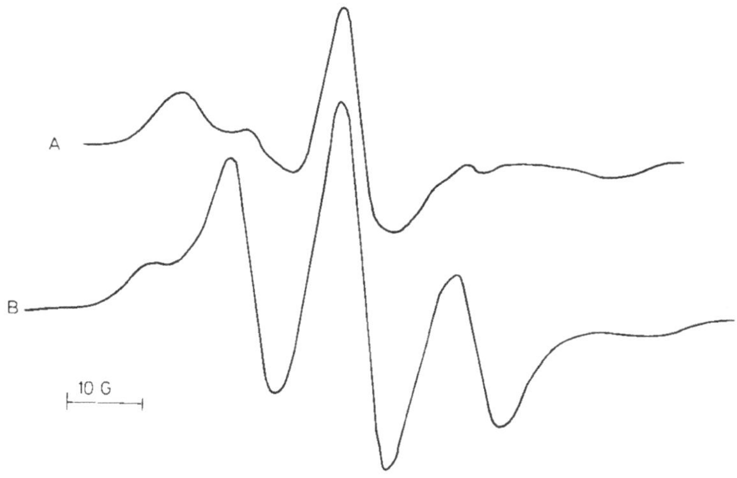

FIGURE 5. ESR spectra obtained with 16-doxylstearic acid (spectrum A) and with a spin-labeled phospholipid (spectrum B) in dispersions of lipid-extracted cytochrome oxidase. The cytochrome oxidase, purified from beef heart mitochondria, was extracted six times with cold 90% acetone and contained 40 mg protein per milliliter and 0.12 mg phospholipid per milligram of protein.

in water.[64] With cytochrome oxidase at the lowest lipid content (below 0.2 mg/mg of protein), i.e., at the level referred to as the boundary lipid, with spin-labeled fatty acids an immobilized spectrum is observed. However, when spin-labeled phospholipids are used under the same conditions, a mobile component is also observed (Figure 5). A quantitative estimation of the spectral components by computer analysis has been performed.[65] It was found that the immobilized component represented 98% in case of spin-labeled fatty acids and only 32% with spin-labeled phospholipids. Consequently, it was concluded that identifying the residual lipid of the cytochrome oxidase complex to tightly bound, immobilized lipid is a simplification and that phospholipids in the vicinity of cytochrome oxidase may exhibit a high degree of mobility.

Jost et al.[51-53] have assumed that the partition of the spin label faithfully reflects the distribution of lipid between the two domains (the boundary lipid and the fluid bilayer). The amount of spin label giving rise to the two components of the ESR spectra has been considered to reflect the size of the two domains, but no theroretical basis has been presented for this approach. Benga et al.[66] have studied purified cytochrome oxidase from beef heart mitochondria without lipid extraction (i.e., a soluble functionally active complex) using phospholipid spin label (PSL) incorporated into the complex. A theory taking into account not only the size of the domains in which the spin-label molecules distribute themselves, but also the different affinities of the label for the domains has been developed. Taking advantage of the variation in spectra obtained with increasing amounts of spin label, computer calculations were made to estimate the extent of the lipid regions or domains in the cytochrome oxidase-lipid complex.

Representative spectra of PSL dispersed in the purified cytochrome oxidase (spectra A-H) and in liposomes prepared from cytochrome oxidase lipids (spectrum I) are shown in Figure 6. The spectra are arranged in order of increasing PSL content of the samples. Several interesting qualitative observations can readily be made from these data. First, a composite spectrum having two components (a mobile and an immobile one) occurs for the PSL dispersed in the purified cytochrome oxidase lipids. This is readily seen in spectrum D where the double arrow indicates the appearance of the mobile component superimposed on the immobilized spectrum (indicated by the single arrow). Second, when different ratios of spin-

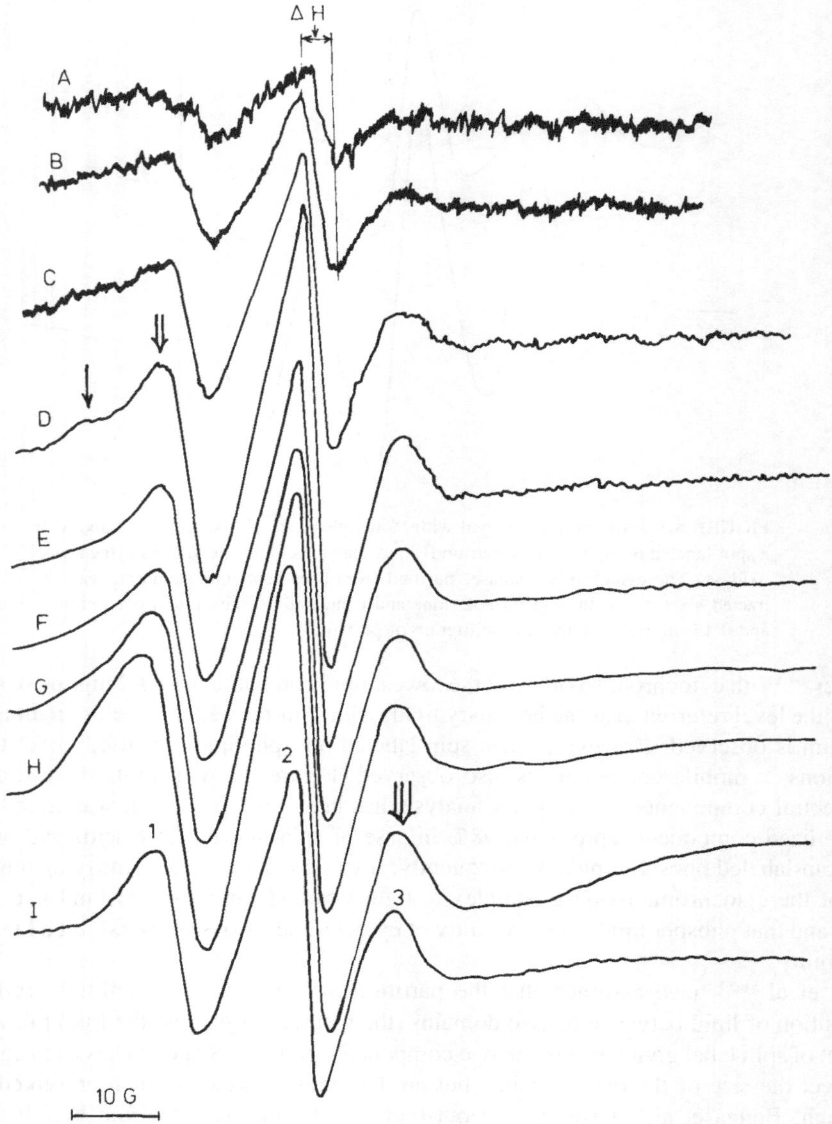

FIGURE 6. ESR spectra obtained with increasing amounts of PSL in a cytochrome oxidase dispersion. (A) 3 nmol PSL per milligram of protein; (B) 15 nmol PSL per milligram protein; (C) 30 nmol PSL per milligram of protein; (D) 60 nmol PSL per milligram of protein; (E) 150 nmol PSL per milligram of protein; (F) 300 nmol PSL per milligram of protein; (G) 600 nmol PSL per milligram of protein; (H) 1500 nmol PSL per milligram of protein; (I) spectrum of PSL in liposomes prepared from the lipids extracted from cytochrome oxidase. (From Benga, Gh., Porumb, T., and Wrigglesworth, J. M., *J. Bioenerg. Biomembr.*, 13, 269, 1981. With permission.)

labeled molecules to protein are used, the proportions of the mobile and immobile components of the spectra vary considerably. At low PSL content of the sample the spectra indicate a relatively high proportion of immobilized spin label (spectra A to C). At higher PSL contents the mobile component of the spectrum increases (spectra D to F). This mobile component bears a close resemblance to the spectrum obtained from the isolated lipids (spectrum I). At very high PSL content (spectra G and H) strong interactions (dipole-dipole) and spin

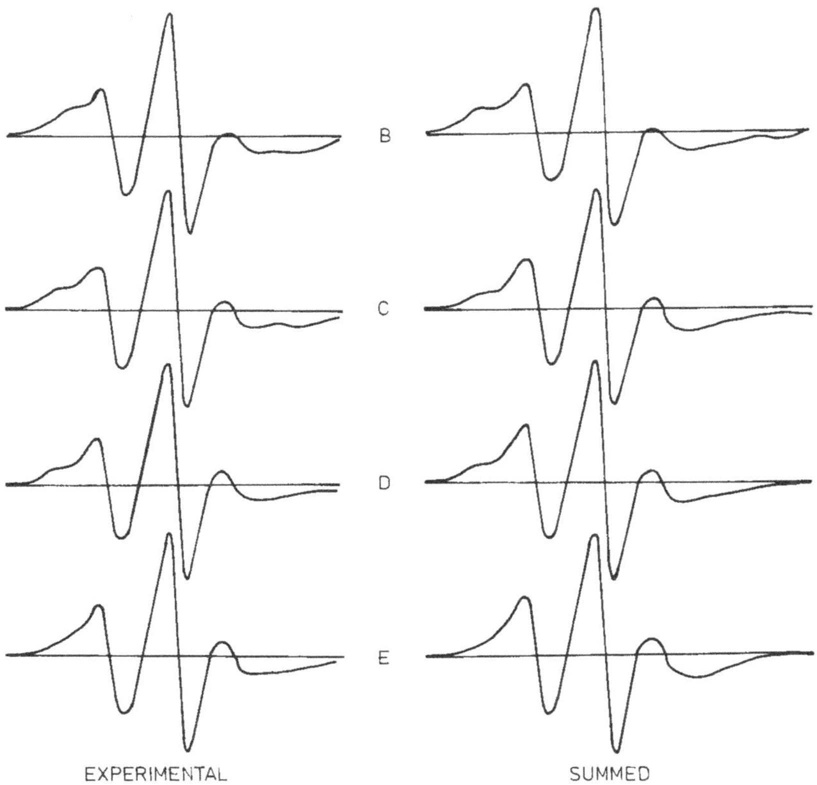

FIGURE 7. The experimental spectra of PSL in cytochrome oxidase dispersion normalized to center peak height and the corresponding synthesized spectra obtained by summation of the mobile and immobile component spectra in different proportions as described in the text. The notation of spectra is that of Figure 6. (From Benga, Gh., Porumb, T., and Wriggles-worth, J. M., *J. Bioenerg. Biomembr.*, 13, 269, 1981. With permission.)

exchange) between the probe molecules occur. This is shown by a characteristic increase in AH (the peak to peak distance of the central band indicated in Figure 6), a decrease in the high-field peak height of the inner hyperfine doublet (triple arrow in Figure 6, spectrum H), and the downward displacement of the high-field baseline. Spectra G and H have not, therefore, been subjected to further analysis by computer. Spectrum A, where the signal-to-noise ratio is very low, has also not been used for quantitative analysis.

Two methods of spectral analysis have been used. One is the spectral titration by subtraction, in which difference spectra are obtained by substraction of one component from the composite experimental spectrum. The endpoint has been judged by the disappearance of that component, by the overall lineshape, and by the appearance of phase reversal of the component. The relative proportion of the mobile and immobile PSL components in the sample B to D (Figure 7) was estimated by carrying out spectral titration against either of the component spectra.

The results were checked using a second method, based on the summation of the mobile and immobile content spectra in different proportions to match with the experimental composite spectra. The sums obtained using different ratios of the components gave synthesized spectra that were good approximations to the experimental spectra. This analysis gives confidence to the interpretation that the spectra of Figure 7 do represent summations of the same two spectral components.

The results of both analytical approaches, subtraction and summation, on the spectra of

Table 2
PROPORTION OF THE IMMOBILE COMPONENT IN THE SPECTRA FROM FIGURE 7

	Immobile component (%)		
Spectrum	Titration against mobile spectrum	Titration against immobile spectrum	Summation of spectral components
B	45 ± 5	60 ± 10	50 ± 5
C	33 ± 8	45 ± 10	36 ± 6
D	29 ± 5	29 ± 5	29 ± 5
E	13 ± 5	16 ± 8	13 ± 1.8

Note: The figure indicated for each procedure corresponds to the best approximation to the experimental spectra.

samples with increasing PSL content are given in Table 2. The agreement in the amount of the immobile component obtained by the two methods of spectral analysis is good, taking into account the difficulties encountered in judging exact endpoints. Although both methods rely on the visual examination of lineshapes, the summation procedure enables more accurate decisions to be made, as the spectra obtained by this method were free of the "noise" sometimes generated by the subtraction procedure. The consistency of the results was also checked by subtractions of the experimental spectra in pairwise fashion.

The results of spectral analysis have shown that, using increasing amounts of spin-labeled phospholipid added to a certain amount of cytochrome oxidase, the proportion of the immobile component of the spectra decreases considerably (from 50% in spectrum B to 13% in spectrum D in Figure 7). If the data are plotted as the reciprocal of the percentage of immobilized PSL vs. the amount of spin label in the sample, the points approximately lie on a straight line (Figure 8) as the theory predicted. An extrapolation to zero concentration of the label gives a proportion of 75 immobilized labeled lipids in the cytochrome oxidase-lipid complex, the remainder being in a mobile phase. The immobile component of the spectrum should not be considered to correspond to the lipid tightly bound to protein, but to the lipid that on the ESR time scale is influenced by the protein. It has been concluded that the number of aliphatic chains influenced by the protein is higher than could coat the hydrophobic surfaces of the protein in a layer that is one aliphatic chain thick.

Knowles et al.[67] have prepared yeast cytochrome oxidase complexes in which ≥99% of the endogenous lipid has been substituted by dimyristoylphosphatidylcholine (DMPC), and the lipid chain immobilization has been studied by spin-labeling ESR. The spectra of a phospholipid spin label consisted of both an immobilized and a fluid lipid bilayer component for all complexes, the proportion of the former increasing with increasing protein/lipid ratio.

Computer difference spectroscopy has been used to obtain the two different spectral components and to determine their relative proportions. Lipid/protein titration of the complexes in the region where the proportion of fluid lipid exceeded that of the immobilized led the authors to conclude that a constant number of lipid molecules (55 ± 5) per protein are immobilized. This was attributed to the first or boundary shell of lipids associated with the protein. The variation of the spectral splittings and line widths of the fluid component difference spectra with lipid/protein ratio was considered to indicate that a second shell of lipids is perturbed by the protein, a third shell is less strongly perturbed, and a further two to three shells may be still weakly perturbed. Thus, the immobilization of the lipid was suggested to extend out to approximately six shells from the protein (Figure 9).

The lipid environment of acetylcholine receptor-rich membranes from *T. marmorata* has also been studied with spin labels.[63] It was concluded that the proportion of lipid in the

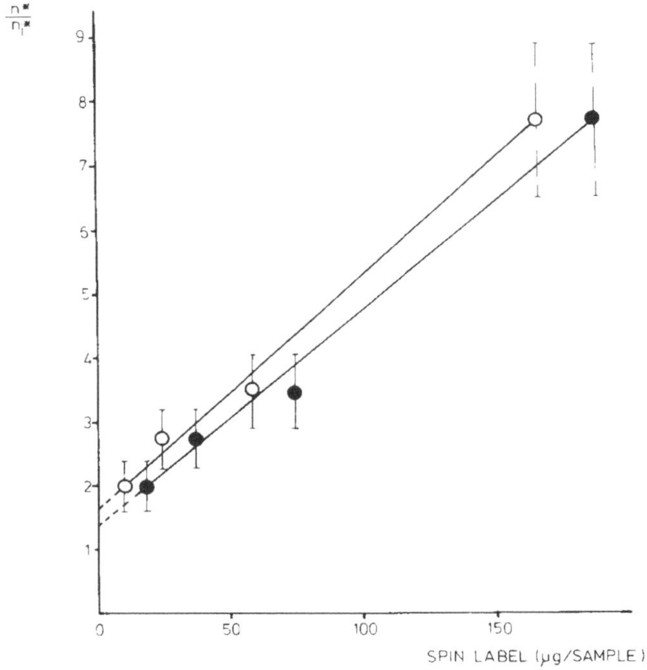

FIGURE 8. Plots of the reciprocal of the fraction of immobilized spin-labeled lipid (n^*/n_i^*) against the amount of spin label added to the sample. The bars correspond to the estimated error from Table 2.

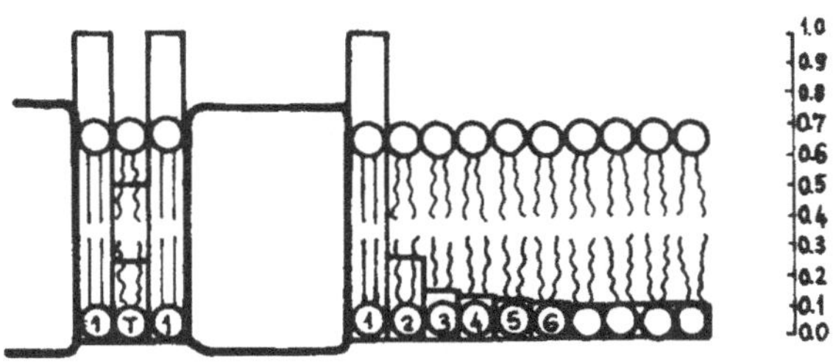

FIGURE 9. Representation of the degree of chain immobilization in the various shells of lipid surrounding cytochrome oxidase in DMPC complexes. (From Knowles, P. F., Watts, A., and Marsh, D., *Biochemistry,* 18, 4480, 1979. With permission.)

immobilized component is greater than that calculated for a single boundary layer around the protein and corresponds more closely to the total interstitial lipid occupying the area between densely packed protein units in these membranes.[68-70]

Devaux and co-workers have synthesized a special class of amphiphilic spin labels: they contain at one end a specific group (e.g., maleimide or isocyanate) that can be attached to a protein while the remaining part of the label is a long chain spin-labeled fatty acid. Such spin labels can be covalently bound to the membrane proteins, however, the alkyl chains will be buried in the hydrophobic lipid bilayer. In this way the spin label will explore the hydrophobic environment of the membrane protein (Figure 10). When such labels are used with rhodopsin and the protein reinserted into PC vesicles, the ESR signal observed at high temperature (37°C) corresponds to a fast motion typical of that expected in the bulk lipid at

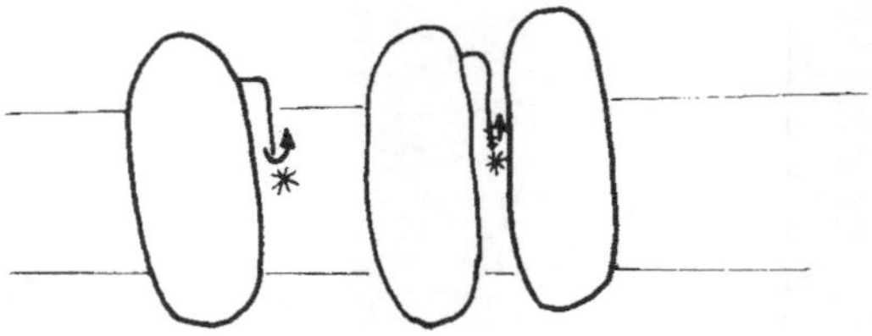

FIGURE 10. A spin-labeled lipid covalently attached to an intrinsic protein can be strongly immobilized only in the case of aggregated proteins. (After Devaux, P. F. and Devoust, J., *Membranes and Transport*, Vol. 1, Martonosi, A. N., Ed., Plenum Press, New York, 1982.)

37°C. This indicated that the probe is not immobilized in a boundary lipid layer. Similar probes have been attached to the ADP carrier in mitochondria and the cholinergic receptor protein and have led to comparable results. However, if aggregation of proteins is provoked, for example, by cross-linking of the proteins by glutaraldehyde, a strong hindrance of the spin label's mobility is observed. Based on such results Devaux[71] presented a view of lipid-protein interactions in membranes different to that of boundary layer concept. According to this model, intrinsic membrane proteins are dissolved in a rather homogenous lipid solvent. But the solvent is near saturation under physiological conditions, and mild modifications such as the removal of small quantities of lipids or temperature variations can result in important protein aggregations. The immobilized lipids detected by ESR would reflect the existence of hydrocarbon chains trapped between adjacent proteins rather than a specific halo of phospholipids surrounding membrane proteins (Figure 10). Similar conclusions were brought about by Chapman et al.[72] from spin-label experiments with gramicidin A considered as a model for intrinsic proteins, and by Swanson et al.[73] from saturation transfer ESR studies on purified cytochrome oxidase incorporated into lipid vesicles. Thus, it is suggested that intrinsic proteins in membranes are not normally surrounded by a layer of immobilized lipid chains, and the immobilized component seen by ESR probably reveals protein-protein interactions, or the existence of specific hydrophobic crevices in the proteins.

Devaux and Davoust[74] pointed out that protein self-associations are not always artifacts, and spin-labeled lipids can be used to detect functional protein oligomers. With Ca^{2+}-ATPase in SR[75] a very large fraction of the spectrum (>80%) of a spin-labeled fatty acid linked by an ester bond to a maleimide residue (and covalently attached to the protein) showed a strong immobilization. It was proposed that the immobilized component arises from lipid chains trapped in protein oligomers, which exist in native SR. Thomas et al.[76] have studied the effects of varying the lipid-to-protein ratio on the dynamics of protein lipid in SR membrane using the same spin-label derivative that attaches covalently to the protein through a maleimide linkage. The observed hydrocarbon immobilization was interpreted as arising, in part, from immobilization at the protein-lipid boundary, but protein-protein interactions that trap hydrocarbon chains have also been accepted. However, when protein aggregation was induced by glutaraldehyde cross-linking, there was no effect on the hydrocarbon probe despite a large immobilizing effect on the protein. The lipid-trapping hypothesis would predict an increase in the immobilized fraction, while the model of an immobilized boundary layer would predict a decrease in the immobilized fraction due to exclusion of boundary lipid at the protein-protein interface. The lack of any effect suggests that hydrocarbon chain immobilization is caused mainly by protein-protein interactions that are not affected by cross-linking, e.g., the interactions of subunits within an oligomeric enzyme.

Recently, Devaux and Davoust[74] pointed out that an immobilized component can be generated in the ESR spectrum of a spin-labeled fatty acid covalently attached to rhodopsin at low temperatures, without implying chain trapping. The motionally restricted component

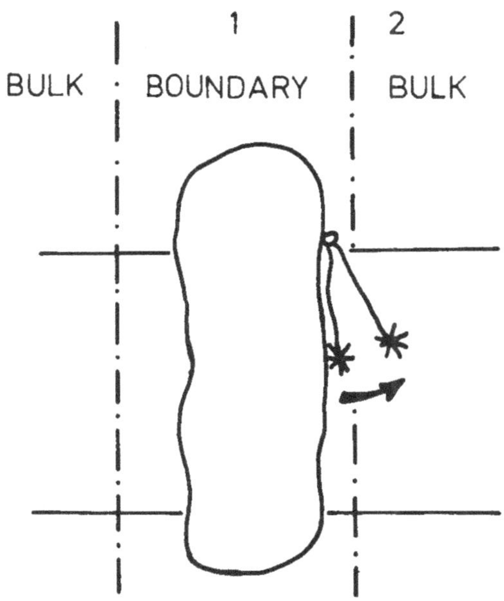

FIGURE 11. The motion of a spin-labeled lipid can be different at the protein-lipid interface (boundary) and in the bulk lipids. If the exchange rate between state 1 and state 2 is rapid enough at the time scale of ESR, only an average mobility is detected. (After Devaux, P. F. and Davoust, J., *Membranes and Transport*, Vol. 1, Martonosi, A. N., Ed., Plenum Press, New York, 1982.)

originates from a reduced exchange rate of the spin label between states 1 (at the protein-lipid interfaces) and 2 (in the bulk lipids), as shown in Figure 11. At high temperatures the motionally restricted component disappears due to rapid exchange between the two states.

Davoust et al.[77] have also studied the collision rates of the lipid chains at the protein boundary and in the bulk lipid phase. For this they have labeled rhodopsin with a covalently attached ^{15}N fatty acid derivative, while a ^{14}N spin-labeled phospholipid was diffused in the bulk phase (Figure 12). The lipid chain collision rates at the protein boundary were found to be of the same order of magnitude as in the bulk lipid phase. In the same time no evidence for a head-group specificity in the direct lipid environment of rhodopsin could be demonstrated.

There are conflicting literature reports with *NMR studies* as to the existence of a boundary layer lipid. Some authors confirmed its existence in cytochrome oxidase using ^{2}H-NMR[78] or in lecithin vesicles containing glycophorin using ^{13}C-NMR.[79] However, it is now considered[80] that these reports are associated with an experimental error or an unwarranted interpretation of the data. Other NMR studies with a variety of proteins including cytochrome oxidase and Ca^{2+}-ATPase did not permit detection of a significant population of heterogenous lipid.[81-84]

NMR studies have brought new ideas in regard to the effects of an intrinsic membrane protein to the surrounding lipids. Seelig and Seelig,[81] using ^{2}H and ^{31}P NMR, have shown that incorporation of cytochrome oxidase into PC bilayers leads to a *more disordered* conformational state of the lipids. They explained this by the following argument. Although in most of the membrane models the membrane proteins are drawn as smooth cylinders or smooth rotation elipsoids, actually the surface of the protein could be uneven, presenting irregularities. The lipids, since they are flexible molecules, will follow the shape of the protein to a certain extent. The NMR data give no indication for a strong, long-lived

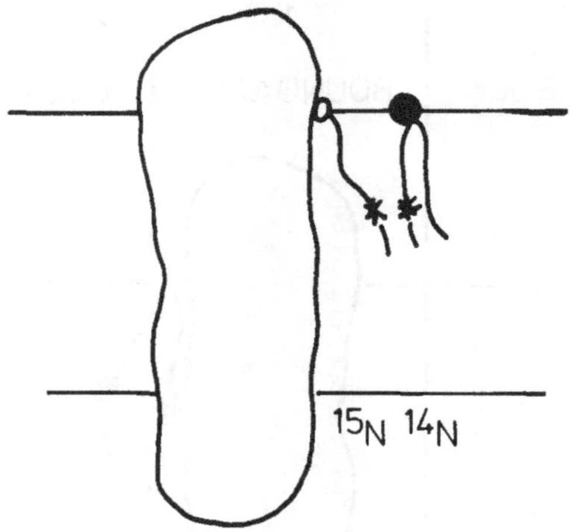

FIGURE 12. Schematic representation of an experiment of double labeling. (After Devaux, P. F. and Davoust, J., *Membranes and Transport*, Vol. 1, Martonosi, A. N., Ed., Plenum Press, New York, 1982.)

interaction between the protein and the lipid. Instead a relatively rapid exchange between those lipids in contact with the protein and those further away from it could be postulated. Within a time interval of 10^{-4} sec, all lipid molecules will touch the rough protein surface, leading to a decrease in the overall phospholipid ordering. Obviously, the protein is always surrounded by a shell of lipids as nearest neighbors (except for protein-protein contacts), but the residence time of this lipid layer at the protein surface is not longer than 10^{-4} sec. Based on ^{13}C NMR studies of rhodopsin-phospholipid interactions Zumbulyadis and O'Brien[85] suggested a model in which the lipid molecules solvate the protein with one hydrocarbon chain (the *sn*-1 chain), which was termed an edge-on orientation rather than with both hydrocarbon chains, termed side-on orientation, or random interactions. The preference for *sn*-1 acyl chain solvation of the hydrophobic region of the protein may be due to the ability of the saturated chain to conform to the geometrical requirements of the protein surface.

Oldfield et al.[82] have also suggested that intrinsic membrane proteins disorder, or at least do not order, the lipid bilayer. The disordering effects have been found to be most pronounced towards the terminal methyl of the hydrocarbon chains. Based on NMR studies these authors aim to present a picture of boundary lipid designed to be consistent with the ESR results.[86-88] It is suggested that proteins cause a large immobilization of polar group of phospholipids with the polar part of membrane protein (due, for example, to electrostatic and hydrogen bond interactions).

This protein-lipid interaction in and near the polar head group region is characterized by line broadening in the ^{2}H and ^{31}P NMR spectra, reflecting increased correlation time.[89] The selective interaction of the lipid head groups with the protein leaves the chains themselves more loosely packed. Thereby, the protein with its irregular or rough surface in the lipid bilayer could facilitate isomerization of the lipid hydrocarbon chain by providing (time-dependent) vacancies at the protein-lipid interface; such vacancies may also arise from conformational fluctuation within the protein molecule itself and from protein rotation in the bilayer. According to this view, the motions of the terminal methyl group would have the largest increase in amplitude (as found), while the center of the hydrocarbon chain would be little affected. A model of protein-lipid interaction was, therefore, proposed in which proteins "immobilize" lipid head groups, while simultaneously disordering somewhat the lipid hydrocarbon chains, due to the rough nature of the protein surface.[90]

A different model of intrinsic protein-lipid interactions was applied by Pink et al.[91] to study ^{2}H NMR resonance of DMPC bilayers containing either gramicidin A or cytochrome oxidase. In this model the lipids are divided into three populations: (1) those that are not adjacent to protein ("free" lipids), (2) those that are adjacent to only one protein ("adjacent" lipids), and (3) those that are "trapped" between two or three proteins. The model proposes that for the lipid molecules adjacent to *one* intrinsic molecule, the methyl group is somewhat more ordered than that in a pure lipid bilayer at the same temperature, while for lipid chains trapped between a cluster of two or three molecules the methyl groups are more disordered statically than in a pure lipid. The model also proposes that there is complete exchange between these three populations of lipids on a time scale of 10^{-5} sec. According to this model the ESR data of Knowles et al.[67] can be understood assuming that as long as one spin-labeled chain is adjacent to a protein, then its motion will be "mechanically" hindered, due to the presence of attached ESR probes, to display an "immobile" component. This would comprise all lipids in populations (2) and (3). Pink et al.[91] emphasized that although they propose that intrinsic molecules can perturb lipids adjacent to them, there is no evidence that supports the concept of a long-lived unbroken annulus of lipid surrounding each intrinsic molecule. On the contrary, they claimed that their data can only be interpreted as showing that this cannot be the case.[91]

Tamm and Seelig[92] have not been able to detect any strong polar interaction between the phosphocholine group and cytochrome *c* oxidase in reconstituted systems, neither in terms of a conformational change of the head group nor in terms of a significant immobilization of individual segments. Similar conclusions have been drawn for the nonpolar interactions in the interior of the lipid bilayer. The motional freedom of the fatty acyl chains was not restricted by cytochrome oxidase in reconstituted systems.

It is interesting to compare the above-mentioned studies with recent NMR studies from other laboratories. Fischer and Levy[93] have shown by proton NMR that rhodopsin incorporation in DMPC bilayers is associated with a domain of approximately 50 lipid molecules which have reduced choline methyl mobilities. Bienvenue et al.[94] have used ^{2}H NMR study rhodopsin-DMPC recombinants. Different molar lipid/protein (L/P) ratios were investigated. Measurements of orientational order parameters showed that the addition of rhodopsin broadened the range of the gel to liquid crystalline transition for $L/P = 150$ and 50. No transition was observed for $L/P = 30$ and 12. Moment analysis and spectral subtraction both showed that the low temperature spectra for $L/P > 30$ had two components. One was a pure phospholipid gel phase spectrum and the other a spectrum attributed to lipids in protein aggregates. The intensity of the second component corresponded to 30 lipids per protein, and its shape was the same as the temperature-independent shape observed for $L/P = 30$ and 12. No such decomposition into two components was possible in the liquid crystalline phase for $L/P > 30$.

Albert and Yeagle[95] have studied by ^{31}P NMR the phospholipid behavior in bovine retinal rod outer segment disk membranes and in PC membranes containing rhodopsin. Two distinguishable resonances have been noticed. One resembles closely the ^{31}P NMR resonance normally obtained from phospholipid bilayers. The other resonance is much broader, and it was considered to arise from the influence of the integral membrane protein rhodopsin on the membrane phospholipid bilayer. This environment was considered to correspond to 23 phospholipids per rhodopsin (30% of the phospholipids) in the native disk membrane and to 40 phospholipids per rhodopsin (50% of the phospholipids) in the reconstituted membrane. ^{31}P NMR studies on other systems have been performed in the same laboratory. Approximately 29 phospholipid molecules have been found immobilized per glycophorin molecule in a recombined system. In SR 32% of the total phospholipid in the membrane are in an immobilized environment. However, the population of this environment decreased to 10% on treatment with 100 m*M* NaCl and could be removed completely by papain proteolysis

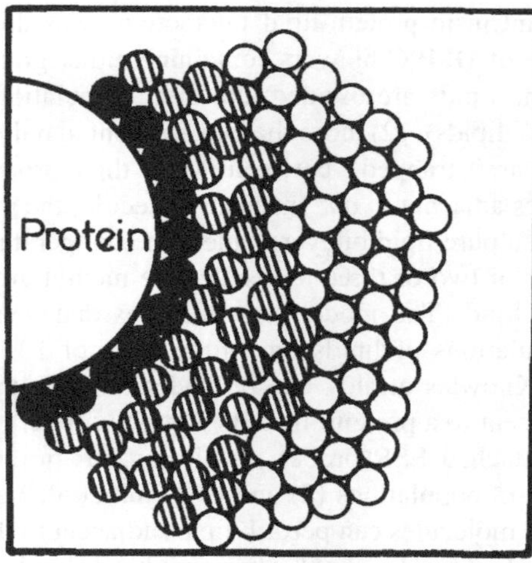

FIGURE 13. Proposed model for the boundary (shaded) and secondary striped lipid layers surrounding the Ca^{2+}-ATPase in SR. A surface view of the membrane is shown. (From Lentz, B. R., Clubb, K. W., Barrow, D. A., and Meissner, G., *Proc. Natl. Acad. Sci. U.S.A.*, 80, 2917, 1983. With permission.)

of the membrane protein. It was concluded that the assignment of the immobilized component to phospholipid interactions with the Ca^{2+}-ATPase is probably not correct. The salt and proteolysis experiments suggested that in the intact membrane the immobilization is caused by the peripheral membrane proteins.[98]

Investigations of ^{31}P NMR[84] and ^{2}H NMR[99] on SR membranes detected a single homogenous phospholipid environment typical of a phospholipid bilayer, even at high protein/lipid ratios. The authors concluded that the Ca^{2+}-ATPase does not perturb the motion of "boundary layer" lipids.

On the contrary, Lentz et al.[100] presented a completely different view for the organization and molecular dynamics of lipid molecules near the Ca^{2+}-ATPase of SR based on calorimetric and fluorometric data. They proposed a model (Figure 13) in which a motionally inhibited lipid annulus (i.e., the boundary domain of enhanced order) is surrounded by a more extensive region of disrupted lipid packing order, which they called secondary lipid domain.

It is obvious from this survey of models of protein-lipid interactions that there are conflicting views on the molecular details on these interactions. The reasons for these discrepancies, as well as some functional implications of intrinsic-protein lipid interactions, will be outlined below.

Some of the discrepancies arise from differences in the molecular properties of time scales that can be sensed by various techniques. This makes it difficult for the experiments with different preparations and different techniques to be compared. The spin-labeling ESR and fluorescent methods are probe techniques, and the presence of the spin label or fluorescent group might perturb the properties of the lipid molecule. For example, a specific interaction between the spin-label group and the protein has sometimes been invoked to explain the difference between spin-label and NMR results, although some authors[101] suggested that the overall features of protein-lipid interactions are independent of any perturbing effect of the probe.

The discrepancy between the NMR and ESR experiments reporting the existence of immobilized lipids around membrane proteins might relate to the differing time scales on which these techniques report; 10^5 and 10^8 sec^{-1}, respectively.[99] An exchange rate of 10^5 sec^{-1} between mobile or immobilized lipids in the boundary region could account for both NMR and ESR observations. An event occurring with a relaxation time of 10^{-5} sec would appear "immobilized" on ESR analysis but not on NMR analysis.[7]

Therefore, the studies in which both NMR and ESR spectroscopy are applied to the same system are of interest. For example, Paddy et al.[102] have studied the same reconstitutions of cytochrome oxidase by ^2H NMR and spin-labeling ESR (using the fatty acid spin label 16 NS). At all temperatures, the ESR spectra show the characteristic "bound" and "free" components, while the ^2H NMR spectra show only a narrow distribution of orientational order parameters. At temperatures near the phase transition of the pure lipid (deuterated 1-palmitoyl-2-palmitoleoyl-sn-glycero-3-phosphocholine), the dependence of the ^2H NMR average orientational order on protein concentration fits a two-state model in which the phospholipid molecules exchange rapidly between two states identified as sites either on or off the protein surface. From this model, the ^2H NMR spectra yielded a value of 0.18 mg of phospholipid per milligram protein as necessary to cover the surface of cytochrome oxidase, which is the same value as derived from the ESR spectra at $-20°C$. Both the ^2H NMR and ESR spectra varied markedly with temperature. At temperatures well above the phase transition of pure lipid, the average orientational parameters derived from the ^2H NMR spectra were independent of protein concentration and were the same for the lipid alone. Analysis of ^2H NMR relaxation rates indicated an additional motion in the presence of protein with a correlation time of 10^{-6} to 10^{-7} sec. If this new motion is associated with exchange between the two states, a minimum value of 10^{-6} to $10^{7 \text{ to } 10}$ sec for the exchange rate was obtained, assuming that lipids on the protein surface are much more motionally restricted than the rest of the lipid. These results were considered to be consistent with short-lived, energetically weak interactions between cytochrome *c* oxidase and the phospholipids used in these studies.

One of the most interesting aspects of the boundary layer concept for its physiological consequences is related to the rate of exchange between the lipids in the vicinity of a intrinsic protein and the lipids in the bulk lipid phase. The limits of this exchange have been put at $10^{-4} > V_{ex} > 10^{-6}$ sec since it can be detected only by spin-labeling ESR and not by NMR. The turnover numbers of enzymes are, in most instances, much slower than this exchange rate. The lipids in the "boundary" layer will change several times during one turnover of the enzyme. Therefore, if these lipids would affect the activity of an intrinsic enzyme they must do this by virtue of its average fluidity.[7] However, the same authors described cases where particular interactions between proteins and lipids appear to take place. β-Hydroxybutyrate dehydrogenase, which depends upon the head group of PC in order to function, segregates this lipid out of the total lipid pool. Rhodopsin and Na^+/K^+-ATPase appeared to segregate acidic phospholipids, while cytochrome oxidase exhibited a preference towards interaction with cardiolipin and phosphatidic acid (PA). This preference would not be simply due to electrostatic interactions, but it was considered that the protein is recognizing the shapes of the molecules. Houslay and Stanley[7] suggested that the molecular geometry of cardiolipin, different from other phospholipids, might well complement the structure of the protein, so forming the basis of the preferential interaction of cytochrome oxidase with this lipid. Cholesterol does not significantly interact with intrinsic membrane proteins because of the inability of the planar sterol ring of cholesterol to adapt to the protein surface. This makes it a poor agent for sealing the protein efficiently into the bilayer. It has been suggested that sealing intrinsic proteins into lipid bilayers such that nonspecific solute leakage at the protein-lipid interface is minimal is of considerable importance. The leakage across the protein-containing plasma and mitochondrial membranes is so small that membrane potentials and solute gradients can be maintained across them. The variety of phospholipids present in biological membranes would allow integral membrane proteins to "select" those that

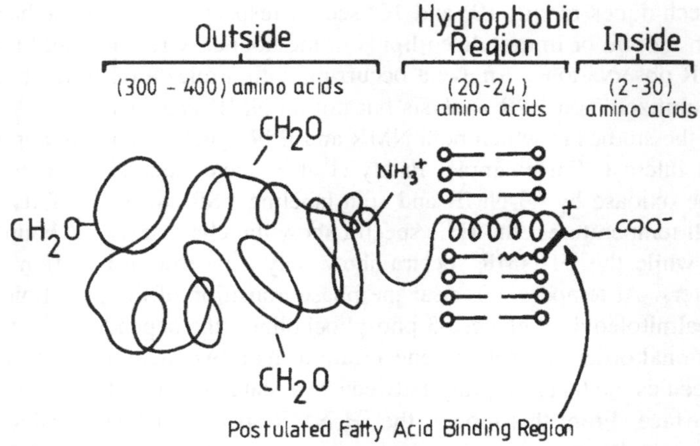

FIGURE 14. Schematic arrangement of vesicular stomatitis virus G protein in the membrane.

complement their structure best, allowing them to be sealed into the bilayer. The extent of the hydrophobic domain on a transmembrane protein may require lipids of specific dimensions to complement these surfaces and seal the protein into the bilayer. It is possible that the fatty acid acylation of membrane proteins (see the review by Schlesinger[103]) can lead to a better "sealing" of transmembrane segments with the surrounding lipids (Figure 14). Recent studies indicated that deacylation of a transmembrane glycoprotein affected its interaction with membrane lipid.[104]

On the other hand, it is also possible that some lipid molecules that exchange slowly (or even do not exchange with the lipid bilayer) are deeply embedded in or totally surrounded by the polypeptide chains of a protein complex. The term of *captive lipid* has suggested, for one case, bacteriochlorophyll *a* in a protein complex from the green photosynthetic bacterium *Chlorobium limicola,* where X-ray crystallography allowed the determination of the arrangement of the lipid and protein. It has been suggested[105] that several molecules of cardiolipin in cytochrome oxidase may fall in the class of captive lipids.

A source of discrepancies between various studies, even when the same protein-lipid system is analyzed by the same technique, is that quantitative estimation on the amount of lipid that is influenced by the protein is dependent upon certain assumptions. If different assumptions are taken into account then different values for this lipid are obtained. This is discussed by McIntyre et al.[106] for their ESR studies of reconstituted SR membrane vesicles. When a certain model was assumed, then about 8 mol of "constrained" phospholipid per mole of calcium pump protein at 25°C was estimated. This amount of constrained lipid is less than one third that estimated to constitute the boundary layer surrounding the calcium pump protein.

On the other hand, even when the same amount of lipid was found to be influenced by the protein, different meanings have been put forward, for example, in the case of calorimetric enthalpy data of protein lipid-reconstituted systems. As the protein concentration within the lipid bilayer increases, the cooperativity of the lipid phase transition is reduced, i.e., the transition is broadened. According to the boundary layer concept, each protein molecule would be surrounded, above the transition temperature, by a lipid layer separating it from the bulk lipid phase. Upon cooling, the protein would segregate laterally but with its boundary lipid; the latter not participating in the thermothropic phase transition. By measuring the enthalpy variation associated with the thermotropic transitions of the pure lipid and lipid-protein recombinants, it is possible to estimate the number of phospholipid molecules that are removed from the bulk lipid phase per molecule of protein incorporated. Given the perimeter of the protein it is possible to estimate the maximum number of phospholipid molecules that can simultaneously contact the protein. It is interesting that this theoretical figure coincides with that estimated from the experimental enthalpy data, as recently shown for a bacteriorhodopsin/lipid system.[107] However, an interpretation different from the boundary layer concept is possible for the reduction of enthalpy which occurs as a function of protein concentration. As the lipid crystallizes below its main transition temperature, proteins are excluded from the crystallizing lipid (but remain within the plane of the bilayer). The proteins in the form of aggregates take with them solvating lipid and form the high protein-lipid patches. The lipids in these patches make no contribution either to the cooperative melting of the lipid chains or to the enthalpy. The remaining enthalpy arises from the residual pure lipid region from which the protein has been excluded. The reduction in the number of the lipid molecules available to form the pure lipid region causes a reduction in its cooperativity and also a reduction in the enthalpy of the melting process.[107]

V. CONCLUSION

In conclusion, it appears that the model of a single immobilized layer coating the hydrophobic surfaces of all intrinsic proteins in membranes is far from being demonstrated, although some recent reviews might give this impression.[101] Further studies on various systems and greater applications of various other techniques should prove valuable as exemplified by recent studies in this field.[108,109]

We will probably have the detailed molecular picture of protein-lipid interactions in biomembranes when we will be able to characterize in membrane proteins the diversities of structure and function that are now recognized for soluble proteins. X-ray diffraction studies of crystalline soluble proteins have resulted in over 100 known tertiary structures. So far, this has not been possible with membrane proteins. There are, however a number of proteins for which structural analysis by X-ray diffraction and electron microscopy has been possible because they form, under certain conditions, two-dimensional crystals.

Henderson and co-workers were the first who successfully performed such structural analyses on bacteriorhodopsin and cytochrome oxidase. The bilayer-intercalated parts of almost all of the intrinsic membrane proteins studied so far are arranged as one or more helices traversing the membrane. The cases of bacteriorhodopsin and rhodopsin[110] are very illustrative. These helices are formed by long stretches of hydrophobic amino acids. Since such stretches are not found in water-soluble proteins, this represents a useful distinction between membrane proteins and other proteins.[5] The α-helical segments that traverse the membrane seem to be predictable from the amino acid sequence through utilization of the physico-chemical characteristics of the residues interacting with the apolar membrane interior.[111] Moreover, a helical "wheel" analysis of the predicted regions indicated the helical faces within the protein interior and in contact with the lipid bilayer.[111] It has been claimed that the specific amino acid sequence of membrane-penetrating segments of transmembrane proteins are less important than the overall hydrophobic character in such segments.[112] It appears, therefore, that although the detailed molecular interactions between lipid and relevant parts of transmembrane proteins are only partially understood, it does seem clear that the "hydrophobic effect",[113] i.e., the thermodynamically favorable removal of hydrophobic segments of the protein from contact with water, plays a dominant role.

On the other hand, the hydrophobic interactions between the intrinsic membrane proteins and lipids must not be overlooked. For example, glycophorin can interact with 300 DMPC molecules in reconstituted systems, and this was explained by hydrophobic protein-lipid interactions involving the large carbohydrate-carrying head group of glycophorin.[114]

We may conclude from this survey that the interaction between proteins and lipids is still far from being understood in molecular terms, and no clear generalization exists apart from the obvious perturbation of the surrounding lipid which must be caused by any intrinsic protein. The proteins are well-defined, three-dimensional structures around which the lipids must conform in a way that leads to a minimum energy configuration of the system. It is possible that we will realize in the future that a diversity of molecular interactions between proteins and lipids in biological membranes exists, and no generalization such as the ''boundary layer concept'' is warranted.

REFERENCES

1. **Jost, P. C. and Griffith, O. H.,** *Lipid-Protein Interactions,* Vol. 1 and 2, John Wiley & Sons, New York, 1982.
2. **Mazliak, P.,** *Les Membranes Protoplasmatiques,* Doin, Paris, 1971.
3. **Rouser, E., Nelson, G. J., Fleischer, S., and Simon, G.,** Lipid composition of animal cell membranes organelles and organs, in *Biological Membranes. Physical Fact and Function,* Vol. 1, Chapman, D., Ed., Academic Press, New York, 1968, 5.
4. **Henderson, R.,** Membrane protein structure, in *Membranes et Communication Intercellulaire,* Ballian, R., Chabre, M., and Devaux, Ph. F., Eds., North-Holland, Amsterdam, 1981, 232.
5. **Benga, Ch. and Holmes, R. P.,** Interactions between components in biological membranes and its implications for cell function, *Prog. Biophys. Mol.,* 43, 195, 1984.
6. **Benga, Gh.,** *Biologia Moleculară a Membranelor cu Aplicatii Medicale,* Dacia, Cluj-Napoca, 1979.
7. **Houslay, M. D. and Stanley, K. K.,** *Dynamics of Biological Membranes. Influence on Synthesis, Structure and Function,* John Wiley & Sons, Chichester, 1982.
8. **Quinn, P. J. and Chapman, D.,** The dynamics of membrane structure, *CRC Crit. Rev. Biochem. 8(1),* 1, 1980.
9. **Doty, P. and Schulman, J. H.,** Formation of lipido-protein monolayers. I. Preliminary investigation on the adsorption of proteins onto lipid monolayers, *Discuss. Faraday Soc.,* 6, 21, 1949.
10. **Eley, D. D. and Hedge, D. G.,** Protein interactions with lecithin and cephalin monolayers, *J. Colloid Sci.,* 11, 445, 1956.
11. **Dervichian, D. G.,** Structural aspect of lipo-protein association, *Discuss. Faraday Soc.,* 6, 7, 1949.
12. **Eley, D. D. and Hedge, D. G.,** The energetics of lipid-protein interactions, *J. Colloid Sci.,* 11, 419, 1957.
13. **Colacicco, G., Rapport, M. M., and Shapiro, D.,** Lipid monolayers: interaction of synthetic dihydroceramide lactosides with proteins, *J. Colloid Interf. Sci.,* 25, 5, 1967.
14. **Phillips, M. C.,** Protein conformation at liquid interfaces and its role in stabilizing emulsions and foams, *Food Technol.,* 50, 1981.
15. **Verger, R. and Pattus, F.,** Lipid-protein interactions in monolayers, *Chem. Phys. Lipids,* 30, 189, 1982.
16. **Dawson, R. M. C.,** The nature of interaction between protein and lipid during the formation of lipoprotein membranes, in *Biological Membranes. Physical Fact and Function,* Vol. 2, Chapman, D. and Wallach, D. F. H., Eds., Academic Press, London, 1973, 203.
17. **Sweet, C. and Zull, J. E.,** Activation of glucose diffusion from egg lecithin liquid crystals by serum albumin, *Biochim. Biophys. Acta,* 173, 94, 1969.
18. **Sogor, B. V. and Zull, J. E.,** Studies of a serum albumin-liposome complex as a model lipoprotein membrane, *Biochim. Biophys. Acta,* 375, 363, 1975.
19. **Papahadjopoulos, D. and Kimelberg, H. K.,** Phospholipid vesicles (liposomes) as models for biological membranes: their properties and interactions with cholesterol and proteins, in *Progress in Surface Science,* Vol. 4, Davison, S. G., Ed., Pergamon Press, Oxford, 1974, 141.
20. **Barrat, M. D., Green, D. K., and Chapman, D.,** Spin-labelled lipid-protein complexes, *Biochim. Biophys. Acta,* 152, 20, 1968.
21. **Butler, K. W., Hanson, A. W., Schneider, H., and Smith, I. C. P.,** The effects of protein on lipid organization in a model membrane system: a spin probe and X-ray study, *Can. J. Biochem.,* 51, 980, 1973.
22. **Benga, Gh. and Chapman, D.,** Protein-lipid interactions in biomembranes. I. Albumin-liposome model system-spin label studies, *Rev. Roum. Biochim.,* 13, 251, 1976.
23. **Mehta, N. G.,** Role of membrane integral proteins in the modulation of red cell shape by albumin, dinitrophenol and the glass effect, *Biochim. Biophys. Acta,* 762, 9, 1983.
24. **Deeley, J. O. T. and Coakley, W. T.,** Interfacial instability and membrane internalization in human erythrocytes heated in the presence of serum albumin, *Biochim. Biophys. Acta,* 727, 293, 1983.

25. **Stremmel, W., Potter, B. J., and Berk, P. D.,** Studies of albumin binding to rat liver plasma membranes. Implications for the albumin receptor hypothesis, *Biochim. Biophys. Acta*, 756, 20, 1983.
26. **Scanu, A. M., Edelstein, C., and Shen, B. W.,** Lipid-protein interactions in plasma lipoproteins. Model: high density lipoproteins, in *Lipid-Protein Interactions*, Vol. 1, Jost, P. C. and Griffith, O. H., Eds., John Wiley & Sons, New York, 1982, 259.
27. **Reijngoud, D. J. and Phillips, M. C.,** Mechanism of dissociation of human apolipoprotein A-I from complexes dimyristoyl-phosphatidylcholine as studied by guanidine hydrochloride denaturation, *Biochemistry*, 21, 2969, 1982.
28. **Reijngoud, D. J., Lund-Katz, S., Hauser, H., and Phillips, M. C.,** Lipid-protein interactions. Effect of apolipoprotein A-I on phosphatidylcholine polar group conformation as studied by proton nuclear magnetic resonance, *Biochemistry*, 21, 2977, 1982.
29. **Williamson, P., Bateman, J., Kozarsky, K., Mattocks, K., Hermanowicz, N., Choe, H. R., and Schlegel, A. R.,** Involvement of spectrin in the maintenance of phase-state asymmetry in the erythrocyte membrane, *Cell*, 30, 725, 1982.
30. **Sweet, C. and Zull, J. E.,** The binding of serum albumin to phospholipid liposomes, *Biochim. Biophys. Acta*, 219, 253, 1970.
31. **Haest, C. W. M.,** Interactions between membrane skeleton proteins and the intrinsic domain of the erythrocyte membrane, *Biochim. Biophys. Acta*, 694, 331, 1982.
32. **Papahadjopoulos, D., Moscarello, M., Eylar, E. H., and Isac, T.,** Effects of proteins on thermotypic phase transitions of phospholipid membranes, *BBA Biochim. Biophys. Acta*, 401, 317, 1975.
33. **Demel, R. A., London, Y., Van Kessel, G., Vossenberg, W. S. M., and Van Deenen, L. L. M.,** The specific interaction of myelin basic protein with lipids at the air-water interface, *Biochim. Biophys. Acta*, 311, 507, 1973.
34. **London, Y., Demel, R. A., Van Kessel, G., Vossenberg, F. G. A., and Van Deenen, L. L. M.,** The protection of A_1 myelin basic protein against the action of proteolytic enzymes after interaction of the protein with lipids at the air-water interface, *Biochim. Biophys. Acta*, 311, 520, 1973.
35. **London, Y. and Vossenberg, F. G. A.,** Specific interaction of central nervous system myelin basic protein with lipids. Specific regions of the protein sequence protected from the proteolytic action of trypsin, *Biochim. Biophys. Acta*, 307, 478, 1973.
36. **Eylar, E. H., Brostoff, S., Hashim, G., Caccam, J., and Burnett, P.,** Basic A_1 protein of the myelin membrane. The complete amino acid sequence, *J. Biol. Chem.*, 246, 5770, 1971.
37. **Boggs, J. M. and Moscarello, M. A.,** Dependence of the boundary lipid fatty acid chain length in phosphatidylcoline vesicles containing a hydrophobic protein from myelin proteolipid, *Biochemistry*, 17, 5734, 1978.
38. **Martenson, R. E.,** The use of gel filtration to follow conformational changes in proteins, *J. Biol. Chem.*, 253, 8887, 1978.
39. **Das, M. L. and Crane, F. L.,** Proteolipids. IV. Formation of complexes between cytochrome *c* and purified phospholipids, *Biochemistry*, 4, 859, 1965.
40. **Gulik-Krzywicky, T., Shechter, E., Luzzati, V., and Faure, M.,** Interactions of proteins and lipids: structure and polymorphism of protein-lipid-water phases, *Nature (London)*, 223, 1116, 1969.
41. **Steinemann, A. A. and Laüger, P.,** Interaction of cytochrome *c* with phospholipid monolayers and bilayer membranes, *J. Membr. Biol.*, 4, 74, 1971.
42. **Van, S. P. and Griffith, O. H.,** Bilayer structure in phospholipid-cytochrome *c* model membranes, *J. Membr. Biol.*, 20, 155, 1975.
43. **Margoliash, E. and Bosshard, H. R.,** Guided by electrostatics, a textbook protein comes of age, *Trends Biochem. Sci.*, 8, 316, 1983.
44. **Steer, C. J., Klausner, R. D. and Blumenthal, R.,** Interaction of liver clathrin coat protein with lipid model membranes, *J. Biol. Chem.*, 257, 8533, 1982.
45. **Ungewickell, E. and Branton, D.,** Triskelions: the building blocks of clathrin coats, *Trends Biochem. Sci.*, 7, 358, 1982.
46. **Moore, B. M., Lentz, B. R., Hoechli, M., and Meissner, G.,** Effect of membrane structure on the adenosine 5′-triphosphate hydrolyzing activity of the calcium-stimulated adenosine-triphosphatase of sarcoplasmic reticulum, *Biochemistry*, 20, 6810, 1981.
47. **Navarro, J., Toivio, L., Kinnucan, M., and Racker, E.,** Effect of lipid compositon on the Ca^{2+}/ATP coupling ratio of the Ca^{2+}-ATPase of sarcoplasmic reticulum, *Biochemistry*, 23, 130, 1984.
48. **Churchill, P., McIntyre, J. O., Eibl, H., and Fleischer, S.,** Activation of D-β-hydroxybutyrate apodehydrogenase using molecular species of mixed fatty acyl phospholipids, *J. Biol. Chem.*, 258, 208, 1983.
49. **Gwak, S. H. and Powell, G. L.,** The association of cardiolipin with detergent-solubilized cytochrome oxidase, *Biophys. J.*, 37, 108, 1982.
50. **Scotto, W. and Swislocki, N. I.,** A solubilized and active adenylate cyclase lipidated with native annular lipids, *Biophys. J.*, 109, 110, 1982.

51. **Jost, P., Griffith, O. H., Capaldi, R. A., and Vanderkooi, G.,** Evidence for boundary lipids in membranes, *Proc. Natl. Acad. Sci. U.S.A.*, 70, 480, 1973.
52. **Jost, P., Griffith, R. A., and Vanderkooi, G.,** Identification and extent of fluid bilayer regions in membranous cytochrome oxidase, *Biochim. Biophys. Acta*, 311, 141, 1973.
53. **Jost, P., Capaldi, R. A., Vanderkooi, G., and Griffith, O. H.,** Lipid-protein and lipid-lipid interactions in cytochrome oxidase model membranes, *J. Supramol. Struct.*, 1, 269, 1973a.
54. **Griffith, O. H., Jost, P., Capaldi, R. A., and Vanderkooi, G.,** Boundary lipid and fluid bilayer regions in cytochrome oxidase model membranes, *Ann. N.Y. Acad. Sci.*, 222, 561, 1973.
55. **Jost, P. C., Nadakavukaren, K. K., and Griffith, O. H.,** Phosphatidylcholine exchange between the boundary lipid and bilayer domains in cytochrome oxidase containing membranes, *Biochemistry*, 16, 3110, 1977.
56. **Warren, G. B., Toon, P. A., Birdsall, N. J. M., Lee, A. G., and Metcalfe, J. C.,** Reconstitution of a calcium pump using defined membrane components, *Proc. Natl. Acad. Sci. U.S.A.*, 71, 622, 1974.
57. **Hesketh, T. R., Smith, G. A., Houslay, M. D., McGill, K. A., Birdsall, N. J. M., Metcalfe, J. C., and Warren, G. B.,** Annular lipids determine the ATPase activity of a calcium transport protein complexed with dipalmitoyllecithin, *Biochemistry*, 15, 4145, 1976.
58. **Stier, A. and Sackman, E.,** Spin labels as enzyme substrates. Heterogenous lipid distribution in liver microsomal membranes, *Biochim. Biophys. Acta*, 311, 400, 1973.
59. **Overath, P. and Träuble, H.,** Phase transitions in cells, membranes and lipids of *Escherichia coli*. Detection by fluorescent probes, light scattering and dilatometry, *Biochemistry*, 12, 2625, 1973.
60. **Marčelja, S.,** Lipid-mediated protein interaction in membranes, *Biochim. Biophys. Acta*, 455, 1, 1976.
61. **Jähnig, F.,** Structural order of lipids and proteins in membranes: evaluation of fluorescence anisotropy data, *Proc. Natl. Acad. Sci. U.S.A.*, 76, 6361, 1979.
62. **Owicki, J. C. and McConnell, H. M.,** Theory of protein-lipid and protein-protein interactions in bilayer membranes, *Proc. Natl. Acad. Sci. U.S.A.*, 76, 4750, 1979.
63. **Marsh, D. and Barrantes, F. J.,** Immobilized lipid in acetylcholine receptor-rich membranes from *Torpedo marmorata*, *Proc. Natl. Acad. Sci. U.S.A.*, 75, 4329, 1978.
64. **Benga, Gh., Popescu, O., and Pop, V.,** Protein-lipid interactions in biological membranes. Cytochrome oxidase-lipid complex: spin label studies, *Rev. Roum. Biochim.*, 16, 175, 1979a.
65. **Benga, Gh., Porumb, R., and Frangopol, P. T.,** Evidence for various degrees of motional freedom of the "boundary" lipid in cytochrome oxidase, *Cell Biol. Int. Rep.*, 3, 651, 1979.
66. **Benga, Gh., Porumb, T., and Wrigglesworth, J. M.,** Estimation of lipid regions in a cytochrome oxidase-lipid complex using spin labelling electron spin resonance: distribution effects on the spin label, *J. Bioenerg. Biomembr.*, 13, 269, 1981.
67. **Knowles, P. F., Watts, A., and Marsh, D.,** Spin-label studies of lipid immobilization in dimyristoylphosphatidylcholine substituted cytochrome oxidase, *Biochemistry*, 18, 4480, 1979.
68. **Bienvenue, A., Rousselet, A., Kato, G., and Devaux, Ph. F.,** Fluidity of the lipids next to the acetylcholine receptor protein of *Torpedo* membrane fragments. Use of amphyhilic reversible spin-labels, *Biochemistry*, 16, 841, 1977.
69. **Baroin, A., Bienvenue, A., and Devaux, Ph. F.,** Spin-label studies of protein-protein interactions in retinal rod outer segment membranes. Saturation transfer electron paramagnetic resonance spectroscopy, *Biochemistry*, 18, 1151, 1979.
70. **Favre, E., Baroin, E., Bienvenue, A., and Devaux, Ph. F.,** Spin-label studies of lipid-protein interactions in retinal rod outer segment membranes. Fluidity of the boundary layer, *Biochemistry*, 18, 1156, 1979.
71. **Devaux, Ph. F.,** Solubility of intrinsic membrane proteins in phospholipid bilayers, in *Membranes and Intercellular Communication*, Ballian, R., Chabre, M., and Devaux, Ph. F., Eds., North-Holland, Amsterdam, 1981, 93.
72. **Chapman, D., Fernandez-Gomez, J. C., and Goni, F. M.,** Intrinsic protein-lipid interactions. Physical and biochemical evidence, *FEBS Lett.*, 98, 211, 1979.
73. **Swanson, M. S., Quintanilha, A. A., and Thomas, D. D.,** Protein rotational mobility and lipid fluidity of purified and reconstituted cytochrome oxidase, *J. Biol. Chem.*, 255, 7494, 1980.
74. **Devaux, P. F. and Davoust, J.,** Current views on boundary lipids deduced from electron spin resonance studies, in *Membranes and Transport*, Vol. 1, Martonosi, A. N., Ed., Plenum Press, New York, 1982.
75. **Andersen, J. P., Fellmann, P., Möller, J. V., and Devaux, P. F.,** Immobilization of a spin-labeled fatty acid chain covalently attached to Ca^{2+}-ATPase from sarcoplasmic reticulum suggests an oligomeric structure, *Biochemistry*, 20, 4928, 1981.
76. **Thomas, D. D., Bigelow, D. J., Squier, T. C., and Hidalgo, C.,** Rotational dynamics of protein and boundary lipid in sarcoplasmic reticulum membrane, *Biophys. J.*, 217, 225, 1982.
77. **Davoust, J., Seigneuret, P., Herve, P., and Devaux, P. F.,** Collisions between nitrogen-14 and spin labels. II. Investigations on the specificity of the lipid environment of rhodopsin, *Biochemistry*, 22, 3146, 1983.

78. **Dahlquist, F. W., Muchmore, D. C., Davis, J. H., and Bloom, M.,** Deuterium magnetic resonance studies of the interaction of lipids with membrane proteins, *Proc. Natl. Acad. Sci. U.S.A.*, 74, 5435, 1977.
79. **Utsumi, H., Tunggal, B. D., and Stoffel, W.,** Carbon-13 nuclear magnetic resonance studies on the interaction of glycophorin with lecithin in reconstituted vesicles, *Biochemistry*, 19, 2385, 1980.
80. **Jardetzky, O.,** NMR in the study of membranes, in *Membranes and Transport*, Vol. 1, Martonosi, A. N., Ed., Plenum Press, New York, 1982, 109.
81. **Seelig, A. and Seelig, J.,** Lipid-protein interaction in reconstituted cytochrome *c* oxidase/phospholipid membranes, *Hoope Seyler's Z. Physiol. Chem.*, 359, 1747, 1978.
82. **Oldfield, E., Gilmore, R., Glaser, M., Gutowsky, H. S., Hshung, J. C., Kung, S. Y., and King, T.,** Deuterium nuclear magnetic resonance investigation of the effects of proteins and polypeptides on hydrocarbon chain order in model membrane systems, *Proc. Natl. Acad. Sci. U.S.A.*, 75, 4657, 1978.
83. **Rice, D. M., Meadows, M. O., Scheinmann, A. O., Goni, F. M., Gomez-Fernandez, J., Moscarello, M. A., Chapman, D., and Oldfield, E.,** Protein-lipid interactions. A nuclear magnetic resonance study of sarcoplasmic reticulum Ca^{2+}, Mg^{2+}-ATPase, lipophilin, and proteolipid apoprotein-lecithin systems and a comparison with the effects of cholesterol, *Biochemistry*, 18, 5893, 1979b.
84. **McLaughlin, A. C., Herbette, L., Blasie, J. K., Wang, C. T., Hymel, L., and Fleischer, S.,** ^{31}P-NMR studies of oriented multilayer formed from isolated sarcoplasmic reticulum. Evidence that "boundary layer" phospholipid is not immobilized, *Biochim. Biophys. Acta*, 643, 1, 1981.
85. **Zumbulyadis, N. and O'Brien, D. F.,** Proton and carbon-13 nuclear magnetic resonance studies of rhodopsin-phospholipid interactions, *Biochemistry*, 24, 5427, 1979.
86. **Kang, S. Y., Gutowsky, H. S., and Oldfield, E.,** Spectroscopic studies of specifically deuterium labeled membrane systems. Nuclear magnetic resonance investigation of protein-lipid interactions, in *Escherichia coli* membranes, *Biochemistry*, 18, 3268, 1979.
87. **Kang, S. Y., Kinsey, R. A., Srinivasan, R., Gutowsky, H. S., Gabridge, M. G., and Oldfield, E.,** Protein-lipid interactions in biological and model membrane systems. Deuterium NMR of *Acholeplasma laidlawii* B, *Escherichia coli*, and cytochrome oxidase systems containing specifically deuterated lipids, *J. Biol. Chem.*, 256, 1155, 1981.
88. **Rajan, S., Kang, S. Y., Gutowsky, H. S., and Oldfield, E.,** Phosphorus nuclear magnetic resonance study of membrane structure. Interactions of lipids with protein, polypeptide, and cholesterol, *J. Biol. Chem.*, 256, 1160, 1981.
89. **Rice, D. M., Hsung, J. C., King, T. E., and Oldfield, E.,** Protein-lipid interactions. High-field deuterium and phosphorus nuclear magnetic resonance spectroscopic investigation of the cytochrome oxidase-phospholipid interaction and the effects of colate, *Biochemistry*, 18, 5885, 1979a.
90. **Oldfield, E.,** NMR of protein-lipid interactions in model and biological systems, in *Membranes and Transport*, Vol. 1, Martonosi, A. N., Ed., Plenum Press, New York, 1982, 115.
91. **Pink, D. A., Georgallas, A., and Chapman, D.,** Intrinsic proteins and their effect upon lipid hydrocarbon chain order, *Biochemistry*, 20, 7152, 1981.
92. **Tamm, L. K. and Seelig, J.,** Lipid solvation of cytochrome c oxidase. Deuterium, nitrogen-14, and phosphorus-31 nuclear magnetic resonance studies on the phosphocholine head group and on cis-unsaturated fatty acyl chains, *Biochemistry*, 22, 1474, 1983.
93. **Fischer, T. H. and Levy, G. C.,** Electron and proton magnetic resonance studies of the effect of rhodopsin incorporation on molecular motion in dimyristoylphosphatidylcholine bilayers, *Chem. Phys. Lipids*, 28, 7, 1981.
94. **Bienvenue, A., Bloom, M., Davis, J. H., and Devaux, P. F.,** Evidence for protein-associated lipids from deuterium nuclear magnetic resonance studies of rhodopsin-dimyristoylphosphatidylcholine recombinants, *J. Biol. Chem.*, 257, 3032, 1982.
95. **Albert, A. D. and Yeagle, P. L.,** Phospholipid domains in bovine retinal rod outer segment disk membranes, *Proc. Natl. Acad. Sci. U.S.A.*, 80, 7188, 1983.
96. **Yeagle, P. L. and Romans, A. Y.,** The glycophorin-phospholipid interface in recombined systems. A ^{31}P-nuclear magnetic resonance study, *Biophys. J.*, 33, 243, 1981.
97. **Yeagle, P. L.,** ^{31}P nuclear magnetic resonance studies of the phospholipid-protein interface in cell membranes, *Biophys. J.*, 37, 227, 1982.
98. **Albert, A. D., Lund, M., and Yeagle, P. L.,** Evidence for the influence of the protein-phospholipid interface on sarcoplasmic reticulum Ca^{++}, Mg^{++} ATPase activity, *Biophys. J.*, 36, 393, 1981.
99. **Seelig, J., Seelig, A., and Tamm, L.,** Nuclear magnetic resonance and lipid-protein interactions, in *Lipid-Protein Interactions*, Vol. 2, Jost, P. C. and Griffith, O. H., Eds., John Wiley & Sons, New York, 1982, 127.
100. **Lentz, B. R., Clubb, K. W., Barrow, D. A., and Meissner, G.,** Ordered and disordered coexist in membranes containing the calcium pump protein of sarcoplasmic reticulum, *Proc. Natl. Acad. Sci. U.S.A.*, 80, 2917, 1983.
101. **Marsh, D.,** Spin-label answers to lipid-protein interactions, *Trends Biochem. Sci.*, 8, 330, 1983.

102. **Paddy, M. R., Dahlquist, F. W., Davis, J. H., and Bloom, M.,** Dynamical and temperature dependent effects of lipid-protein interactions. Application of deuterium nuclear magnetic resonance and electron paramagnetic resonance spectroscopy to the same reconstitutions of cytochrome *c* oxidase, *Biochemistry*, 20, 3152, 1981.
103. **Schlesinger, M. J.,** Proteolipids, *Annu. Rev. Biochem.*, 50, 193, 1981.
104. **Schlesinger, M. J. and Magee, A. I.,** Fatty acid acylation of membrane proteins, *Biophys. J.*, 37, 126, 1982.
105. **Griffith, O. H. and Jost, P. C.,** Lipid-protein associations, in *Molecular Specialization and Symmetry in Membrane Function*, Solomon, A. K. and Karnovsky, M., Eds., Harvard University Press, Cambridge, 1978, 31.
106. **McIntyre, J. O., Samson, P., Brenner, S. C., Dalton, L., and Fleischer, S.,** EPR studies of the motional characteristics of the phospholipid in functional reconstituted sarcoplasmic reticulum membrane vesicles, *Biophys. J.*, 37, 53, 1982.
107. **Alonso, A., Restall, C. J., Turner, M., Gomez-Fernandez, J. C., Goñi, F. M., and Chapman, D.,** Protein-lipid interactions and differential scanning calorimetric studies of bacteriorhodopsin reconstituted lipid-water systems, *Biochim. Biophys. Acta*, 689, 283, 1982.
108. **Cortijo, M., Alonso, A., Gomez-Fernandez, J. C. and Chapman, D.,** Intrinsic protein-lipid interactions. Infrared spectroscopic studies of gramicidin A, bacteriorhodopsin and Ca^{2+}-ATPase in biomembranes and reconstituted systems, *J. Mol. Biol.*, 157, 597, 1982.
109. **Cornell, B. A., Hiller, R. G., Raison, J., Separovic, F., Smith, R., Vary, J. C., and Morris, C.,** Biological membranes are rich in low-frequency motion, *Biochim. Biophys. Acta*, 732, 1983.
110. **Ovchinnikov, Yu. A.,** Rhodopsin and bacteriorhodopsin: structure-function relationships, *FEBS Lett.*, 148, 179, 1982.
111. **Argos, P., Rao, J. K., and Hargrave, P. A.,** Structural prediction of membrane-bound proteins, *Eur. J. Biochem.*, 128, 565, 1982.
112. **Von Heijne, G.,** Membrane proteins. The amino acid composition of membrane-penetrating segments, *Eur. J. Biochem.*, 120, 275, 1981.
113. **Tanford, C.,** *The Hydrophobic Effect*, 2nd ed., John Wiley & Sons, New York, 1980.
114. **Rüppel, D., Kapitza, H. G., Galla, H. J., Sixl, F., and Sackmann, E.,** On the microstructure and phase diagram of dimyristoylphosphatidylcholine-glycophorin bilayers. The role of defects and the hydrophylic lipid-protein interaction, *Biochim. Biophys. Acta*, 692, 1, 1982.

Chapter 9

BASEMENT MEMBRANE STRUCTURE, FUNCTION, AND ALTERATION IN DISEASE

Lewis D. Johnson

TABLE OF CONTENTS

I. INTRODUCTION

Basement membranes are ubiquitous extracellular structures that are composed of several macromolecules arranged in a highly organized complex. They form the boundary between the layers of different types of cells that make up various tissues. Alterations in basement membranes have been observed in a number of diseases, particularly diabetes mellitus. Thus, it is not surprising that the literature describing studies of the structure and function of basement membranes has steadily expanded during the past 15 years. Basement membranes lie immediately adjacent to the plasma membranes of cells and the component macromolecules of basement membranes appear to function in concert with some molecular components of the plasma membrane; thus, it is appropriate that basement membranes be included in a review of cell membranes.

A number of reviews have been published that deal with the biochemical composition and function of basement membranes.[1-5] In addition, the alterations of the basement membrane that are present in human and animal diseases have been the focus of other reviews.[6-8] The purpose of this review is to indicate how new observations have increased our understanding of the structural bases for basement membrane function. The use of electron microscopic techniques, such as immunohistochemistry and rotary shadowing of molecules, in conjunction with advances in connective tissue biochemistry in general, permit insight into the nature of basement membranes.

Basement membranes are composed of three classes of macromolecules: unique types of collagen, noncollagenous glycoproteins, and proteoglycans. These components appear to be restricted to specific ultrastructural divisions of the basement membrane, either the lamina densa, the electron dense layer, or the lamina rara, the electron lucent layer. Modifications of this general structure are occasionally present in specific organs. For example, the lens capsule and Descemet's membrane in the eye have no lamina rara, and the glomerular basement membrane in the kidney is composed of three layers: lamina rara externa, lamina densa, and lamina rara interna.

II. EARLY STUDIES

The investigations into the biochemical composition of basement membranes began in the mid 1960s. Attention focused on collagenous components because the presence of hydroxyproline made identification relatively easy within purified basement membranes. The studies of a mucoprotein basement membrane component of neoplastic cell origin by Mukerjee et al.[9] indicate the difficulties in separating collagenous and noncollagenous components. The component they described resembled a noncollagenous protein in many respects but contained hydroxyproline, an amino acid found almost exclusively in collagen. The renal glomerular basement membrane (GBM) received early attention because it is involved in a variety of human diseases. The studies of Spiro[10] focused on the carbohydrate unit of GBM protein solubilized by digestion of isolated GBM with relatively crude collagenase. Two types of carbohydrate chain were found, a glucosylgalactose disaccharide linked to hydroxylysine and a heterosaccharide containing *n*-acetylhexosamine, mannose, galactose, fucose, and *n*-acetylneuraminic acid in a molar ratio of 5:3:4:1:3, presumably linked to asparagine. This was the first indication that basement membrane collagen was more heavily glycosylated than previously studied collagen types. Early studies by Huang and Kalant[11] on rat GMB identified six components after treatment with collagenase, urea, and gel filtration. Amino acid analysis indicated a mixture of collagenous and noncollagenous proteins ranging from 7.5 to 18.9% carbohydrate. The collagenous components, those containing hydroproline and hydroxylysine, never contained more than 17% glycine, indicating a mixture of proteins. The two problems of greatest concern to investigators were the insolubility of basement

Table 1
AMINO ACID COMPOSITION OF THE COLLAGEN FROM SHEEP ANTERIOR LENS CAPSULE AND GUT

	Residues/1000 residues		
Amino acids[a]	After pronase	After pepsin	Gut
Hydroxylysine	56.5	52.0	8.4
Lysine	6.8	9.0	25.7
Histidine	6.4	8.0	6.0
Arginine	29.6	30.6	51.0
3-Hydroxyproline	24.0	25.0	—
4-Hydroxyproline	134.7	129.0	102.7
Aspartic	47.6	45.7	46.7
Threonine	20.6	18.7	19.0
Serine	30.7	30.5	36.0
Glutamic	83.6	78.0	81.3
Proline	65.0	65.0	113.6
Glycine	317.0	338.0	337.6
Alanine	32.0	38.0	101.3
Half-cystine	6.0	5.0	—
Valine	26.6	24.3	17.2
Methionine	7.0	4.7	2.4
Isoleucine	24.0	22.4	12.0
Leucine	54.9	52.0	26.0
Tyrosine	4.3	4.0	1.0
Phenylalanine	27.0	23.7	11.4

[a] Average of two runs.

Taken from Denduchis, B., Kefalides, N. A., and Bezkorovainy, A., *Arch. Biochem. Biophys.*, 138, 582, 1970. With permission.

membranes and the presence of collagen, which could have resulted from connective tissue contamination. In order to avoid contamination, the bovine lens capsule was studied[12] and the isolated components were found to resemble those previously described, but the collagenous components still had a relatively low glycine content of 17 to 19% rather than the 33%, typical of tendon or dermal collagen. Analysis by X-ray diffraction of basement membrane material synthesized by epithelial cells also indicated the presence of a collagenous component, with intensity maxima corresponding to Bragg spacings of 10.5, 7.0, 4.6, 3.35, 3.14, and 2.9 Å, strongly resembling the X-ray diffraction pattern of freeze-dried tendon collagen.[13] Several other early studies resulted in similar findings with variations probably reflecting different isolation and solubilization techniques, in addition to different tissues and different animal species.[14-20] Denduchis et al.[21] isolated a collagenous component from sheep lens capsule basement membrane, after limited pronase or pepsin digestion at low temperature, that strongly resembled a heavily glycosylated collagen, as shown in Table 1. Additional characteristics of this collagen were a molecular weight of 284,000 and an intrinsic viscosity of 13.1 dℓ/g. Examination by electron microscopy revealed two morphologically distinct components, thin strands measuring 2 to 3 nm in diameter and globules measuring 5 to 20 nm in diameter.[22] Pepsin treatment destroyed the globular component leaving the filaments intact, suggesting a nonhelical structure.

Some studies yielded numerous components during solubilization of basement membranes, making interpretation of data very difficult. Nonspecific enzymatic degradation, either as the result of contaminants in commercial collagenase preparations[23] or as the result of endogenous enzymes, had to be prevented before precise characterization of basement membrane components could be accomplished. The results of a number of studies performed without protease inhibitors are difficult to interpret.[24-34] A consistent observation of such studies was that basement membranes were composed of multiple peptides, ranging in molecular weight from 30,000 to 700,000. An early study by Freytag et al.[35] failed to support the idea that small molecular weight peptides were the result of proteolysis. They found little qualitative difference when components separated by electrophoresis on polyacrylamide gels were compared in isolates with or without protease inhibitors during homogenization of tissue and fractionation.

Some investigators reported phospholipids as components of the GBM.[36-38] Such findings probably indicated the difficulty in separating basement membranes from cell membranes. This particular problem was first addressed by Mohos,[39] with regard to sialic acid rather than lipids. Discrepancies in the biochemical composition of basement membranes prompted extensive investigation into methods of isolation of basement membranes.[40-45] Isolation of a noncollagenous component, probably laminin, by buffer extraction avoided contamination but not fragmentation because of failure to use protease inhibitors.[31-32] Presently, isolation of basement membrane components is performed routinely in the presence of inhibitors of four classes of proteases. Single molecules obtained are in excess of 150,000 molecular weight. The conditions best suited for extraction from tissues and separation of component molecules are described in recent papers by Kleinman et al.[46] and Duhamel et al.[47] The most widely used source of basement membrane components is the murine Engelbreth-Holm-Swarm (EHS) sarcoma, first described by Orkin et al.[48] A recently described rat tumor also appears to be a suitable source[49] as does Reichert's membrane, the extraembryonic membrane of both the rat and mouse.[50] Although many sources have been used, the establishment of the EHS sarcoma, a source of abundant basement membrane material that is antigenically very similar to normal basement membrane of most species, has paved the way for extensive analytical studies of component molecules, the most widely studied being the collagenous components.

III. COLLAGEN

Collagen was first identified as a component of basement membranes in the early 1950s. Since that time, there has been considerable disagreement as to the number and biochemical nature of collagenous components. Kefalides[51] first reported that basement membranes treated with pepsin yielded a collagenous protein that closely resembled interstitial collagens, in many respects. The molecular weight of 110,000 for α-chains and glycine content of 310 residues per 1000 were characteristic of other known collagens, but the sum of lysine + hydroxylysine as well as proline + hydroxyproline was greater than found in interstitial collagens. In addition, alanine content was 25% of that found in interstitial collagens, there were 8 residues of half-cystine per 1000 residues of amino acid, and the 3-hydroxyproline isomer was present in relative abundance. This unusual collagen was more heavily glycosylated than interstitial collagens. Cleavage of α-chains with CNBr yielded 12 peptides, one of which contained all of the half-cystine residues and a heterosaccharide, suggesting that a noncollagenous peptide was linked to the collagen chain. Glycine comprised one third of the amino acids in each peptide; 4-hydroxyproline was found in every peptide and 3-hydroxyproline in seven peptides. Hydroxylysine was present in nine peptides; most residues were glycosylated with the galactosylglucose disaccharide but only one with the galactose monosaccharide.[52] The new molecule was designated α_1 (IV). The molecular conformation of the collagen molecule was designated $[\alpha_1\ (IV)]_3$. For several years, it was thought to contain only one type of α-chain.[53] In 1973, Tanzer and Kefalides[54] reported that basement membranes contained aldehyde-derived cross-links as did interstitial collagens. Man and

Adams[55] devised an assay for 3- and 4-hydroxyproline in tissues and reported that 18 to 25 residues of 3-hydroxyproline per 1000 residues were present in basement membrane collagen, in comparison to 1 per 1000 residues in interstitial collagen. The high hydroxyproline content correlates with the increased thermal stability (Tm = 40.0 ± 1.0°C). The presence of disulfide bonds in the helical region may also contribute.[56] Gryder et al.[57] found that 3-hydroxyproline was always in the -x- position of the sequence Gly-x-y and 4-hydroxyproline was predominantly, if not exclusively, in the -y position. Daniels and Chu[58] first suggested the possibility of more than one type of α-chain in studies of peptides obtained from reduced denatured steer GBM. Chung et al.[59] isolated three collagenous components from basement membranes in several tissues. They designated the chains isolated from skin as A and B (later found to be type V collagen). A smaller chain (55,000 molecular weight) was present only in the aorta and highly vascular organs. They concluded that A chain may be of epithelial cell origin and B chain of smooth muscle cell origin. Burgeson et al.[60] isolated similar collagen chains from fetal membranes that coprecipitated in 4.0 *M* NaCl in a constant ratio, suggesting the molecular structure $\alpha A(\alpha B)_2$. Bentz et al.[61] support this molecular conformation. In addition, denaturation studies indicate that only the B chain is able to form a stable triple helical structure. The observation that pepsin-solubilized human GBM collagen was composed of a heterogenous mixture of polypeptides further supported the idea that basement membranes contain more than one type of collagenous chain.[62] Dixit[63] isolated two distinct collagen α-chains, designated C and D (type IV), from human and porcine GBM, each with an apparent molecular weight of 95,000. There have been numerous reports, subsequently, that clearly indicate the existence of two genetically distinct polypeptide chains of basement membrane collagens.[64-71] These are now conventionally referred to as α_1(IV) and α_2(IV).

The first report of analysis of EHS sarcoma basement membrane, by Timpl et al.,[72] described the extraction of type IV collagen from the tumor with dilute acetic acid. The mice in which the sarcoma was grown were made lathrytic to facilitate extraction. Characteristics of the intact protein were (1) chromatographic mobility on CM-cellulose between the α-chains and β-components of type I collagen; (2) linkage by disulfide bonds; (3) circular dichroism spectra that resembled those obtained from other types of collagen; (4) a glycine content of only 280 residues per 1000 residues. After reduction of disulfide bonds and treatment with either pepsin or trypsin, a resistant fragment was obtained that contained one third glycine residues. The molecular weight of this resistant fragment was 140,000, a reduction of 15 to 20% from the size of the original collagen chain of approximately 160,000 molecular weight. Degradation of this fragment yielded two disulfide-linked, triple helical segments of 72,000 and a cystine-free fragment of 55,000 molecular weight.[73] The larger disulfide cross-linked chain was extremely heat stable (Tm = 70°C) and resistant to collagenase digestion.[74]

Levine and Spiro[75] described a small fragment of type IV collagen from GBM of apparent molecular weight 3820 that contains both disaccharide and complex heterosaccharide and is thought to be transitional between polar and helical positions of the collagen chain. Dixit[76] and Dixit and Kang[77] reported that anterior lens capsule contained two collagenous components and that the C-chain of anterior lens capsule yielded 12 peptides after cleavage with CNBr, ranging in molecular weight from 604 to 19,025, as calculated by amino acid analysis (see Table 2).

Type IV collagen contains large N- and C-terminal extensions, recognized by electron microscopy as nonbanded globular masses at each end of segment-long-spacing crystallites.[78] These end regions may impart the strong interactive properties between chains as well as salt precipitability and heat gelation characteristics of type IV collagen.[79] In addition, the noncollagenous end regions may play a role in inhibiting D-periodic striated fibril formation in basement membranes in vivo. The extent of intermolecular interaction is such that in vivo

Table 2
AMINO ACID COMPOSITION[a] OF CNBr PEPTIDES FROM THE C CHAIN OF BOVINE ANTERIOR LENS CAPSULE

	1	2	3	4	5	6	7	8	9	10	11	12	Total
3-Hyp	0	0	0	0	0	1(0.7)	0	1(1.1)	0	0	1(0.8)	1(0.9)	4
4-Hyp	1(1.0)	2(1.6)	4(3.7)	6(5.5)	6(5.6)	6(5.8)	8(8.1)	12	10(9.7)	12	19	23	109
Asp	1(1.1)	1(1.0)	1(1.0)	1(1.0)	2(2.1)	1(1.1)	4(4.2)	4(3.8)	4(4.0)	4(3.7)	8(8.2)	9(8.9)	40
Thr	0	1(0.9)	0	2(1.8)	1(0.9)	1(1.0)	1(1.0)	2(2.0)	2(1.9)	1(0.8)	2(2.0)	2(2.0)	15
Ser	0	1(0.9)	1(1.1)	2(1.9)	0	0	4(3.7)	3(3.0)	2(2.0)	6(6.0)	8(7.6)	7(7.0)	34
Glu	0	0	2(1.8)	5(5.1)	3(3.2)	5(4.7)	3(2.8)	10(10.2)	7(6.9)	7(7.0)	16	17	75
Pro	0	1(1.1)	2(1.8)	2(2.3)	3(3.2)	5(5.0)	2(2.2)	10(9.8)	4(3.8)	4(4.1)	12	14	59
Gly	2(2.2)	3(2.8)	8(7.9)	12	13	13	20	39	26	31	63	71	301
Ala	0	0	0	0	1(1.0)	1(1.0)	2(2.0)	3(2.8)	4(4.0)	2(2.0)	6(6.0)	7(6.8)	26
Val	0	1(0.9)	0	0	0	1(0.9)	2(1.8)	3(3.0)	2(1.7)	2(1.8)	5(4.7)	5(5.2)	21
Ile	0	0	0	1(1.0)	0	0	0	2(2.1)	3(2.7)	3(2.8)	6(6.0)	7(6.8)	22
Leu	1(1.0)	0	1(1.0)	3(3.2)	3(2.8)	2(2.0)	3(2.7)	6(6.1)	3(3.1)	5(5.0)	9(9.2)	10	46
Tyr	0	0	0	0	0	0	0	0	0	1(0.9)	0	0	1
Phe	0	1(0.8)	1(1.1)	0	1(0.9)	0	2 (2.1)	5(5.0)	1(0.8)	3(3.0)	5(5.1)	6(6.2)	25
Hyl	0	0	1(1.2)	3(2.7)	2(1.8)	1(1.1)	4(3.8)	7(6.6)	7(6.6)	7(6.7)	11	10	53
Lys	0	0	0	0	0	0	1(1.1)	0(0.2)	1(0.7)	1(1.3)	2(1.6)	2(1.7)	7
His	0	1(0.8)	0	0	0	1(0.8)	1(1.0)	1(0.8)	1(0.8)	0	1(0.9)	1(1.1)	7
Arg	0	0	0	0	1(1.0)	2(1.9)	1(1.1)	1(1.0)	1(1.1)	1(1.1)	6(6.0)	7(7.2)	20
Hse	1(1.0)	0	1(0.8)	1(0.8)	1(1.0)	1(0.9)	1(0.9)	1(0.8)	1(0.8)	1(0.8)	1(0.8)	1(0.9)	11
Total	6	12	22	38	37	41	59	110	79	91	181	200	876
Mol wt by aa analysis	604	1,198	2,115	3,725	3,576	4,017	5,515	10,474	7,624	9,329	17,310	19,025	
Mol wt by PAGE	604[b]	1,198[b]	2,850	4,420	3,800	4,230	5,820	13,800	8,400	12,300	19,200	20,400	

[a] Residues per peptide to the nearest whole number. Actual values are given where less than ten residues occur.
[b] Molecular weight calculated by amino acid analysis.

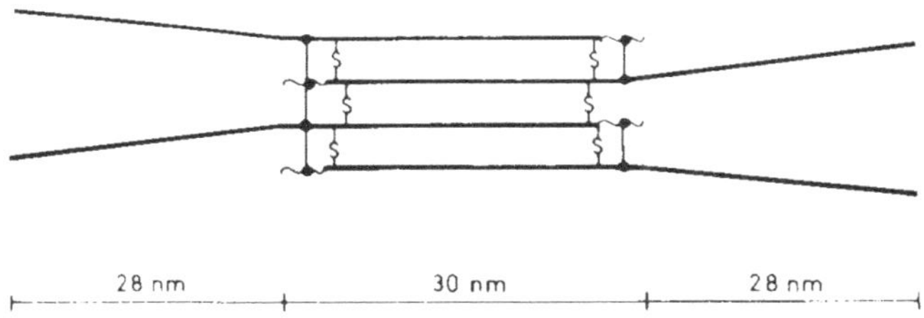

FIGURE 1. Model of 7S collagen (long form) representing the terminal cross-linking region of basement membrane collagen type IV. Each thick line represents a triple helical segment, the thin lines indicate nonhelical sequences at the end of the helix. The molecules are connected by disulfide bridges (-S-) and by nonreducible cross-links (•———•), presumably derived from oxidized lysine residues. The number and positions of cross-links are tentative. The four triple helices are, presumably, arranged in a tetragonal array. (Taken from Kuhn, K., Wiedemann, H., Timpl, R., Risteli, J., Dieringer, H., Voss, T., and Glanville, R. W., *FEBS Lett.*, 125, 123, 1981. With permission.)

aggregates as large as 30 polypeptide chains of α-size exist with disulfide and aldehyde-derived cross-links.[80]

The unusual sensitivity of type IV collagen to pepsin and other noncollagenolytic proteases is best explained by the presence of discontinuities in the amino acid sequence, Gly-X-Y. Shupper et al.[81] have demonstrated two long stretches of seven to eight amino acids located at both the N and C termini of a pepsin-derived fragment, P1, as well as a smaller interruption, identified as a single deletion of glycine, 30 residues from the N terminal of P1. Shorter interruptions have also been found in smaller pepsin fragments of both α_1(IV) and α_2(IV) chains.[82]

In 1980, Risteli et al.[83] reported the isolation of an unusual collagen by pepsinization and trypsinization of basement membranes from a number of tissues. The molecule is triple helical and exhibits a biphasic melting profile, indicating two conformationally distinct domains. This collagen is resistant to bacterial collagenase, contains 22% carbohydrate (predominantly disaccharide), and has a molecular weight of 360,000. This collagen, tentatively named 7-S collagen because of its hydrodynamic properties, was examined electron microscopically, using the technique of rotary shadowing, and found to consist of two rod-like structures. The dimensions were 3 × 95 nm (long form) and 3.4 × 40 to 50 nm (short form). Subsequent studies indicate that 7-S collagen is the cross-linking domain of type IV collagen.[84] The molecular weight of the short form is 200,000. The schematic reporesentation of 7-S collagen is shown in Figure 1. The four-domain structure of type IV collagen is shown in Figure 2. The four domains are two triple helical segments, a discontinuous major triple helix, the 7-S domain, and two noncollagenous segments. Intermolecular interactions occur at the ends of the molecule.[84] Basement membrane collagen isolated from bovine lung exhibits the same properties, with disulfide bonds contributing to thermal stability and resistance to enzymatic cleavage.[85]

Further analysis of type IV collagen pinpoints the site of an interruption in the triple helix within the α1 chain and suggest that an interruption at a similar point in the α2 chain would imply the structure $[\alpha 1(IV)_2]\alpha 2(IV)$.[86] This conformation is suggested by other studies.[87,88] The position of the two 3-hydroxyproline residues in the C-terminal peptide of α1(IV), in both cases, is in the homologous sequence, Gly-Phe-Xaa-Gly-3-Hyp-4-Hyp-Gly-Pro. Both residues are located close to an interruption in the triple helical sequence, thus facilitating hydroxylation.[85] The use of monoclonal antibodies has greatly facilitated studies of the localization of domains within the type IV collagen molecule.[89] Using immunological tech-

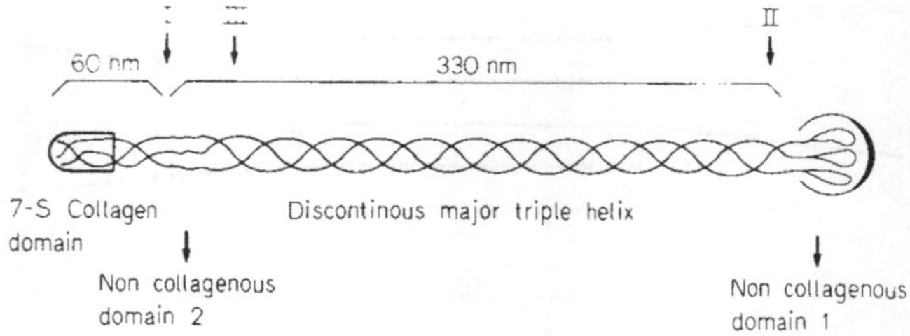

FIGURE 2. Four-domain structure of type IV collagen including two triple-helical segments and two noncollagenous segments NC1 and NC2. The position of segment NC2 is still tenative. Thick lines at both ends of the molecule identify the sites involved in interaction with other type IV collagen molecules. The dimensions are arbitrary except for the relative length of the triple-helical segments. Arrows on top denote major cleavage sites during solubilization of type IV collagen: (I) cleavage by unknown agents during prolonged extraction with acids; (II) cleavage by pepsin at 6 to 8°C (3 to 4 hr); (III) additional cleavage by pepsin at 15°C (6 hr). (Taken from Timpl, R., Wiedemann, H., Van Delden, V., Furthmayr, H., and Kuhn, K., *Eur. J. Biochem.*, 120, 203, 1981. With permission.)

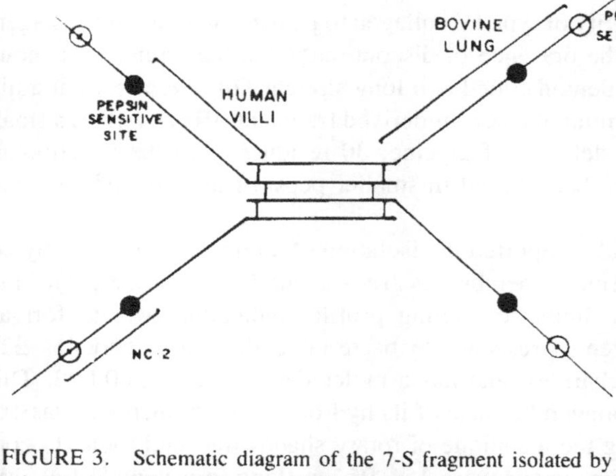

FIGURE 3. Schematic diagram of the 7-S fragment isolated by pepsin solubilization. The differences in length of the peripheral arms noted in bovine lung, murine tumor, and human placenta preparations are thought to be due to differences in the relative susceptibilities of the noncollagenous domains located on the peripheral arms of the spiderlike molecules. NC2 = noncollagenous domain 2; (⊙) pepsin-sensitive site for bovine lung and murine tumor matrix type IV collagen; (·) pepsin-sensitive site for human placental type IV collagen. (Reprinted with permission from Reference 90. Copyright 1983 American Chemical Society.)

niques, Madri et al.[90] concluded that the structural differences in type IV collagens from various species are the result of pepsin cleavage at different sites in noncollagenous regions adjacent to the 7-S domain (Figure 3).

Recently, a collagenous fragment, first described by Chung et al.[59] and called intima collagen[91] because of its original isolation from aortic intima, has been studied more extensively. This distinct collagen type resembles basement membrane collagens more closely than interstitial collagens.[91-93]

Studies of the synthesis and subsequent processing of basement membrane collagen have demonstrated that unlike interstitial collagen, type IV collagen have demonstrated that unlike interstitial collagen, type IV collagen exists in the same form extracellularly as intracellularly, i.e., as type IV procollagen. However, early studies suggested posttranslational processing.[94-96] Later studies made it clear that conversion did not take place,[97-101] and the molecular weight of the α-size chain was approximately 140,000 rather than 110,000 as originally reported. A number of studies describe biosynthetic events, such as the delay in the appearance of ^{14}C-hydroxyproline in extracellular protein,[102,103] the delay of 4 hr between ^{14}C-proline incorporation and the formation of cross-links.[104] The findings of studies of biosynthesis in vitro indicate that the procollagen IV chains have a molecular weight of approximately 185,000 for the pro-α1 chain and 175,000 for the pro-α2 chain which is larger than the α1 and α2 chains obtained by controlled proteolysis of basement membranes.[105-112] Studies by Laurie et al.[113] indicate that collagenous components are secreted from the synthesizing cells by the same route as other glycoproteins.

IV. NONCOLLAGENOUS GLYCOPROTEINS

In 1972, Boesken and Hammer[114] reported the characterization of a basement membrane glycoprotein found in the urine of nephritic rabbits. This high molecular weight molecule was collagen-free, as determined by amino acid analysis, and contained approximately 12% carbohydrate of the composition to suggest a complex type oligosaccharide chain. At about the same time, Johnson and Starcher[31] described the characteristics of a low molecular weight glycopeptide secreted by murine parietal yolk sac carcinoma cells in vitro. A similar molecule of 32,000 molecular weight was also extracted from murine kidney.[32] More recent studies in several other laboratories indicate that these glycopeptides are probably protease-resistant fragments of laminin. Collagenase digests of glomerular basement membrane contain two noncollagenous glycopeptides of molecular weights of approximately 1,000,000 and <200,000 according to Wieslander et al.[115]

A. Laminin

The most abundant and best studied of the noncollagenous glycoprotein components of basement membrane is laminin, first described by Timpl et al.[116] Laminin consists of at least two polypeptide chains of molecular weights 220,000 and 440,000, joined to each other by disulfide bonds. Pepsin digestion results in a cystine-rich fragment that retains most of the antigenicity of the original glycoprotein. No data concerning the carbohydrate portion of the glycoprotein were given. Chung et al.[117,118] described a glycoprotein of similar composition synthesized in vitro by a mouse embryonal carcinoma-derived cell line. Rohde et al.[119] extracted laminin from the EHS sarcoma under nondenaturing conditions and determined the antigenic sites within the glycoprotein. Antibodies against native laminin or a large pepsin-resistant fragment cross-reacted strongly, but failed to react with reduced, alkylated laminin. Major antigenic determinants were located in a disulfide knot comprising one third of the molecule that was resistant to cleavage by pepsin or CNBr. Minor determinants shared by native and reduced laminin were not found. These findings suggest that laminin consists of conformationally rigid domains and more flexible domains. Immunofluorescent studies indicated that laminin was present in the basement membrane of most mammalian tissues. Hogan reported that mouse extraembryonic membranes contained two collagenase-resistant glycoproteins of molecular weights 350,000 and 220,000 that were probably laminin polypeptides.[120] Rohde et al.[121] reported that pepsin treatment of native laminin released two homogeneous fragments of large size that contained 90% of the disulfide bonds and accounted for one third of the mass of laminin. An additional source of laminin, free of other basement membrane components, is the rat yolk sac tumor described by Wewer et al.[122] Risteli and Timpl[123] examined a large pepsin-resistant fragment of 290,000 molecular weight, from human placenta and renal basement membranes. Although the similar fragment of mouse laminin is smaller, immunologic data and amino acid analysis indicate strong similarities.

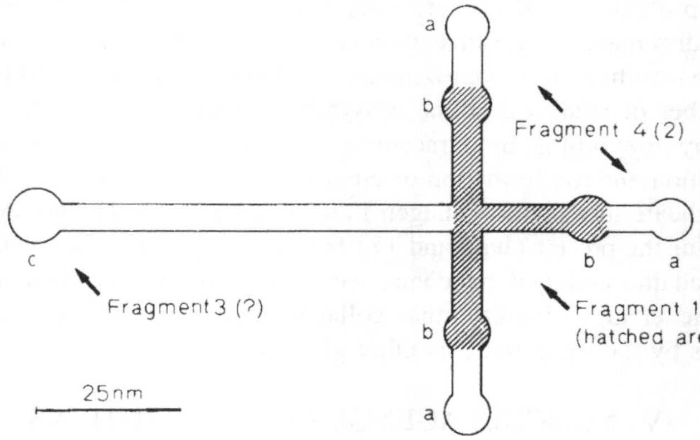

FIGURE 4. Scheme of the cross-shaped structure of laminin (9) and tentative assignment of its protease-resistant fragments; a, b, and c denote domains in the molecule possessing a globular morphology. The hatched area outlines the region occupied by fragment 1. the bar indicates a length of 25 nm. (Taken from Ott, U., Odermatt, E., Engel, J., Furthmayr, H., and Timpl, R., *Eur. J. Biochem.*, 123, 63, 1982. With permission.)

The technique of rotary shadowing of isolated laminin molecules permits visualization of this molecule.[124,125] Laminin, as depicted in Figure 4, is a rigid asymmetric cross-consisting of a long arm (77 nM) and three apparently identical short arms (36 nM). These rod-like arms terminate in globular units, 5 to 7 nm in diameter. The large pepsin-resistant fragment, described earlier, appeared to be a rigid structure with three short arms, 26 nm long, lying at a preferred angle of 90°. These shapes are consistent with the molecular weights and hydrodynamic properties noted when laminin is in solution. Rao et al.[126] cleaved laminin with the enzyme α-thrombin and isolated the α-subunit (200,000 mol wt). By electron microscopy, only a "stump" of the long arm remained attached, indicating the site of enzymatic attack to be near the point where the arms cross. Shibata et al.[127] used an immobilized galactose-binding lectin to isolate laminin from digests of EHS sarcoma and found that galactosyl end groups were constituents of both polypeptide chains. Engvall et al.[128] isolated laminin from rat yolk sac tumor and compared it immunologically with mouse laminin. Similarities were noted in that both molecules were recognized by rabbit antibodies against either. However, mouse antiserum to rat laminin reacted only weakly, and mice immunized with mouse laminin did not produce antibodies. Ohno et al.[129] very recently reported the isolation of laminin components from human placenta and found four major components: laminin A, B1, and B2 (350,000, 195,000, and 185,000 mol wt, respectively) that correspond to the laminin components of other basement membranes, and laminin M (240,000 mol wt).

Studies of laminin synthesis further indicate a more complicated picture than originally painted when digests of basement membranes were analyzed. Mouse parietal endoderm cells synthesize four glycoproteins of molecular weights 450,000 (PYSA), 230,000 and 240,000 (PYSB1 and B2), and 150,000 (PYSC).[130,131] The first three components represent those described by others and PYSC is probably entactin, described below. These findings have been confirmed by Clark and Kefalides.[109] The synthesis of laminin components appears to be sequential.[132] Studies in mouse oocytes and embryos indicate that eggs and oocytes synthesize only the B1 chain, 4- to 8-cell embryos synthesize both B1 and B2, and 16-cell embryos synthesize A1, B1, and B2. Data from one study indicate that the route of synthesis is the same as for other glycoproteins,[133] with basement membrane material first appearing in the basal portions of the secreting cells of the embryonic mouse and later outside the cell, accompanied by fine filaments.[134] Studies of the synthesis of basement membrane components and deposition into the rat parietal yolk sac strongly support the idea that fibronectin is not an intrinsic component of the basement membrane.[135]

B. Entactin

In 1981, Carlin et al. described a novel sulfated glycoprotein associated with basement membranes.[136] The glycoprotein had a molecular weight of 158,000, but was not degraded by chondroitinase ABC; therefore, it was not chondroitin sulfate or dermatan sulfate. They called this newly described component entactin. Electron microscopic studies immunologically localized entactin within the murine renal glomerulus, predominantly along the epithelial cell processes and to a lesser extent in the basement membrane proper. The possibility that entactin represented the peptide core of heparan sulfate was considered. Examination of entactin localization in several other murine tissues confirmed these findings that the lamina rara externa was the predominant location.[137] Parietal endoderm cells, derived from teratocarcinomas of the mouse, also synthesize a peptide of approximately 150,000 mol wt that is precipitated by antibodies against entactin.[138]

C. Fibronectin

Fibronectin is a large extracellular glycoprotein, composed of two identical 220,000-mol wt peptide chains held together by disulfide bonds.[139-141] Immunofluorescent microscopic studies have described fibronectin in the GBM, in greatest quantity along the cell surface where it abuts the basement membrane,[142] both in the lamina rara interna and the lamina rara externa.[143] Later studies at the level of the light microscope fail to support the earlier observations that fibronectin is a basement membrane component.[144] Although there is still some controversy, the studies by Martinez-Hernandez et al.[145] at the electron microscopic level clearly show that fibronectin is not an integral basement membrane component. Instead, it may be trapped in the GBM during plasma filtration and subsequently appear in the mesangial region of the glomerulus, as described,[142] during the process of cleansing the GBM of trapped proteins.

D. Others

Several reports of glycoproteins that are associated with basement membranes or the extracellular matrix are too limited in scope to determine whether these are laminin or entactin or represent additional basement membrane glycoproteins. Hogan et al.[146] described a sulfated glycoprotein of 180,000 mol wt in Reichert's membrane of the mouse that does not represent a precursor of entactin. Howe and Salter[147] reported isolation of a 200,000-mol wt glycoprotein from extraembryonic membranes, but it did not precipitate with anti-GBM antisera. Stanley et al.[148] isolated a 220,000-mol wt sulfated glycoprotein unique to stratified squamous epithelium, but did not discuss the relationship to entactin or the similar-sized sulfated glycoprotein described by Hogan.[146] Hunt et al.[149] isolated three glycoprotein antigens from human GBM, using affinity chromatography and collagenase digestion. These glycoproteins are stated to be noncollagenous, but amino acid analysis indicates the presence of 7 to 15% collagen-type amino acids. A 130,000 mol wt glycoprotein component of the muscle cell surface complex was described by Marton and Arnason.[150] This glycoprotein appears to be a general basement membrane glycoprotein by immunologic studies, and does not appear to be a fragment of any known basement membrane components.

At the present time, there are only two well-defined basement membrane glycoproteins, laminin and entactin, and there is still a great deal to learn about the localization and function of entactin.

V. HEPARAN SULFATE PROTEOGLYCANS

The presence of anionic sites within the basement membrane prompted the search for the component carrying these sites.[151] Heparan sulfate proteoglycan was isolated by several groups from the GBM[152-154] and from the EHS sarcoma.[155] The characteristics of the proteoglycan were 750,000 mol wt consisting of nearly equal amounts of protein and covalently linked heparan sulfate. Subsequent studies also indicate the presence of heparan sulfate proteoglycan in retinal capillary basement membrane.[156] Studies of sulfated proteoglycan synthesis by isolated glomeruli indicate that the proteoglycans have a molecular weight of 130,000, with 85% being heparan sulfate proteoglycan and the remainder chrondroitin and dermatan sulfates. The glycosaminoglycan released by nitrous acid (heparan sulfate) has a molecular weight of approximately 26,000, thus the proteoglycan contains four to five glycosaminoglycan chains. Data of Kanwar indicate that the heparan sulfate and chondroitin sulfate(s) were located on different core proteins.[157] Brown et al.[158] and Lemkin and Farquhar[159] reported similar findings that confirmed those of Kanwar et al.[157] A basement membrane-producing cell line, called PYS-2, synthesizes a heparan sulfate proteoglycan of 400,000 mol wt[160] as do murine parietal endoderm cells,[161] a cell type very similar to PYS-2. Unlike other basement membrane components, heparan sulfate has a short half-life, approximately 7 days in the GBM.[162] The ultrastructural architecture in the GBM, after staining with polyanionic dyes, is a network of tiny filamentous structures about 100 to 160 nm in length, with a mesh of about 60-nm width. Lateral branches appear along the filaments about 20 nm apart and 25 nm in length, projecting within the meshwork. The filamentous structure probably represents the protein core and the branches are the glycosaminoglycan chains.[163]

Recent immunochemical studies describe the localization of laminin, type IV collagen, and type V collagen in the basement membrane area at the epidermal-dermal junction of a basal cell carcinoma,[164] the kidney,[165-168] lung,[169] muscle capillaries and skeletal muscle,[170] and several other organs.[171] Studies of basement membrane synthesized by PYS-2 cells in vitro show a different distribution than that described in vivo.[172] The explanation for this difference is unclear. A schematic representation of the molecular organization of basement membrane molecules is shown in Figure 5, taken from Martinez-Hernandez and Amenta.[7] The charge properties of the basement membrane are contributed primarily by heparan sulfate, the bulk of which is located adjacent to cell surfaces. Laminin is predominantly localized within, the laminae rarae, and type IV collagen is dispersed throughout the basement membrane. Type V collagen, not shown in this scheme, is probably extrinsic to the basement membrane proper and functions by anchoring the basement membrane to adjacent cellular or extracellular structures, such as components of the interstitium. The precise location of entactin is not clearly understood at present. The distribution of components within various basement membranes in the same organ is not constant and probably reflects the differences in the primary function of the basement membrane at particular tissue sites.[173-175] The techniques required to extract basement membrane components from the EHS sarcoma provide additional clues to the molecular organization of the basement membrane. Results of studies by Kleinman et al.[176] suggest that lysine-derived cross-links and disulfide bonds stabilize collagen molecules, but that some other type of bond occurs between laminin, heparan sulfate proteoglycan, and collagen. Sakashita et al.[177] reported that laminin binds to heparan sulfate that has been immobilized on a column of Sepharose.

This binding is reversible and is disrupted by 1.0 M NaCl. Similarly, Del Rosso et al.,[178] using an affinity chromatography system, demonstrated that all sulfated glycosaminoglycans bind to laminin, but hyaluronic acid, a nonsulfated glycosaminoglycan, does not. Competitive-release experiments indicated that the glycosaminoglycan shared a common binding site on the laminin molecule. Zimmermann et al.[179] noted that the basement membrane of the EHS sarcoma is multilamellar. Treatment with trypsin widened the spaces between lamellae and resulted in loss of granular structures, presumably laminin. Treatment with testicular hyaluronidase widened only the outer lamellae, and treatment with collagenase left an extensively folded sheet composed predominantly of granules. These studies provide

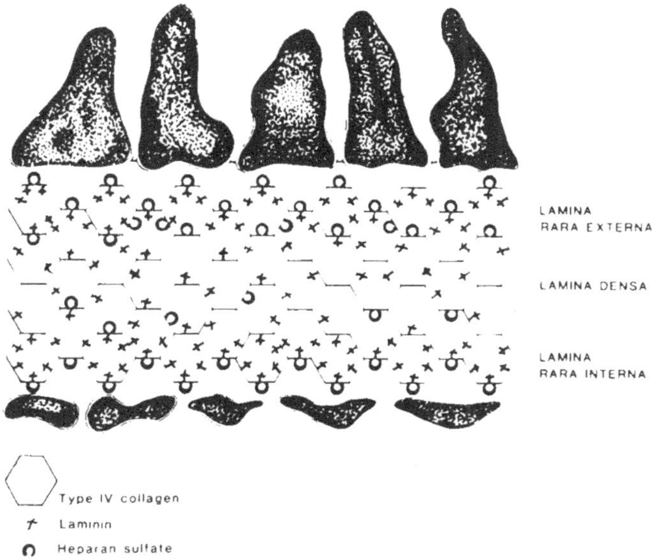

FIGURE 5. Schematic representation of the proposed supramolecular organization of glomerular basement membrane based on electron immunohistochemical studies. Type IV collagen is distributed throughout all layers of the GBM, forming the structural backbone upon which other components are attached. Laminin and heparan sulfate proteoglycan, although present in all layers, seem to be predominant in the laminae rarae. Entactin is not illustrated because further studies are required before its exact localization can be ascertained. (Taken from Martinez-Hernandez, A. and Amenta, P. S., *Lab. Invest.*, 48, 656, 1983. With permission.)

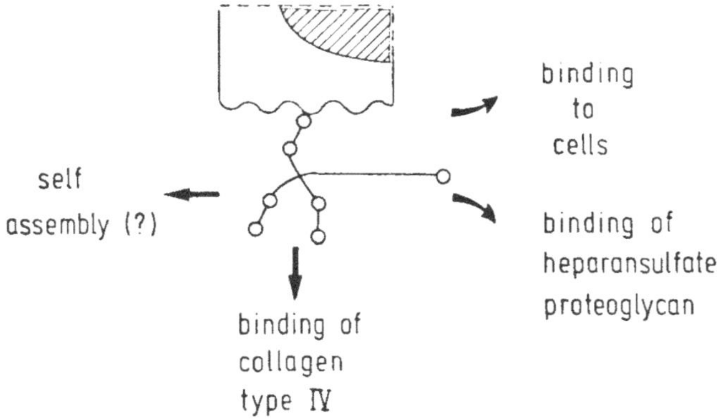

FIGURE 6. Hypothetical scheme of the interactions of laminin within basement membranes. (Taken from Timpl, R., Engel, J., and Martin, G. R., *Trends Biochem. Sci.*, 8, 207, 1983. With permission.)

valuable insights into the molecular organization of EHS sarcoma basement membrane, but it is not clear how this relates to the structure of normal basement membrane in vivo.

Laminin probably demonstrates the greatest diversity of interactions of any of the molecular species of the basement membrane.[180] As shown in Figure 6, the short arms of the molecule bind to plasma membranes, type IV collagen, and probably other laminin molecules to form high molecular weight aggregates. The long arm binds to heparan sulfate proteoglycan.

The molecular organization of the basement membrane accounts for the diverse functions of this important extracellular structure. These are described in the following section.

VI. BASEMENT MEMBRANE FUNCTIONS

A major function of basement membranes is selective filtration, best demonstrated by the renal GBM. This topic was reviewed by Farquhar.[181] Early studies by Caulfield and Farquhar[182] indicated that the rat GBM readily permitted passage of dextran molecules of average molecular weight 32,000 and Einstein-Stokes radius (ESR) of 38 Å, but markedly restricted passage of larger molecular weight dextrans. They concluded that the GMB was the major permeability barrier to proteins from 40,000 to 200,000 mol wt. Rennke et al.[183] subsequently showed that charge, as well as size, was important. They found that cationized ferritin was trapped in the laminae rarae of the GBM, whereas neutral and anionized ferritins of the same size were not. The filtration pores appear to distinguish between the shapes of molecules, as studies have shown that the fractional clearance of dextrans (prolate ellipsoid) is greater than that of ficoll (sphere), particularly in the ESR range from 24 to 44 Å.[184] The glycoprotein component of the GBM was originally thought to be the filtration barrier.[185] The discovery that glycosaminoglycans, heparan sulfate, in particular, were components of the laminae rarae portion of the basement membrane provided the basis for several studies into the role of polyanions in GBM filtration. Treatment of rat kidney with heparinase resulted in penetration of the GBM by native ferritin and failure of cationized ferritin to bind along the laminae rarae.[186] If rat kidney was perfused with high molarity buffers in order to neutralize sulfated glycosaminoglycans, native ferritin and bovine serum albumin, which were then perfused into the kidney, gradually clogged the basement membrane.[187] Studies by Simionescu et al.[188] showed that polyanions were in microdomains within the laminae rarae, and Charonis and Wissig[189] showed that phosphate and carboxyl groups, as well as sulfate groups, are present in the laminae rarae with sulfate groups predominating in renal capillary basement membrane and carboxyl and phosphate groups in muscle capillary basement membranes. These data indicate the role of laminin and heparan sulfate in basement membrane filtration. That laminin or some other, as yet undefined, protein plays a role is supported by the observation that trypsin or leukocyte proteinase digestion of isolated basement membranes increases the permeability of the basement membrane to plasma proteins.[190]

Because of the close proximity of basement membranes to the cells they subtend, investigators have sought the mechanisms of interaction. Terranova et al.[191] found that PAM 212 (epithelial) cells preferentially attach to type IV collagen via laminin but not to serum proteins, including fibronectin. Studies of regenerating liver cells[192] and embryonal carcinoma cells[193] indicate that laminin facilitates attachment to growth surfaces or collagen. The cell attachment activity of laminin is not heat sensitive (95°C for 5 min) but is almost completely inhibited by reduction alkylation or by periodate oxidation.[194,195] In 1982, Giavazzi et al.[196] reported that a murine tumor cell line of macrophage origin did not adhere to growth surfaces in the presence of laminin. Later studies indicate that the reason for failure of macrophages to adhere is the presence of laminin on the cell surface.[197] Cammarata and Spiro[198] found that lens cell attachment to laminin was energy dependent, required Ca^{++} but not Mg^{++}, and was blocked by sulfhydryl inhibitors. There was no apparent polarity of the cells to the basement membrane. The domain of laminin that is involved in cellular attachment is an α-subunit of 200,000 mol wt. Pepsin treatment removed the globular end and abolished attachment, indicating the role of the globular domain of the short arm in cellular attachment.[199] A high binding affinity (Kd = 2 nM) receptor on the cell surface to which laminin attaches was first described on human breast carcinoma cells.[200] Studies of a melanoma cell line indicated approximately 110,000 receptors per cell.[201] The receptor has a molecular weight of 67,000; binding is abolished by trypsinization of the cell membrane and binding affinity is altered slightly in isolated plasma membranes. This receptor would appear to be a glycolipid as laminin-mediated agglutination of aldehyde-fixed sheep red blood cells is inhibited by the gangliosides G_{M1}, G_{D1a}, and G_{T2b} in the order $G_D > G_T > G_M$. Monosaccharides or ceramide alone were not inhibitory.[202]

The role of the basement membrane in providing a substratum for cell attachment is essential in the maintenance of tissue morphology. Bernfield et al.[203] described the role of proteoglycan in the basement membrane in maintenance of salivary gland morphology during organogenesis. Vracko and Benditt[204] described the role of basement membrane in orderly replacement of skeletal muscle and capillaries following injury, and Vracko[205] reviewed basement membrane significance in maintenance of morphology. Later studies of the role of the basement membrane in salivary gland morphogenesis further indicate that the morphology is maintained by an intact basement membrane, synthesized by the epithelium, and that changes in morphology result from selective degradation of the basement membrane by mesenchymal cells.[206] These observations were made prior to an understanding of basement membrane biochemistry; therefore, no specific components were identified as playing a unique role. Gerfaux et al.[207] postulated that maintenance of tissue differentiation, or morphology, is governed by interaction of cells with sites along the basement membrane that are protease resistant and is the result of interaction between domains of type IV collagen with lectin-like activity and glycoprotein. Based on immunohistochemical studies, Ekblom et al.[208] reported that the presence of laminin was temporally related to renal tubular epithelial cell aggregation during the genesis of the embryonic kidney. These studies were expanded to demonstrate that heparan sulfate proteoglycan and type IV collagen were present, along with laminin, during the inductive phase of renal tubular morphogenesis.[209] Brownell et al.[210] studied molar tooth organ development in embryonic mice and concluded that basement membrane organization resulted from interaction between the epithelium-derived molecules (type IV collagen, glycoprotein, and proteoglycan) and the mesenchyme-derived cell surface molecule (fibronectin). The role of the basement membrane in development of the brain is thought to be one of guidance and support.[211] Studies of early rat astrocytes in vitro showed that laminin was deposited pericellularly, on top of the astrocytes, in contact with neuronal elements.

The appearance of type IV collagen during the growth of myoblasts but prior to the formation of myotubes suggests a role of the basement membrane in muscle development.[212] The role of basement membranes in developmental biology has been reviewed recently.[213]

The importance of the basement membrane to mammalian birth and development was most dramatically demonstrated by the effects of intravenously injected antibodies to type IV collagen and laminin into pregnant mice.[214] Pregnancies were aborted due to disruption of the extraembryonic fetal membranes and separation of maternal and fetal tissues. The greatest effect was seen with antitype IV collagen antibodies, suggesting that type IV collagen plays a more critical role than laminin in controlling the functional integrity of placental structures.

Changes in the relative quantities of basement membrane components appear to occur during development. Clark et al.[215] reported that during development of the rat parietal yolk sac, the ratio of collagen to protein in Reichert's membrane declined from 12.5 to 17.5 days gestation. Several studies show that the sequence of synthesis of components is initiated by a noncollagenous glycoprotein. Hintner et al.[216] noted that the sequence of appearance of basement membrane zone antigens at the dermal-epidermal junction in cultured human skin was bullous pemphigoid antigen prior to type IV collagen and laminin. Laminin is secreted by cells of the early mouse embryo prior to the appearance of type IV collagen.[217] Laminin was detected immunologically in the stage XIII embryonic chick, a time very early in development of the chick.[218] However, these findings are not universal. Laurie and Leblond[219] reported that rat embryonic tissues synthesize type IV collagen prior to laminin and heparan sulfate proteoglycan.

Studies of glycosaminoglycans in developing mouse salivary glands show that during branching, degradation proceeds at a more rapid rate than synthesis.[220] Studies of tissues of the eye indicate that changes in the distribution or composition of basement membrane components may occur during development.[221] Gulati et al.[222] studied the distribution of laminin during limb regeneration in the newt over a period of 42 days. Laminin was not detected until day 42, suggesting that it played no role in the regeneration process but only during the maintenance of architecture. Based on these results, it would appear that laminin may function somewhat differently in lower animals than in mammals, since it first appears quite late in the sequence of synthetic events.

VII. ALTERATIONS IN DISEASE

The human disease that has been most extensively studied with regard to changes in the composition of the basement membrane, in particular, the glomerular basement membrane of the kidney, is diabetes mellitus. The structural change is thickening of the GBM, which is associated with the functional change of impaired selective filtration. There has always been disagreement among investigators with regard to the composition of diabetic basement membranes.

The earliest report indicated no difference in the amount of specific amino acids or sugars, but simply an increase in the total amount of basement membrane.[223] Nearly 10 years later, the first report of compositional differences indicated a defect in the posttranslational hydroxylation of lysine and subsequent glycosylation of hydroxylysine. The isolated GBM from diabetic human kidney contained increased amounts of hydroxylysine and hydroxylysine-linked glycosides.[224] However, there was no change in the heterosaccharide content.[225] Westberg and Michael[226] reported that human GBM from diabetics was minimally enriched in collagen. Kefalides[227] also found a slight increase in the amount of collagen but no changes that would indicate an alteration in hydroxylation of amino acids or glycosylation. The conclusion that the changes reported by others probably represents contamination by non-GBM collagen appears most likely. However, reports continued to indicate changes in the composition of collagen. Khalifa and Cohen[228] found an increase in protocollagen lysylhydroxylase activity in the glomeruli of rats made diabetic by treatment with streptozotocin. Sato et al.[229] and Klein et al.[230] found that basement membranes obtained from human diabetic kidneys were enriched in collagen. Beisswenger[231] and Wahl et al.[232] confirmed the earlier findings of Kefalides[227] who reviewed the field in 1981 and pointed out that the discrepant results reflected the difficulties of obtaining adequate material for chemical analysis early in the disease and obtaining pure basement membrane for study late in the disease.[233]

Studies of basement membrane collagen synthesis in alloxan-diabetic rats indicated that collagen synthesis and proline hydroxylation were accelerated in glomeruli from diabetic rats.[234] However, it is not clear that only type IV collagen synthesis was being studied. A phenomenon associated with diabetic mellitus is nonenzymatic glycosylation of proteins, serum albumin and hemoglobin, for example. Similarly, Cohen et al.[235] studied GBM collagen from nondiabetic rats and streptozotocin-diabetic rats and reported that diabetic GBM collagen contained over twice the amount of glucose and/or mannose in ketoamine linkage. Later studies indicated lysine and hydroxylysine were the amino acids in hyperglycemic rats that were subject to nonenzymatic glycosylation.[236] These changes may affect the cross-links between collagen chains. LePape et al.[237] reported decreased amounts of hydroxylysinonorleucine and dihydroxylysinonorleucine concomitant with an increase in nonenzymatically derived hexosyllysine in experimentally diabetic rats. Uitto et al.[238] reported their findings in a similar study. The GBM from patients with diabetes contained collagen that was more extensively glycosylated, nonenzymatically. They concluded that this would result in decreased turnover since collagens with a high carbohydrate content are more resistant to proteolysis. In contrast to these reports, Wahl et al.[232] noted a decrease in the amount of hexose, both glucose and mannose, in human GBM. In addition, the half-cystine content was lower, an alteration noted in many studies. Another possible mechanism to explain decreased basement membrane turnover was proposed by Sternberg et al.[239] They reported decreased α-glucosidase activity in renal cortex and brain tissue of experimentally diabetic rats concomitant with GBM and brain capillary basement thickening. The α-glucosidase is specific for the hydroxylysine-linked disaccharide units of collagen. Nonenzymatic glycosylation of collagen in relation to development of diabetic complications has been recently reviewed.[240] There have been only isolated reports of changes in noncollagenous glycoproteins. These observations provide an explanation for the thickening of basement membranes in diabetes but do not address the diminished filtering capacity. The most recent studies of diabetic basement membranes have been directed at heparan sulfate proteoglycan. Cohen and Surma[241] first demonstrated decreased synthesis of heparan sulfate by diabetic tissue as measured by decreased sulfate incorporation. Rohrbach et al.[242] and Brown et al.[243] noted the same phenomenon. In addition, increased degradation was noted in the latter studies. Rohrbach et al.[242] found that the proteoglycan was heterogeneous in size, which could be the result of either diminished synthesis or degradation at an increased rate. They suggest that GBM thickening might be the result of compensatory synthesis of other components at an increased rate (see Figure 7). These results provide an explanation for the altered filtering selectivity of the GBM and an insight into the resultant thickening. Recent reviews have been prepared by Rohrbach and Martin[244] and Williamson and Kilo.[245]

Some of the compositional changes noted in association with diabetes mellitus could reflect changes resulting from aging. Blue and Lange[246] noted increased sugar content in the GBM of adult in comparison with neonatal mice. The total sugar increase was the result of increases in hexoses, sialic acids, hexosamine, and fucose. Kalant et al.[247] demonstrated an increase in GBM collagen with increasing age, as well as decreasing solubility. Deyl et al.[248] used the ratio of 3-hydroxyproline to 4-hydroxyproline to estimate the content of basement membrane collagen, in relation to total collagen, in several tissues from cows, rats, hamsters, and humans. They noted an increasing type IV collagen content in all tissues of all species studied with increasing age, probably as the result of decreased turnover. Langeveld et al.[249] noted age-related increases in hydroxyproline and hydroxylysine and glycosylated hydroxylysine in cattle. Similar findings were found when human GBM and tubular basement membranes were analyzed, suggesting a change in the ratio of collagenous to noncollageneous basement membrane proteins with age.[250] The ratio of 3-hydroxyproline to 4-hydroxyproline increased considerably during fetal and neonatal development. Similar data resulted when rat GBM was analyzed.[251] Other studies of human GBM indicated a decrease in the hydroxyproline isomer ratio, which could reflect a greater amount of non-basement membrane collagen in the GBM preparations.[252] In spite of some data suggesting altered GBM solubility, Wu and Cohen[253] showed no difference in the number of reducible cross-links in relation to aging in humans. The aging of our population will probably stimulate further studies of the age-related changes of basement membranes, because such changes potentially affect the function of virtually every organ in the body.

Another form of renal disease characterized by impaired glomerular filtration is nephrosis, which can be produced experimentally by injection of the aminonucleoside of puromycin. Several investigators have studied the composition of the renal GBM in this model. Blau and Michael[254] reported a decrease in the sialic acid content of rat GBM glycoprotein in aminonucleoside nephrosis. Lui and Kalant[255] reported a similar change and, in addition, a decrease in hexosamine. Bohrer et al.[256] reported that the defect in nephrosis was a reduction in the ratio of total pore surface area to pore length in the GBM rather than a change in the effective pore radius. Caulfield and Farquhar[257] observed a decrease in the total fixed negative charge of the laminae rarae of the GBM which correlates with the findings of Bohrer et

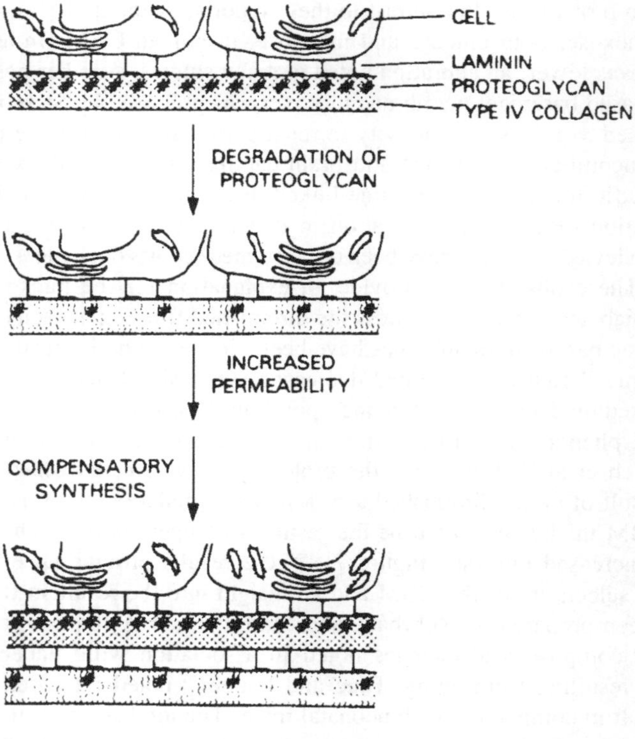

FIGURE 7. Schematic representation of possible events leading to the thickening of basement membrane in diabetes. At the top of the figure, synthesis of the common constituents of basement membrane (laminin, type IV collagen, and the heparan sulfate proteoglycan) has been completed and the functional basement membrane is intact. In the diabetic state, the level of heparan sulfate proteoglycan decreases due either to increased degradation (middle) or inadequate synthesis (not depicted). As a result of the loss of proteoglycan, the basement membrane becomes more porous. We propose that a high porosity of the basement membrane triggers the compensatory synthesis of more basement membrane components which leads to an increase in type IV collagen and laminin as shown at the bottom. (Taken from Rohrbach, D. H., Hassell, J. R., Kleinman, H. K., and Martin, G. R., *Diabetes*, 31, 185, 1982. Reproduced with permission from the American Diabetes Association, Inc.)

al.[256] Mynderse et al.[258] reported a loss of heparan sulfate proteoglycan from the GBM, as expected in light of our current knowledge of the molecular basis for the GBM filter. However, a recent report by Abrahamson et al.[259] indicates a prominent role for laminin in maintaining the selective filtering capacity. They injected antilaminin immunolgobulin into aminonucleoside nephrotic rats and noted an increased proteinuria, presumably, due to detachment of glomerular epithelial cells from the GBM rather than a decreased amount of laminin.

Other diseases have been associated with changes in the composition of basement membranes. Alport's syndrome, an inherited disorder characterized by nephritis, hearing loss, and defects of the optic lens and macula, may be, in part, due to a defect in the composition of the basement membranes of the glomerulus, lens, and inner ear.[260] Fibrosis of the liver has been associated with increased amounts of type IV collagen synthesis resulting in an altered ratio of collagen types in the liver.[261] Interaction of microbial pathogens with basement membrane components may be important in the initiation of infectious disease. An interesting observation by Speziale et al.[262] was that strains of uropathic *Escherichia coli* bound to laminin. This binding may be an important step in the development of pyelonephritis.

The association of cancer cells with the basement membrane has attracted considerable attention recently.[200,201] Varani et al.[263] reported that a line of mouse fibrosarcoma cells, that demonstrate the ability to metastasize, easily attach to type IV collagen. Cells with little tendency to metastasize do not attach unless laminin is added to the system. This difference between relatively aggressive and nonaggressive tumor cells may be an important characteristic with respect to the tendency for metastasis.

Another aspect of the relationship between tumor cells and the basement membrane deals with the quality of the basement membrane components synthesized by transformed or neoplastic cells. David and Van den Berghe[264] reported that transformed mouse mammary epithelial cells synthesize an undersulfated heparan sulfate proteoglycan which may explain the disrupted basement membrane associated with tumors. The interaction of neoplastic cells with the basement membrane, as regards metastasis, is not restricted to the type of basement membrane formed by these cells.

Liotta et al.[265] found that a metastatic murine tumor secreted an enzyme that specifically cleaved type IV collagen. The site of attack appeared to be within a largely helical domain of the collagen molecule. This collagenase has been characterized further. It has a molecular weight of 70,000 to 80,000 and requires trypsin activation and divalent cations. It produces specific cleavage products of both chains of type IV collagen and functions at neutral pH.[266] Basement membrane-degrading collagenase is secreted by several transformed cell lines and may play an important role in metastasis.[267,268] Other enzymes have been described that cleave basement membrane glycosaminoglycans.[269] It has been reported recently that migrating endothelial cells degrade type IV collagen.[270] Such local disruption may facilitate the outgrowth of capillary buds toward the angiogenesis stimulus elaborated by tumor cells.

The volume of knowledge about the composition and function of the basement membranes has expanded exponentially in the past 5 years. However, our understanding of all of the molecular and cellular interactions is just beginning to develop.

ACKNOWLEDGMENTS

I want to extend my thanks to J. B. Caulfield and S. D. Fowler for their helpful comments and to Pat Hutto for preparing this manuscript.

REFERENCES

1. **Kefalides, N. A.,** The chemistry and structure of basement membranes, *Arthritis Rheum.*, 12, 427, 1969.
2. **Kefalides, N. A.,** Structure and biosynthesis of basement membranes, in *International Review of Connective Tissue Research*, Hall, D. A. and Jackson, D. S., Eds., Academic Press, New York, 1973, 63.
3. **Kefalides, N. A.,** Basement membranes: structural and biosynthetic considerations, *J. Invest. Dermatol.*, 65, 85, 1975.
4. **Kefalides, N. A., Alper, R., and Clark, C. C.,** Biochemistry and metabolism of basement membranes, in *International Review of Cytology*, Vol. 61, Bourne, G. H. and Danielli, A. F., Eds., Academic Press, New York, 1979, 167.
5. **Grant, M. E., Heathcote, J. G., and Orkin, R. W.,** Current concepts of basement membrane structure and function, *Biosci. Rep.*, 1, 819, 1981.
6. **Johnson, L. D.,** The biochemical properties of basement membrane components in health and disease, *Clin. Biochem.*, 13, 204, 1980.
7. **Martinez-Hernandez, A. and Amenta, P. S.,** The basement membrane in pathology, *Lab. Invest.*, 48, 656, 1983.

8. **Martin, G. R., Rohrbach, D. H., Terranova, V. P., and Liotta, L. A.,** Structure, function and pathology of basement membranes, in *Connective Tissue Diseases*, Wagner, B. M. and Fleischmajer, R., Eds., Williams & Wilkins, Baltimore, 1983.
9. **Mukerjee, H., Sri Ram, J., and Pierce, G. B., Jr.,** Basement membranes. V. Chemical composition of neoplastic basement membrane mucoprotein, *Am. J. Pathol.*, 46, 49, 1965.
10. **Spiro, R. G.,** Studies on the renal glomerular basement membrane-nature of the carbohydrate units and their attachment to the peptide portion, *J. Biol. Chem.*, 242, 1923, 1967.
11. **Huang, F. and Kalant, N.,** Isolation and characterization of antigenic components of rat glomerular basement membrane, *Can. J. Biochem.*, 46, 1523, 1968.
12. **Fukushi, S. and Spiro, R. G.,** The lens capsule-sugar and amino acid composition, *J. Biol. Chem.*, 244, 2041, 1969.
13. **Lee, P. A., Blasey, K., Goldstein, I. J., and Pierce, G. B.,** Basement membrane: carbohydrates and X-ray diffraction, *Exp. Mol. Pathol.*, 10, 323, 1969.
14. **Spiro, R. G.,** Studies on the renal glomerular basement membrane — preparation and chemical composition, *J. Biol. Chem.*, 242, 1915, 1967.
15. **Spiro, R. G. and Fukushi, S.,** The lens capsule — studies on the carbohydrate units, *J. Biol. Chem.*, 244, 2049, 1969.
16. **Westberg, N. G. and Michael, A. F.,** Human glomerular basement membrane-preparation and composition, *Biochemistry*, 9, 3837, 1970.
17. **Mahieu, P. and Winand, R. J.,** Chemical structure of tubular and glomerular basement membranes of human kidney: isolation, purification, carbohydrate and amino acid composition, *Eur. J. Biochem.*, 12, 410, 1970.
18. **Mahieu, P. and Winand, R. J.,** Chemical structure of tubular and glomerular basement membranes of human kidneys. Isolation and characterization of the carbohydrate units, *Eur. J. Biochem.*, 15, 520, 1970.
19. **Mahieu, P., Lambert, P. H., and Miescher, P. A.,** Detection of antiglomerular basement membrane antibodies by a radioimmunological technique, *J. Clin. Invest.*, 54, 128, 1974.
20. **Hudson, B. G. and Spiro, R. G.,** Studies on the nature of reduced alkylated renal glomerular basement membrane. Solubility, subunit size and reaction with cyanogen bromide, *J. Biol. Chem.*, 247, 4229, 1972.
21. **Denduchis, B., Kefalides, N. A., and Bezkorovainy, A.,** The chemistry of sheep anterior lens capsule, *Arch. Biochem. Biophys.*, 138, 582, 1970.
22. **Olsen, B. R., Alper, R., and Kefalides, N. A.,** Structural characterization of a soluble fraction from lens-capsule basement membrane, *Eur. J. Biochem.*, 38, 220, 1973.
23. **Skoza, L. and Mohos, S. C.,** Enzymatic solubilization and separation of glomerular collagen and sialo-protein-effect of contaminating enzymes present in commercial collagenase preparations, *Lab. Invest.*, 30, 93, 1974.
24. **Ferwerda, W., Meijer, J. F. M., van den Eijnden, D. H., and Van Dijk, W.,** Epithelial basement membrane of bovine renal tubuli-isolation and chemical characterization, *Hoppe-Seyler's Z. Physiol. Chem.*, 355, 976, 1974.
25. **Peczon, B. D., Venable, J. H., Beams, C. G., Jr., and Hudson, B. G.,** Intestinal basement membrane of Ascaris suum — preparation, morphology, and composition, *Biochemistry*, 14, 4069, 1975.
26. **Sato, T. and Spiro, R. G.,** Studies on the subunit composition of the renal glomerular basement membrane, *J. Biol. Chem.*, 251, 4062, 1976.
27. **Krisko, I., DeBernardo, E., and Sato, C. S.,** Isolation and characterization of rat tubular basement membrane, *Kidney Int.*, 12, 238, 1977.
28. **Hung, C. H., Ohno, M., Freytag, J. W., and Hudson, B. G.,** Intestinal basement membrane of *Ascaris suum* — analysis of polypeptide components, *J. Biol. Chem.*, 252, 3995, 1977.
29. **Munakata, H., Sato, I., and Yosizawa, Z.,** Study on the porcine tubular basement membrane, *Tohoku J. Exp. Med.*, 125, 77, 1978.
30. **Sato, T., Munakata, H., Yoshinaga, K., and Yoshizawa, Z.,** Chemical composition of glomerular and tubular basement membranes of human kidney, *Tokoku J. Exp. Med.*, 115, 299, 1975.
31. **Johnson, L. D. and Starcher, B. C.,** Epithelial basement membranes: the isolation and identification of a soluble component, *Biochim. Biophys. Acta*, 290, 158, 1972.
32. **Johnson, L. D. and Warfel, J.,** Isolation and characterization of an epithelial basement membrane glycoprotein from murine kidney and further characterization of an epithelial basement membrane glycoprotein secreted by murine teratocarcinoma cells *in vitro*, *Biochim. Biosphys. Acta*, 455, 538, 1976.
33. **Megaw, J. M., Priest, J. H., Priest, R. E., and Johnson, L. D.,** Differentiation in human amniotic fluid cell cultures. II. Secretion of an epithelial basement membrane glycoprotein, *J. Med. Genet.*, 14, 163, 1977.
34. **Kefalides, N. A.,** A collagen of unusual composition and a glycoprotein isolated from canine glomerular basement membrane, *Biochem. Biophys. Res. Commun.*, 22, 26, 1966.
35. **Freytag, J. W., Ohno, M., and Hudson, B. G.,** Bovine renal glomerular basement membrane-assessment of proteolysis during isolation, *Biochem. Biophys. Res. Commun.*, 72, 796, 1976.

36. **Misra, R. P. and Berman, L. B.,** Glomerular basement membrane: insights from molecular models, *Am. J. Med.*, 47, 337, 1969.
37. **Fung, K. K. and Kalant, N.,** Phospholipid of the rat glomerular basement membrane in experimental nephrosis, *Biochem. J.*, 129, 733, 1972.
38. **Kibel, G., Heilhecker, A., and von Bruchhausen, F.,** Lipids associated with bovine kidney glomerular basement membranes, *Biochem. J.*, 155, 535, 1976.
39. **Mohos, S. C.,** Glomerular sialoprotein, *Science*, 164, 1519, 1969.
40. **Paulini, K. and Beneke, G.,** Untersuchungen an der isolierten Descemetschen membran-ein beitrag zum problem der struktur der basalmembran, *Virchows Arch. Abt. B Zellpath.*, 4, 208, 1970.
41. **Myers, C. and Bartlett, P.,** Separation of glomerular basement membrane substances by sodium dodecylsulfate disc gel electrophoresis and gel filtration, *Biochim. Biophys. Acta*, 290, 150, 1972.
42. **Marquardt, H., Wilson, C. B., and Dixon, F. J.,** Human glomerular basement membrane-selective solubilization with chaotropes and chemical and immunologic characterization of the components, *Biochemistry*, 12, 3260, 1973.
43. **Mahieu, P. M. and Winand, R. J.,** Carbohydrate and amino acid composition of human glomerular basement fractions purified by affinity chromatography, *Eur. J. Biochem.*, 37, 157, 1973.
44. **Ligler, F. S. and Robinson, G. B.,** A new method for the isolation of renal basement membranes, *Biochim. Biophys. Acta*, 468, 327, 1977.
45. **Carlson, E. C., Brendel, K., Hjelle, J. T., and Meezan, E.,** Ultrastructural and biochemical analysis of isolated basement membranes from kidney glomeruli and tubules and brain and retinal microvessels, *J. Ultrastruct. Res.*, 62, 26, 1978.
46. **Kleinman, H. K., McGarvey, M. L., Liotta, L .A., Robey, P. G., Tryggvason, K., and Martin, G. R.,** Isolation and characterization of type IV procollagen, laminin and heparan sulfate proteoglycans from the EHS sarcoma, *Biochemistry*, 24, 6188, 1982.
47. **Duhamel, R. C., Meezan, E., and Brendel, K.,** Selective solubilization of two populations of polypeptides from bovine retinal basement membranes, *Exp. Eye Res.*, 36, 257, 1983.
48. **Orkin, R. W., Gehron, P., McGoodwin, E. B., Martin, G. R., Valentine, T., and Swarm, R.,** A murine tumor producing a matrix of basement membrane, *J. Exp. Med.*, 145, 204, 1977.
49. **Wewer, U.,** Characterization of a rat yolk sac carcinoma cell line, *Dev. Biol.*, 93, 416, 1982.
50. **Smith, K. K. and Strickland, S.,** Structural components and characteristics of Reichert's membrane, an extra-embryonic basement membrane, *J. Biol. Chem.*, 256, 4654, 1981.
51. **Kefalides, N. A.,** Isolation of a collagen from basement membranes containing three identical α-chains, *Biochem. Biophys. Res. Commun.*, 45, 226, 1971.
52. **Kefalides, N. A.,** Isolation and characterization of cyanogen bromide peptides from basement membrane collagen, *Biochem. Biophys. Res. Commun.*, 47, 1151, 1972.
53. **Dehm, P. and Kefalides, N. A.,** The collagenous component of lens basement membrane — the isolation and characterization of an α chain size collagenous peptide and its relationship to newly synthesized lens components, *J. Biol. Chem.*, 253, 6680, 1978.
54. **Tanzer, M. L. and Kefalides, N. A.,** Collagen crosslinks — occurrence in basement membrane collagen, *Biochem. Biophys. Res. Commun.*, 51, 775, 1973.
55. **Man, M. and Adams, E.,** Basement membrane and interstitial collagen content of whole animals and tissues, *Biochem. Biophys. Res. Commun.*, 66, 9, 1975.
56. **Gelman, R. A., Blackwell, J., Kefalides, N. A., and Tomichek, E.,** Thermal stability of basement membrane collagen, *Biochim. Biophys. Acta*, 427, 492, 1976.
57. **Gryder, R. M., Lamon, M., and Adams, E.,** Sequence position of 3-hydroxyproline in basement membrane collagen-isolation of glygly-3-hydroxyprolyl-4-hydroxyproline from swine kidney, *J. Biol. Chem.*, 250, 2470, 1975.
58. **Daniels, J. R. and Chu, G. H.,** Basement membrane collagen of renal glomerulus, *J. Biol. Chem.*, 250, 3531, 1975.
59. **Chung, E., Rhodes, R. K., and Miller, E. J.,** Isolation of three collagenous components of probable basement membrane origin from several tissues, *Biochem. Biophys. Res. Commun.*, 71, 1167, 1976.
60. **Burgeson, R. E., El Adli, F. A., Kaitile, I. I., and Hollister, D. W.,** Fetal membrane collagens: identification of two new collagen alpha chains, *Proc. Natl. Acad. Sci. U.S.A.*, 73, 2579, 1976.
61. **Bentz, H., Bachinger, H. P., Glanville, R., and Kuhn, K.,** Physical evidence for the assembly of A and B chains of human placental collagen in a single triple helix, *Eur. J. Biochem.*, 92, 563, 1978.
62. **Tryggvason, K. and Kivirikko, K. I.,** Heterogeneity of a pepsin-solubilized human glomerular basement membrane collagen, *Nephron*, 21, 230, 1978.
63. **Dixit, S. N.,** Isolation and characterization of two α-chain size collagenous polypeptide chains, C and D, from glomerular basement membranes, *FEBS Lett.*, 106, 379, 1979.
64. **Glanville, R. W., Rauter, A., and Fietzek, P. P.,** Isolation and characterization of a native placental basement membrane collagen and its component α-chain, *Eur. J. Biochem.*, 95, 383, 1979.

65. **Bailey, A. J., Sims, T. J., Duance, V. C., and Light, N. D.,** Partial characterization of a second basement membrane collagen in human placenta — evidence for the existence of two type IV collagen molecules, *FEBS Lett.*, 99, 361, 1979.
66. **Sage, H., Woodbury, R. G., and Bornstein, P.,** Structural studies on human type IV collagen, *J. Biol. Chem.*, 254, 9893, 1979.
67. **Gay, S. and Miller, E. J.,** Characterization of lens capsule collagen — evidence for the presence of two unique chains in molecules derived from major basement membrane structures, *Arch. Biochem. Biophys.*, 198, 370, 1979.
68. **Kresina, T. F. and Miller, E. J.,** Isolation and characterization of basement membrane collagen from human placental tissue — evidence for the presence of two genetically distinct collagen chains, *Biochemistry*, 18, 3089, 1979.
69. **Dixit, S. N.,** Type IV collagen, isolation and characterization of two structurally distinct collagen chains from bovine kidney cortices, *Eur. J. Biochem.*, 106, 563, 1980.
70. **Crouch, E., Sage, H., and Bornstein, P.,** Structural basis for an apparent heterogeneity of collagens in human basement membrane: type IV procollagen contains two distinct chains, *Proc. Natl. Acad. Sci. U.S.A.*, 77, 745, 1980.
71. **Robey, P. G. and Martin, G. R.,** Type IV collagen contains two distinct chains in separate molecules, *Coll. Relat. Res.*, 1, 27, 1981.
72. **Timpl, R., Martin, G. R., Bruckner, P., Wick, G., and Wiedemann, H.,** Nature of the collagenous protein in a tumor basement membrane, *Eur. J. Biochem.*, 84, 43, 1978.
73. **Timpl, R., Bruckner, P., and Fietzek, P.,** Characterization of pepsin fragments of basement membrane collagen obtained from a mouse tumor, *Eur. J. Biochem.*, 95, 255, 1979.
74. **Timpl, R., Risteli, J., and Bachinger, H. P.,** Identification of a new basement membrane collagen by the aid of a large fragment resistant to bacterial collagenase, *FEBS Lett.*, 101, 265, 1979.
75. **Levine, M. J. and Spiro, R. G.,** Isolation from glomerular basement membrane of a glycopeptide containing both asparagine-linked and hydroxylysine-linked carbohydrate units, *J. Biol. Chem.*, 254, 8121, 1979.
76. **Dixit, S. N.,** Isolation and characterization of two collagenous components from anterior lens capsule, *FEBS Lett.*, 85, 153, 1978.
77. **Dixit, S. N. and Kang, A. H.,** Anterior lens capsule collagens: cyanogen bromide peptides of the C chain, *Biochemistry*, 18, 5686, 1979.
78. **Schwartz, D. and Veis, A.,** Characterization of bovine anterior lens capsule basement membrane collagen. II. Segment-long-spacing precipitates — further evidence for large N-terminal and C-terminal extensions, *Eur. J. Biochem.*, 103, 29, 1980.
79. **Schwartz, D., Chin-Quee, T., and Veis, A.,** Characterization of bovine anterior lens capsule basement membrane collagen. I. Pepsin susceptibility, salt precipitation and thermal gelatin — a property of non-collagen component integrity, *Eur. J. Biochem.*, 103, 21, 1980.
80. **West, T. W., Fox, J. W., Jodlowski, M., Freytag, J. W., and Hudson, B. G.,** Bovine glomerular basement membrane — properties of the collagenous domain, *J. Biol. Chem.*, 255, 10451, 1980.
81. **Shuppan, D., Timpl, R., and Glanville, R. W.,** Discontinuities in the triple helical sequence, Gly-X-Y, of basement membrane (type IV) collagen, *FEBS Lett.*, 115, 297, 1980.
82. **Glanville, R. W. and Rauter, A.,** Pepsin fragments of human placental basement membrane collagens showing interrupted triple helical amino acid sequences, *Hoppe-Seyler's Z. Physiol. Chem.*, 362, 943, 1981.
83. **Risteli, J., Bachinger, H. P., Engel, J., Furthmayr, H., and Timpl, R.,** 7-S collagen: characterization of an unusual basement membrane structure, *Eur. J. Biochem.*, 108, 239, 1980.
84. **Kuhn, K., Wiedemann, H., Timpl, R., Risteli, J., Dieringer, H., Voss, T., and Glanville, R. W.,** Macromolecular structure of basement membrane collagens — identification of 7S collagen as a crosslinking domain of type IV collagen, *FEBS Lett.*, 125, 123, 1981.
85. **Timpl, R., Wiedemann, H., Van Delden, V., Furthmayr, H., and Kuhn, K.,** A network model for the organization of type IV collagen molecules in basement membranes, *Eur. J. Biochem.*, 120, 203, 1981.
86. **Fujiwara, H. and Nagai, Y.,** Basement membrane collagen from bovine lung — its chain associations as observed by two-dimensional electrophoresis, *Coll. Relat. Res.*, 1, 491, 1981.
87. **Trueb, V., Grobli, B., Spiess, M., Odermatt, B. F., and Winterhalter, K. H.,** Basement membrane (type IV) collagen is a heteropolymer, *J. Biol. Chem.*, 257, 5239, 1982.
88. **Sakai, L. Y., Engvall, E., Hollister, D. W., and Burgeson, R. E.,** Production and characterization of a monoclonal antibody to human type IV collagen, *Am. J. Pathol.*, 108, 310, 1982.
89. **Mayne, R., Sanderson, R. D., Wiedemann, H., Fitch, J. M., and Lisenmayer, T. F.,** The use of monoclonal antibodies to fragments of chicken type IV collagen in structural and localization studies, *J. Biol. Chem.*, 258, 5794, 1983.
90. **Madri, J. A., Foellmer, H. G., and Furthmayr, H.,** Ultrastructural morphology and domain structure of a unique collagenous component of basement membranes, *Biochemistry*, 22, 2797, 1983.

91. **Odermatt, E., Risteli, J., van Delden, V., and Timpl, R.,** Structural diversity and domain composition of a unique collagenous fragment (intima collagen) obtained from human placenta, *Biochem. J.*, 211, 295, 1983.
92. **Furthmayr, H., Wiedemann, H., Timpl, R., Odermatt, E., and Engel, J.,** Electron microscopical approach to a structural model of intima collagen, *Biochem. J.*, 211, 303, 1983.
93. **Jander, R., Rauterberg, J., and Glanville, R. W.,** Further characterization of the three polypeptide chains of bovine and human short-chain collagen (intima collagen), *Eur. J. Biochem.*, 133, 39, 1983.
94. **Grant, M. E., Kefalides, N. A., and Prockop, D. J.,** The biosynthesis of basement membrane collagen in embryonic chick lens. I. Delay between the synthesis of polypeptide chains and the secretion of collagen by matrix-free cells, *J. Biol. Chem.*, 247, 3539, 1972.
95. **Grant, M. E., Kefalides, N. A., and Prockop, D. J.,** The biosynthesis of basement membrane collagen in embryonic chick lens. II. Synthesis of a precursor form in matrix-free cells and a time-dependent conversion to α-chains in intact lens, *J. Biol. Chem.*, 247, 3545, 1972.
96. **Kefalides, N. A., Cameron, J. D., Tomichek, E. A., and Yanoff, M.,** Biosynthesis of basement membrane collagen by rabbit corneal endothelium, *in vitro*, *J. Biol. Chem.*, 251, 730, 1976.
97. **Minor, R. R., Clark, C. C., Strause, E. L., Koszalka, T. R., Brent, R. L., and Kefalides, N. A.,** Basement membrane procollagen is not converted to collagen in organ cultures of parietal yolk sac endoderm, *J. Biol. Chem.*, 251, 1789, 1976.
98. **Heathcote, J. G., Sears, C. H. J., and Grant, M. E.,** Studies on the assembly of the rat lens capsule — biosynthesis and partial characterization of the collagen components, *Biochem. J.*, 176, 283, 1978.
99. **Ko, C. Y., Johnson, L. D., and Priest, R. E.,** Isolation and characterization of hydroxyproline-containing proteins secreted by a murine carcinoma cell line, *Biochim. Biophys. Acta*, 581, 252, 1979.
100. **Tryggvason, K., Robey, P. G., and Martin, G. R.,** Biosynthesis of type IV procollagens, *Biochemistry*, 19, 1284, 1980.
101. **Alitalo, K., Vaheri, A., Krieg, T., and Timpl, R.,** Biosynthesis of two subunits of type IV procollagen and of other basement membrane proteins by a human tumor cell line, *Eur. J. Biochem.*, 109, 247, 1980.
102. **Clark, C. C., Tomichek, E. A., Koszalka, T. R., Minor, R. R., and Kefalides, N. A.,** The embryonic rat parietal yolk sac — the role of the parietal endoderm in the biosynthesis of basement membrane collagen and glycoprotein, *in vitro*, *J. Biol. Chem.*, 250, 5259, 1975.
103. **Karakashian, M. W. and Kefalides, N. A.,** Studies on the underhydroxylated basement membrane procollagen synthesized by rat parietal yolk sacs in the presence of α, α_1-dipyridyl, *Connect. Tissue Res.*, 10, 247, 1982.
104. **Williams, I. F., Harwood, R., and Grant, M. E.,** Triple helix formation and disulphide bonding during the biosynthesis of glomerular basement membrane collagen, *Biochem. Biophys. Res. Commun.*, 70, 200, 1976.
105. **Clark, C. C. and Kefalides, N. A.,** Partial characterization of newly synthesized basement membrane procollagen, in *Cellular and Biochemical Aspects in Diabetic Retinopathy*, INSERM Symp. No. 7, Regnault, F. and Duhault, J., Eds., Elsevier/North-Holland, Amsterdam, 1978, 21.
106. **Howard, B. V., Macarak, E. J., Gunsan, D., and Kefalides, N. A.,** Characterization of the collagen synthesized by endothelial cells, *Proc. Natl. Acad. Sci. U.S.A.*, 73, 2361, 1976.
107. **Crouch, E. and Bornstein, P.,** Characterization of a type IV procollagen synthesized by human amniotic fluid cells in culture, *J. Biol. Chem.*, 254, 4197, 1979.
108. **Killen, P. D. and Striker, G. E.,** Human glomerular visceral epithelial cells synthesize a basal lamina collagen *in vitro*, *Proc. Natl. Acad. Sci. U.S.A.*, 96, 3518, 1979.
109. **Clark, C. C. and Kefalides, N. A.,** Partial characterization of collagenous and noncollagenous basement membrane proteins synthesized by the 14.5-day rat embryo parietal yolk sac *in vitro*, *Connect. Tissue Res.*, 10, 303, 1982.
110. **Karakashian, M. W., Dehm, P., Gramling, T. S., and LeRoy, E. C.,** Precursor-size components are the basic collagenous subunits of murine tumor basement membrane, *Coll. Relat. Res.*, 2, 3, 1982.
111. **Sundar-Raj, C. V. and Freeman, I. L.,** Structure and biosynthesis of rabbit lens capsule collagen, *Invest. Ophthalmol. Vis. Sci.*, 23, 743, 1982.
112. **Dean, D. C., Barr, J. F., Freytag, J. W., and Hudson, B. G.,** Isolation of type IV procollagen-like polypeptides from glomerular basement membrane-characterization of pro-α_1, (IV), *J. Biol. Chem.*, 258, 590, 1983.
113. **Laurie, G. W., Leblond, C. P., and Martin, G. R.,** Intracellular localization of basement membrane precursors in the endodermal cells of the rat parietal yolk sac. II. Immunostaining for type IV collagen and its precursors, *J. Histochem. Cytochem.*, 30, 983, 1982.
114. **Boesken, W. H. and Hammer, D. K.,** Purification and chemical characterization of a basement membrane glycoprotein in the urine of nephritic rabbits, *Hoppe-Seyler's Z. Physiol. Chem.*, 353, 1429, 1972.
115. **Wieslander, J., Bygren, P., and Heinegard, D.,** Human glomerular basement membrane heterogeneity of antigenic determinents, *Biochim. Biophys. Acta*, 553, 244, 1979.

116. **Timpl, R., Rohde, H., Robey, P. G., Rennard, S. I., Foidart, J. M., and Martin, G. R.,** Laminin — a glycoprotein from basement membranes, *J. Biol. Chem.*, 254, 9933, 1979.
117. **Chung, A. E., Jaffe, R., Freeman, I. L., Vergnes, J. P., Braginski, J. E., and Carlin, B.,** Properties of a basement membrane related glycoprotein synthesized in culture by a mouse embryonal carcinoma-derived cell line, *Cell*, 16, 277, 1979.
118. **Chung, A. E., Freeman, I. L., and Braginski, J. E.,** A novel extracellular membrane elaborated by a mouse embryonal carcinoma-derived cell line, *Biochem. Biophys. Res. Commun.*, 79, 859, 1977.
119. **Rohde, H., Wick, G., and Timpl, R.,** Immunochemical characterization of the basement membrane glycoprotein laminin, *Eur. J. Biochem.*, 102, 195, 1979.
120. **Hogan, B. L. M.,** High molecular weight extracellular proteins synthesized by endoderm cells derived from mouse teratocarcinoma cells and extraembryonic membranes, *Dev. Biol.*, 76, 275, 1980.
121. **Rohde, H., Bachinger, H. P., and Timpl, R.,** Characterization of a pepsin fragment of laminin in a tumor basement membrane, *Hoppe-Seylers Z. Physiol. Chem.*, 361, 1651, 1980.
122. **Wewer, U., Albrechtsen, R., and Ruoslahti, E.,** Laminin, a noncollagenous component of epithelial basement membrane synthesized by a rat yolk sac tumor, *Cancer Res.*, 41, 1518, 1981.
123. **Risteli, L. and Timpl, R.,** Isolation and characterization of pepsin fragments of laminin from human placental and renal basement membranes, *Biochem. J.*, 193, 749, 1981.
124. **Engel, J., Odermatt, E., Engel, A., Madri, J. A., Furthmayr, H., Rohde, H., and Timpl, R.,** Shapes, domain organization and flexibility of laminin and fibronectin, two multifunctional proteins of the extracellular matrix, *J. Mol. Biol.*, 150, 97, 1981.
125. **Ott, U., Odermatt, E., Engel, J., Furthmayr, H., and Timpl, R.,** Protease resistance and conformation of laminin, *Eur. J. Biochem.*, 123, 63, 1982.
126. **Rao, C. N., Margulies, I. M. K., Tralka, T. D., Terranova, V. P., Madri, J. A., and Liotta, L. A.,** Isolation of a subunit of laminin and its role in molecular structure and tumor cell attachment, *J. Biol. Chem.*, 257, 9740, 1982.
127. **Shibata, S., Peters, B. P., Roberts, W. D., Goldstein, I. J., and Liotta, L. A.,** Isolation of laminin by affinity chromatography on immobilized *Griffonia simplicifolia*. I. Lectin, *FEBS Lett.*, 142, 194, 1982.
128. **Engvall, E., Krusius, T., Wewer, U., and Ruoslahti, E.,** Laminin from rat yolk sac tumor: isolation, partial characterization and comparison with mouse laminin, *Arch. Biochem. Biophys.*, 222, 649, 1983.
129. **Ohno, M., Martinez-Hernandez, A., Ohno, N., and Kefalides, N. A.,** Isolation of laminin from human placental basement membranes: amnion, chorion and chorionic microvessels, *Biochem. Biophys. Res. Commun.*, 112, 1091, 1983.
130. **Hogan, B. L. M., Cooper, A. R., and Kurkinen, M.,** Incorporation into Reichert's membrane of laminin-like extracellular proteins synthesized by parietal endoderm cells of the mouse embryo, *Dev. Biol.*, 80, 289, 1980.
131. **Cooper, A. R., Kurkinen, M., Taylor, A., and Hogan, B. L. M.,** Studies on the biosynthesis of laminin by murine parietal endoderm cells, *Eur. J. Biochem.*, 119, 189, 1981.
132. **Cooper, A. R. and MacQueen, H. A.,** Subunits of laminin are differently synthesized in mouse eggs and early embryos, *Dev. Biol.*, 96, 467, 1983.
133. **Laurie, G. W., Leblond, C. P., Martin, G. R., and Silver, M. H.,** Intracellular localization of basement membrane precursors in the endodermal cells of the rat parietal yolk sac. III. Immunostaining for laminin and its precursors, *J. Histochem. Cytochem.*, 30, 991, 1982.
134. **Csato, W. and Merker, H. J.,** Production and formation of the basement membrane in embryonic tissues of the mouse, *Cell Tissue Res.*, 228, 85, 1983.
135. **Amenta, P. S., Clark, C. C., and Martinez-Hernandez, A.,** Deposition of fibronectin and laminin in the basement membrane of the rat parietal yolk sac: immunochemical and biosynthetic studies, *J. Cell Biol.*, 96, 104, 1983.
136. **Carlin, B., Jaffe, R., Bender, B., and Chung, A. E.,** Entactin, A novel basal lamina-associated sulfated glycoprotein, *J. Biol. Chem.*, 256, 5209, 1981.
137. **Bender, B. L., Jaffe, R., Carlin, B., and Chung, A. E.,** Immunolocalization of entactin, a sulfated basement membrane component in rodent tissues, and comparison with GP-2 (laminin), *Am. J. Pathol.*, 103, 419, 1981.
138. **Kurkinen, M., Barlow, D., Jenkins, J. R., and Hogan, B. L.,** *In vitro* synthesis of laminin and entactin polypeptides, *J. Biol. Chem.*, 258, 6543, 1983.
139. **Hynes, R. O.,** Fibronectin and its relation to cellular structure and behavior, in *Cell Biology of Extracellular Matrix*, Hay, E. D., Ed., Plenum Press, New York, 1981, 295.
140. **Ruoslahti, E., Hayman, E. G., Pierschbacher, M., and Engvall, E.,** Fibronectin: purification, immunochemical properties and biological activities, *Methods Enzymol.*, 82, 803, 1982.
141. **Ruoslahti, E., Valeri, A., Kinsela, P., and Linder, C.,** Fibroblast surface antigen: a new serum protein, *Biochim. Biophys. Acta*, 322, 352, 1973.
142. **Oberley, T. D., Mosher, D. F., and Mills, M. D.,** Localization of fibronectin within the renal glomerulus and its production by cultured glomerular cells, *Am. J. Pathol.*, 96, 651, 1979.

143. **Courtoy, P. J., Kanwar, Y. S., Hynes, R. O., and Farquhar, M. G.,** Fibronectin localization in the rat glomerulus, *J. Cell. Biol.*, 87, 691, 1980.
144. **Boselli, J. M., Macarak, E. J., Clark, C. C., Brownell, A. G., and Martinez-Hernandez, A.,** Fibronectin: its relationship to basement membranes. I. Light microscopic studies, *Coll. Res.*, 5, 391, 1981.
145. **Martinez-Hernandez, A., Marsh, C. A., Clark, C. C., Macarak, E. J., and Brownell, A. G.,** Fibronectin: its relationship to basement membranes. II. Ultrastructural studies in rat kidneys, *Coll. Res.*, 5, 405, 1981.
146. **Hogan, B. C. M., Taylor, A., Kurkinen, M., and Couchman, J. R.,** Synthesis and localization of two sulfated glycoproteins associated with basement membranes and the extracellular matrix, *J. Cell Biol.*, 95, 197, 1982.
147. **Howe, C. C. and Salter, D.,** Identification of noncollagenous basement membrane glycopeptides synthesized by mouse parietal endoderm and an endodermal cell line, *Dev. Biol.*, 77, 480, 1980.
148. **Stanley, J. R., Hanley-Nelson, P., Yuspa, S. H., Shevach, E. M., and Katz, S. I.,** Characterization of bullous pemphigoid antigen: a unique basement membrane protein of stratified squamous epithelia, *Cell*, 24, 897, 1981.
149. **Hunt, J. S., Macdonald, P. R., and McGiven, A. R.,** Characterization of human glomerular basement membrane antigenic fractions isolated by affinity chromatography utilizing anti-glomerular basement antibodies, *Biochem. Biophys. Res. Commun.*, 104, 1025, 1982.
150. **Marton, L. S. G. and Arnason, B. G. W.,** A basement membrane-associated glycoprotein from skeletal muscle, *J. Cell Biochem.*, 19, 363, 1982.
151. **Kanwar, Y. S. and Farquhar, M. G.,** Anionic sites in the glomerular basement membrane, *in vivo* and *in vitro* localization to the laminae rarae by cationic probes, *J. Cell Biol.*, 81, 137, 1979.
152. **Kanwar, Y. S. and Farquhar, M. G.,** Presence of heparan sulfate in the glomerular basement membrane, *Proc. Natl. Acad. Sci. U.S.A.*, 76, 1303, 1979.
153. **Parthasarathy, N. and Spiro, R. G.,** Characterization of the glycosaminoglycan component of the renal glomerular basement membrane and its relationship to the peptide portion, *J. Biol. Chem.*, 256, 507, 1981.
154. **Cohen, M. P., Wu, V. Y., and Surma, M. L.,** Non-collagen protein and proteoglycan in renal glomerular basement membrane, *Biochim. Biophys. Acta*, 678, 322, 1981.
155. **Hassell, J. R., Robey, P. G., Barrach, H. J., Wilczek, J., Rennard, S. I., and Martin, G. R.,** Isolation of a heparan sulfate-containing proteoglycan from basement membrane, *Proc. Natl. Acad. Sci. U.S.A.*, 77, 4494, 1980.
156. **Cohen, M. P. and Ciborowski, C. J.,** Presence of glycosaminoglycans in retinal capillary basement membrane, *Biochim. Biophys. Acta*, 674, 400, 1981.
157. **Kanwar, Y. S., Hascall, V. C., and Farquhar, M. G.,** Partial characterization of newly synthesized proteoglycans isolated from the glomerular basement membrane, *J. Cell. Biol.*, 90, 527, 1981.
158. **Brown, D. M., Michael, A. F., and Oegma, T. R.,** Glycosaminoglycan synthesis by glomeruli *in vivo* and *in vitro*, *Biochim. Biophys. Acta*, 674, 96, 1981.
159. **Lemkin, M. C. and Farquhar, M. G.,** Sulfated and nonsulfated glycosaminoglycans and glycopeptides are synthesized by kidney *in vivo* and incorporated into glomerular basement membranes, *Proc. Natl. Acad. Sci. U.S.A.*, 78, 1726, 1981.
160. **Oohira, A., Wight, T. N., McPherson, J., and Bornstein, P.,** Biochemical and ultrastructural studies proteoheparan sulfate synthesized by PYS-2, a basement membrane synthesizing cell line, *J. Cell Biol.*, 92, 357, 1982.
161. **Hogan, B. L. M., Taylor, A., and Cooper, A. R.,** Murine parietal endoderm cells synthesize heparan sulfate and 170K and 145K sulfated glycoproteins as components of Reichert's membrane, *Dev. Biol.*, 90, 210, 1982.
162. **Cohen, M. P. and Surma, M. L.,** *In vivo* biosynthesis and turnover of 35S-labeled glomerular basement membrane, *Biochim. Biophys. Acta*, 716, 337, 1982.
163. **Reale, E., Luciano L., and Kuhn, K. W.,** Ultrastructural architecture of proteoglycans in the glomerular basement membrane: a cytochemical approach, *J. Histochem. Cytochem.*, 31, 662, 1983.
164. **Van Cauwenberge, D., Pierard, G. E., Foidart, J. M., and Lapiere, C. M.,** Immunochemical localization of laminin, type IV and type V collagen in basal cell carcinoma, *Br. J. Dermatol.*, 108, 163, 1983.
165. **Roll, F. J., Madri, J. A., Albert, J., and Furthmayr, H.,** Codistribution of collagen types IV and AB2 in basement membrane and mesangium of the kidney — an immunoferritin study of ultrathin frozen sections, *J. Cell Biol.*, 85, 597, 1980.
166. **Scheinman, J. I., Foidart, J. M., Gehron-Robey, P., Fish, A. J., and Michael, A. F.,** The immunohistology of glomerular antigens. IV. Laminin, a defined non-collagenous basement membrane glycoprotein, *Clin. Immunol. Immunopathol.*, 15, 175, 1980.
167. **Scheinman, J. I., Foidart, J. M., and Michael, A. F.,** The immunohistology of glomerular antigens. V. The collagenous antigens of the glomerulus, *Lab. Invest.*, 43, 373, 1980.

168. **Oberley, T. D., Chung, A. E., Murphy-Ullrich, J. E., and Mosher, D. F.,** Studies on the localization of the glycoprotein, GP-2, within the renal glomerulus *in vivo* and in cultured kidney cell strains *in vitro, J. Histochem. Cytochem.*, 29, 1237, 1981.
169. **Vaccaro, C. A. and Brody, J. S.,** Structural features of alveolar wall basement membrane in the rat lung, *J. Cell Biol.*, 91, 427, 1981.
170. **Charnois, A. S., Tsilibary, E. C., Kramer, R. H., and Wissig, S. L.,** Localization of heparan sulfate and laminin in the basement membrane of muscle capillaries and skeletal muscle cells, *J. Cell Biol.*, 91, 165a, 1981.
171. **Foidart, J. M., Bere, E. W., Jr., Yaar, M., Rennard, S. I., Gullino, M., Martin, G. R., and Katz, S. I.,** Distribution and immunoelectron microscopic localization of laminin, a non-collagenous basement membrane glycoprotein, *Lab. Invest.*, 42, 336, 1980.
172. **Leivo, I.,** Basement membrane-like matrix of teratocarcinoma-derived endoderm cells, *J. Histochem. Cytochem.*, 31, 35, 1983.
173. **Madri, J. A., Roll, F. J., Furthmayr, H., and Foidart, J. M.,** Ultrastructural localization of fibronectin and laminin in the basement membrane of the murine kidney, *J. Cell Biol.*, 86, 682, 1980.
174. **Martinez-Hernandez, A., Miller, E. J., Damjanov, I., and Gay, S.,** Laminin secreting yolk sac carcinoma of the rat. Biochemical and electron immunohistochemical studies, *Lab. Invest.*, 47, 247, 1982.
175. **Monaghan, P., Warburton, M. J., Perusinghe, N., and Rudland, P. S.,** Topographical arrangement of basement membrane proteins in lactating rat mammary gland: comparison of the distribution of type IV collagen, laminin, fibronectin and Thy-1 at the ultrastructural level, *Proc. Natl. Acad. Sci. U.S.A.*, 80, 3344, 1983.
176. **Kleinman, H. K., McGarvey, M. C., Liotta, L. A., Robey, P. G., Tryggvason, K., and Martin, G. R.,** Isolation and characterization of type IV procollagen, laminin and heparan sulfate proteoglycan from the EHS sarcoma, *Biochemistry*, 24, 6188, 1982.
177. **Sakashita, S., Engvall, E., and Ruoslahti, E.,** Basement membrane glycoprotein, laminin, binds to heparin, *FEBS Lett.*, 116, 243, 1980.
178. **Del Rosso, M., Cappelletti, R., Viti, M., Vannucchi, S., and Chiarugi, V.,** Binding of the basement membrane glycoprotein laminin to glycosaminoglycansan affinity chromatography study, *Biochem. J.*, 199, 699, 1981.
179. **Zimmermann, B., Merker, H. J., and Barrach, H. J.,** Basement membrane alteration after treatment with trypsin, hyaluronidase and collagenase, *Virchow's Arch. (Cell Pathol.)*, 40, 9, 1982.
180. **Timpl, R., Engel, J., and Martin, G. R.,** Laminin-a multifunctional protein of basement membranes, *Trends Biochem. Sci.*, 8, 207, 1983.
181. **Farquhar, M. G.,** The glomerular basement membrane, a selective macromolecular filter, in *Cell Biology of Extracellular Matrix,* Hay, E. D., Ed., Plenum Press, New York, 1981, 335.
182. **Caulfield, J. B. and Farquhar, M. G.,** The permeability of glomerular capillaries to graded dextrans-identification of the basement membrane as the primary filtration barrier, *J. Cell Biol.*, 63, 883, 1974.
183. **Rennke, H. G., Cotran, R. S., and Venkatachalam, M. A.,** Role of molecular charge in glomerular permeability-tracer studies with cationized ferritins, *J. Cell Biol.*, 67, 638, 1975.
184. **Bohrer, M. P., Deen, W. H., Robertson, C. R., Troy, J. L., and Brenner, B. M.,** Influence of glomerular configuration on the passage of macromolecules across the glomerular capillary wall, *J. Gen. Physiol.*, 74, 583, 1979.
185. **Latta, H. and Johnston, W. H.,** The glycoprotein inner layer of glomerular capillary basement membrane as a filtration barrier, *J. Ultrastruct. Res.*, 57, 65, 1976.
186. **Kanwar, Y. S., Linker, A., and Farquhar, M. G.,** Increased permeability of the glomerular basement membrane to ferritin after removal of glycosaminoglycans (heparan sulfate) by enzyme digestion, *J. Cell Biol.*, 86, 688, 1980.
187. **Kanwar, Y. S. and Rosenzweig, L. J.,** Clogging of the glomerular basement membrane, *J. Cell Biol.*, 93, 489, 1982.
188. **Simionescu, M., Simionescu, N., and Palade, G. E.,** Preferential distribution of anionic sites on the basement membrane and the abluminal aspect of the endothelium in fenestrated capillaries, *J. Cell Biol.*, 95, 425, 1982.
189. **Charnois, A. S. and Wissig, S. L.,** Anionic sites in basement membranes — differences in their electrostatic properties in continuous and fenestrated capillaries, *Microvasc. Res.*, 25, 265, 1983.
190. **Cotter, T. G. and Robinson, G. B.,** The effects of proteinases on the filtration properties of isolated basement membrane, *Int. J. Biochem.*, 12, 191, 1980.
191. **Terranova, V. P., Rohrbach, D. H., and Martin, G. R.,** Role of laminin in the attachment of PAM 212 (epithelial) cells to basement membrane collagen, *Cell*, 22, 719, 1980.
192. **Carlsson, R., Engvall, E., Freeman, A., and Ruoslahti, E.,** Laminin and fibronectin in cell adhesion: enhanced adhesion of cells from regenerating liver to laminin, *Proc. Natl. Acad. Sci. U.S.A.*, 78, 2403, 1981.

193. **Darmon, M.Y.,** Laminin provides a better substrate than fibronectin for attachment, growth and differentiation of 1003 embryonal carcinoma cells, *In Vitro*, 18, 997, 1982.
194. **Johansson, S., Kjellen, L., Hook, M., and Timpl, R.,** Substrate adhesion of rat hepatocytes: a comparison of laminin and fibronectin as attachment proteins, *J. Cell Biol.*, 90, 260, 1981.
195. **Palotie, A., Peltonen, L., Risteli, L.,and Risteli, J.,** Effect of the structural components of basement membranes on the attachment of teratocarcinoma-derived cells, *Exp. Cell Res.*, 144, 31, 1983.
196. **Giavazzi, R., Liotta, L., and Hart, I.,** Laminin inhibits the adhesion of a murine tumor of macrophage origin, *Exp. Cell Res.*, 140, 315, 1982.
197. **Wicha, M. S. and Huard, T. K.,** Macrophages express cell surface laminin, *Exp. Cell Res.*, 143, 475, 1983.
198. **Cammarata, P. R. and Spiro, R. G.,** Lens epithelial cell adhesion to lens capsule: a model system for cell-basement membrane interaction, *J. Cell Physiol.*, 113, 273, 1982.
199. **Rao, C. N., Margulies, I. M. K., Tralka, T. S., Terranova, V. P., Madri, J. A., and Liotta, L. A.,** Isolation of a subunit of laminin and its role in molecular structure and tumor cell attachment, *J. Biol. Chem.*, 257, 9740, 1982.
200. **Terranova, V. P., Rao, C. N., Kabelic, T., Margulies, I. M., and Liotta, L. A.,** Laminin receptor on human breast carcinoma cells, *Proc. Natl. Acad. Sci. U.S.A.*, 80, 444, 1983.
201. **Rao, C. N., Barkley, H., Terranova, V. P., and Liotta, L. A.,** Isolation of a tumor cell laminin receptor, *Biochem. Biophys. Res. Commun.*, 111, 804, 1983.
202. **Kennedy, D. W., Rohrbach, D. H., Martin, G. R., Momoi, T., and Yamada, K. M.,** The adhesive glycoprotein laminin is an agglutinin, *J. Cell. Physiol.*, 114, 257, 1983.
203. **Bernfield, M. R., Banerjee, S. D., and Cohn, R. H.,** Dependence of salivary epithelial morphology and branching morphogenesis upon acid mucopolysaccharide-protein (proteoglycan) at the epithelial surface, *J. Cell Biol.*, 52, 674, 1972.
204. **Vracko, R. and Benditt, E. P.,** Basal lamina: the scaffold for orderly cell replacement — observations on regeneration of injured skeletal muscle fibers and capillaries, *J. Cell Biol.*, 55, 406, 1972.
205. **Vracko, R.,** Basal lamina scaffold — anatomy and significance for maintenance of orderly tissue structure, *Am. J. Pathol.*, 77, 313, 1974.
206. **Banerjee, S. D., Cohn, R. H., and Bernfield, M. R.,** Basal lamina of embryonic salivary epithelia — production by the epithelium and role in maintaining lobular morphology, *J. Cell Biol.*, 73, 445, 1977.
207. **Gerfaux, J., Chany-Fournier, F., Bardos, P., Muh, J. P., and Chany, C.,** Lectin-like activity of components extracted from human glomerular basement membrane, *Proc. Natl. Acad. Sci. U.S.A.*, 76, 5129, 1979.
208. **Ekblom, P., Alitalo, K., Vaheri, A., Timpl, R., and Saxen, L.,** Induction of a basement membrane glycoprotein in embryonic kidney — possible role of laminin in morphogenesis, *Proc. Natl. Acad. Sci. U.S.A.*, 77, 485, 1980.
209. **Ekblom, P.,** Formation of basement membranes in the embryonic kidney — an immunohistochemical study, *J. Cell Biol.*, 91, 1, 1981.
210. **Brownell, A. G., Bessem, C. C., and Slavkin, H. C.,** Possible functions of mesenchyme cell-derived fibronectin during formation of basal lamina, *Proc. Natl. Acad. Sci. U.S.A.*, 78, 3711, 1981.
211. **Liesi, P., Dahl, D., and Vaheri, A.,** Laminin is produced by early rat astrocytes in primary culture, *J. Cell Biol.*, 96, 920, 1983.
212. **Bailey, A. J., Shellworth, G. B., and Duance, V. C.,** Identification and change of collagen types in differentiating myoblasts and developing chick muscles, *Nature (London)*, 278, 67, 1979.
213. **Leivo, I.,** Strucutre and composition of early basement membranes: studies with early embryos and teratocarcinoma cells, *Med. Biol.*, 61, 1, 1983.
214. **Foidart, J. M., Yaar, M., Figueroa, A., Wilk, A., Brown, K. S., and Liotta, L.,** Abortion in mice induced by intravenous injections of antibodies to type IV collagen or laminin, *Am. J. Pathol.*, 110, 346, 1983.
215. **Clark, C. C., Minor, R. R., Koszalka, T. R., Brent, R. L., and Kefalides, N. A.,** The embryonic rat parietal yolk sac — changes in the morphology and composition of the basement membrane during development, *Dev. Biol.*, 46, 243, 1975.
216. **Hintner, H., Fritsch, P. O., Foidart, J. M., Stingl, G., Schuler, G., and Katz, S. I.,** Expression of basement membrane zone antigens at the dermoepibolic junction in organ cultures of human skin, *J. Invest. Dermatol.*, 74, 200, 1980.
217. **Leivo, I., Vaheri, A., Timpl, R., and Wartiovaara, J.,** Appearance and distribution of collagens and laminin in the early mouse embryo, *Dev. Biol.*, 76, 100, 1980.
218. **Mitrani, E.,** Primitive streak-forming cells of the chick invaginate through a basement membrane, *Wilhelm Roux's Arch.*, 191, 320, 1982.
219. **Laurie, G. W. and Leblond, C.P.,** What is known of the production of basement membrane components, *J. Histochem. Cytochem.*, 31, 159, 1983.

220. **Bernfield, M. and Banerjee, S. D.,** The turnover of basal lamina glycosaminoglycan correlates with epithelial morphogenesis, *Dev. Biol.*, 90, 291, 1982.
221. **Fitch, J. M. and Linsenmayer, T. F.,** Monoclonal antibody analysis of ocular basement membranes during development, *Dev. Biol.*, 95, 137, 1983.
222. **Gulati, A. K., Zalewski, A. A., and Reddi, A. H.,** An immunofluorescent study of the distribution of fibronectin and laminin during limb regeneration in the adult newt, *Dev. Biol.*, 96, 355, 1983.
223. **Lazarow, A. and Speidel, E.,** The chemical composition of the glomerular basement membrane and its relationship to the production of diabetic complications, in *Small Blood Vessel Involvement in Diabetes Mellitus*, Siperstein, M. D., Colwell, A. R., Sr., and Meyer, K., Eds., American Institute of Biological Sciences, Washington, D.C., 1964, 127.
224. **Beisswenger, P. J. and Spiro, R. G.,** Human glomerular basement membrane — chemical alteration in diabetes mellitus, *Science*, 168, 596, 1970.
225. **Beisswenger, P. J. and Spiro, R. G.,** Studies on the human glomerular basement membrane: composition, nature of the carbohydrate units and chemical changes in diabetes mellitus, *Diabetes*, 22, 180, 1973.
226. **Westberg, N. G. and Michael, A. F.,** Human glomerular basement membrane — chemical composition in diabetes mellitus, *Acta Med. Scand.*, 194, 39, 1973.
227. **Kefalides, N. A.,** Biochemical properties of human glomerular basement membranes in normal and diabetic kidneys, *J. Clin. Invest.*, 53, 403, 1974.
228. **Khalifa, A. and Cohen, M. P.,** Glomerular protocollagen lysyl oxidase activity in streptozotocin diabetes, *Biochim. Biophys. Acta*, 386, 332, 1975.
229. **Sato, T., Munakata, H., Yoshinaga, K., and Yoshizawa, Z.,** Comparison of the chemical composition of glomerular and tubular basement membranes obtained from human kidneys of diabetics and non-diabetics, *Clin. Chim. Acta*, 61, 145, 1975.
230. **Klein, L., Butcher, D. L., Sudilovsky, O., Kikkawa, R., and Miller, M.,** Quantification of collagen in renal glomeruli isolated from human non-diabetic and diabetic kidneys, *Diabetes*, 24, 1057, 1975.
231. **Beisswenger, P. J.,** Glomerular basement membrane — biosynthesis and chemical composition in the streptozotocin diabetic rat, *J. Clin. Invest.*, 58, 844, 1974.
232. **Wahl, P., Deppermann, D., and Hasslacher, C.,** Biochemistry of glomerular basement membranes of the normal and diabetic human, *Kidney Int.*, 21, 744, 1982.
233. **Kefalides, N. A.,** Basement membrane research in diabetes mellitus, *Coll. Relat. Res.*, 1, 295, 1981.
234. **Brownlee, M. and Spiro, R. G.,** Glomerular basement membrane metabolism in the diabetic rat — in vivo studies, *Diabetes*, 28, 121, 1979.
235. **Cohen, M. P. and Urdanivia, E., Surma, M., and Wu, V. Y.,** Increased glycosylation of glomerular basement membrane collagen in diabetes, *Biochem. Biophys. Res. Commun.*, 95, 765, 1980.
236. **Cohen, M. P. and Wu, V. Y.,** Identification of specific amino acids in diabetic glomerular basement membrane collagen subject to nonenzymatic glycosylation *in vivo*, *Biochem. Biophys. Res. Commun.*, 100, 1549, 1981.
237. **LePape, A., Guitton, J. D., and Muh, J. P.,** Modifications of glomerular basement membrane cross-links in experimental diabetic rats, *Biochem. Biophys. Res. Commun.*, 100, 1214, 1981.
238. **Uitto, J., Perejda, A. J., Grant, G. A., Rowold, E. A., Kilo, C., and Williamson, J. R.,** Glycosylation of human glomerular basement membrane collagen — increased content of hexose in ketoamine linkage and unaltered hydroxylysine-o-glycosides in patients with diabetes, *Connect. Tissue Res.*, 10, 287, 1982.
239. **Sternberg, M., Grochulski, A., Peyroux, J., Hirbec, G., and Poirier, J.,** Studies on the α-glucoside specific for collagen disaccharide units: variations associated with capillary basement membrane thickening in kidney and brain of diabetic and aged rats, *Coll. Relat. Res.*, 2, 495, 1982.
240. **Perejda, A. J. and Uitto, J.,** Nonenzymatic glycosylation of collagen and other proteins: relationship to development of diabetic complications, *Coll. Relat. Res.*, 2, 81, 1982.
241. **Cohen, M. P. and Surma, M. L.,** ^{35}S-sulfate incorporation into glomerular basement membrane glycosaminoglycans is decreased in experimental diabetes, *J. Lab. Clin. Med.*, 98, 715, 1981.
242. **Rohrbach, D. H., Hassell, J. R., Kleinman, H. K., and Martin, G. R.,** Alterations in the basement membrane (heparan sulfate) proteoglycan in diabetic mice, *Diabetes*, 31, 185, 1982.
243. **Brown, D. M., Klein, D. J., Michael, A. F., and Oegema, T. R.,** ^{35}S-glycosaminoglycan and ^{35}S-glycopeptide metabolism by diabetic glomeruli and aorta, *Diabetes*, 21, 418, 1982.
244. **Rohrbach, H. and Martin, G. R.,** Structure of basement membrane in normal and diabetic tissues, *Ann. N.Y. Acad. Sci.*, 401, 203, 1982.
245. **Williamson, J. R. and Kilo, C.,** Capillary basement membranes in diabetes, *Diabetes*, 32 (Suppl. 2), 96, 1983.
246. **Blue, W. T. and Lange, C. F.,** Age-related carbohydrate content of mouse kidney glomerular basement membrane and its reactivity of antistreptococcal membranes antisera, *Immunochemistry*, 13, 295, 1976.
247. **Kalant, N., Satomi, S., White, R., and Tel, E.,** Changes in renal glomerular basement membrane with age and nephritis, *Can. J. Biochem.*, 55, 1197, 1977.

248. **Deyl, Z., Macek, K., and Adam, M.,** Changes in the proportion of collagen type IV with age — possible role in transport processes, *Exp. Gerontol.*, 13, 263, 1978.
249. **Langeveld, J. P. M., Veerkamp, J. H., Monnens, L. A. H., and Van Haelst, U. J. G.,** Chemical characterization of glomerular and tubular basement membranes of cattle of different ages, *Biochim. Biophys. Acta*, 514, 225, 1978.
250. **Langeveld, J. P. M., Veerkamp, J. H., Duyf, C. M. P., and Monnens, L. A. H.,** Chemical characterization of glomerular and tubular basement membrane of men of different ages, *Kidney Int.*, 20, 104, 1981.
251. **Hoyer, J. R. and Spiro, R. G.,** Studies on the rat glomerular basement membrane: age related changes in composition, *Arch. Biochem. Biophys.*, 185, 496, 1978.
252. **Smalley, J. W.,** Age-related changes in the amino acid composition of human glomerular basement membrane, *Exp. Gerontol.*, 15, 43, 1980.
253. **Wu, V. Y. and Cohen, M. P.,** Reducible cross-links in human glomerular basement membrane, *Biochem. Biophys. Res. Commun.*, 104, 911, 1982.
254. **Blau, E. B. and Michael, A. F.,** Rat glomerular glycoprotein composition and metabolism in aminonucleoside nephrosis, *Proc. Soc. Exp. Biol. Med.*, 141, 164, 1972.
255. **Lui, S. and Kalant, N.,** Carbohydrate of the glomerular basement membrane in normal and nephrotic rats, *Exp. Mol. Pathol.*, 21, 52, 1974.
256. **Bohrer, M. P., Baylis, C., Robertson, C. R., and Brenner, B. M.,** Mechanism of the puromycin induced defect in the transglomerular passage of water and macromolecules, *J. Clin. Invest.*, 60, 152, 1977.
257. **Caulfield, J. P. and Farquhar, M. G.,** Loss of anionic sites from the glomerular basement membrane in aminonucleoside nephrosis, *Lab. Invest.*, 29, 505, 1978.
258. **Mynderse, L. A., Hassell, J. R., Kleinman, H. K., Martin, G. R., and Martinez-Hernandez, A.,** Loss of heparan sulfate proteoglycan from glomerular basement membranes of nephrotic rats, *Lab. Invest.*, 48, 292, 1983.
259. **Abrahamson, D. R., Hein, A., and Caulfield, J. P.,** Laminin in glomerular basement membranes of aminonucleoside nephrotic rats — increased proteinuria induced by antilaminin immunoglobulin G, *Lab. Invest.*, 49, 38, 1983.
260. **DiBona, G. F.,** Alport's syndrome: a genetic defect in biochemical composition of basement membranes of glomerulus, lens and inner ear?, *J. Lab. Clin. Med.*, 101, 817, 1983.
261. **Hahn, E., Wick, G., Pencev, D., and Timpl, R.,** Distribution of basement membrane proteins in normal and fibrotic liver: collagen type IV, laminin, and fibronectin, *Gut*, 21, 63, 1980.
262. **Speziale, P., Hook, M., Wadstrom, T., and Timpl, R.,** Binding of the basement membrane protein laminin to *Escherichia coli*, *FEBS Lett.*, 146, 55, 1982.
263. **Varani, J., Lovett, E. J., III, McCoy, J. P., Jr., Shibata, S., Maddox, D. E., Goldstein, I. J., and Wicha, M.,** Differential expression of a laminin-like substance by high- and low-metastatic tumor cells, *Am. J. Pathol.*, 111, 27, 1983.
264. **David, G. and Van den Berghe, H.,** Transformed mouse mammary epithelial cells synthesize undersulfated basement membrane proteoglycan, *J. Biol. Chem.*, 258, 7338, 1983.
265. **Liotta, L. A., Abe, S., Robey, P. G., and Martin, G. R.,** Preferential digestion of basement membrane collagen by an enzyme derived from a metastatic murine tumor, *Proc. Natl. Acad. Sci. U.S.A.*, 76, 2268, 1979.
266. **Liotta, L. A., Tryggvason, K., Garbisa, S., Robey, P. G., and Abe, S.,** Partial purification and characterization of a neutral protease which cleaves type IV collagen, *Biochemistry*, 20, 100, 1981.
267. **Salo, T., Liotta, L. A., Keski-Oja, J., Turpeenneimi-Hujanen, T., and Tryggvason, K.,** Secretion of basement membrane collagen degrading enzyme and plasminogen activator by transformed cells — role in metastasis, *Int. J. Cancer*, 30, 669, 1982.
268. **Salo, T., Liotta, L. A., and Tryggvason, K.,** Purification and characterization of a murine basement membrane collagen-degrading enzyme secreted by metastatic tumor cells, *J. Biol. Chem.*, 258, 3058, 1983.
269. **Smith, R. L. and Bernfield, M.,** Mesenchyme cells degrade epithelial basal lamina glycosaminoglycan, *Dev. Biol.*, 94, 378, 1982.
270. **Kalebic, T., Garbisa, S., Glaser, B., and Liotta, L. A.,** Basement membrane collagens: degradation by migrating endothelial cells, *Science*, 22, 281, 1983.

INDEX

A

B

C

F

G

M

N

O

P

Q

R

S

T

U

V

W

X

For Product Safety Concerns and Information please contact our
EU representative GPSR@taylorandfrancis.com Taylor & Francis
Verlag GmbH, Kaufingerstraße 24, 80331 München, Germany